ELECTRODES AND THE MEASUREMENT
OF BIOELECTRIC EVENTS

ELECTRODES AND THE MEASUREMENT OF BIOELECTRIC EVENTS

L. A. GEDDES

Baylor College of Medicine
Houston, Texas

WILEY-INTERSCIENCE, a Division of John Wiley & Sons, Inc.
New York • London • Sydney • Toronto

Library of Congress Catalog Card Number: 75-180270

ISBN 0-471-29490-X

Printed in the United States of America.

10 9 8 7 6 5 4 3 2

Dedicated to
MY ALMA MATER
McGill University

Wonderful as are the laws and phenomena of electricity when made evident to us in inorganic or dead matter, their interest can bear scarcely any comparison with that which attaches to the same force when connected with the nervous system and with life. . . .

Michael Faraday
Philosophical Transactions of the Royal Society of London, 1839

Preface

This book is written both for the life scientist and the physical scientist who are faced with the problems associated with the measurement of bioelectric events. Selecting the correct types of electrodes and knowing their characteristics is surely the starting point. Contained in the chapters that follow is a large amount of practical information relating to the fabrication of electrodes and their electrical characteristics. The electrodes described range in size from those used to record the electrocardiogram of a submerged whale to those small enough to record the transmembrane potential of a single living cell. In the application of theoretical principles, mathematical derivations are held to a minimum. Wherever possible, the stories told by mathematical expressions are presented graphically — a method that is most familiar to the life scientist and indeed is the method used by the physical scientist when dealing with complex systems in which it is difficult to define or measure all the underlying components.

There seems to be an aura of mystery surrounding the construction and electrical properties of electrodes; the author has sought to remove some of it by describing the properties of electrodes in quantitative terms wherever possible. Thus the book opens with a discussion of the properties of the common denominator in virtually all electrode systems, the electrode-electrolyte interface. Following this there is a description of the skin; then the various types of electrodes are discussed in detail. Insofar as possible, measured data are given for each type. Some of these data were already available to the reader and the author; where gaps were found to exist, the author has made an effort to fill them by measurements in his own laboratory. For the reader who wishes to consult original sources, an extensive bibliography has been included.

This book constitutes the first compilation of data on the electrodes presently used to measure bioelectric events. From the data presented, the reader should be able to construct electrodes of his own design and predict, with fair accuracy, their characteristics and properties.

Because there are many who desire to use and understand the properties of electrodes, the author has included six easily conducted experiments in an appendix. These experiments were selected on the basis of their fundamental importance to the measurement of bioelectric events. To the ingenious instructor they can serve as the starting point in the design of additional experiments.

This book derives its origin from the early and mid-1940s when the fields of clinical electromyography and electroencephalography were being developed; at this time the first transmembrane potentials were being measured with micropipets. Also at this time the author was fortunate enough to be working with Wilder Penfield and Herbert Jasper, two of the pioneers of applied and basic electrophysiology. This fortunate association, which lasted for a period of eight years, placed the author in contact with the many fascinating problems in electrophysiology and gave him ample opportunity to test his ideas for the solution of some of them. It also permitted the author to initiate courses for the fellows and residents who were conducting research in this field. From this initial exposure and that which followed during the years from 1952 to the present, when the author came under the guidance of Hebbel Hoff, the nature of bioelectric events and the absorbing problems connected with their measurement became clear. From these fortunate contacts, and because so many life and physical scientists now desire to measure bioelectric events, the author has collected the information and experience he has accumulated and has tried to present a logical and practical analysis of the problems encountered in the measurement of bioelectric events.

The electrodes from which data were obtained were fabricated by the author's team. Therefore, for all phases of measurement and data processing, special thanks and appreciation must go to A. G. Moore and J. Bourland who direct the activities of the mechanical and electronics laboratories in the Division of Biomedical Engineering. Supporting them with their considerable talents were M. Hinds, G. Cantrell, T. Coulter, J. Vasku, C. Martinez, and E. Arriaga. Special thanks must go to Carter Jordan and Narvin Foster who were always ready to prepare the animals (no matter how large or small, docile or dangerous) needed for the studies. Recognition must also be accorded to Dr. Lee E. Baker, the author's collaborator in many of the studies and to Gary Wise and Dr. L. E. Geddes, who verified the experiments that appear in the appendix. Dr. G. Greeley of Texas A & M College of Veterinary Medicine is commended for his patience and skill in composing the frog drawings, made while the author demonstrated the various techniques. Dr. J. Bear of the Department of Chemistry of the University of Houston and Dr. Max Valentinuzzi of Baylor are hereby thanked for their reading of selected parts of the book. Special thanks must go to Miss Lucia Bonno, secretary to the Division of Biomedical Engineering, and to Diane Moffitt, her successor, who converted almost illegible notes and rough drafts into manuscript form which was considered worthy of publication.

Perhaps the greatest appreciation should go to the many students with genuinely inquiring minds with whom the author has been fortunate enough to be associated; for it was mainly their desire for knowledge that prompted the author to seek out pertinent information wherever it could be found. Without the contributions of all the persons mentioned, there would have been no book and no experiments.

Houston, Texas   L. A. GEDDES
September 1971

Contents

Tables

ELECTRODES AND THE MEASUREMENT
OF BIOELECTRIC EVENTS

Introduction

Interest in electrodes originated with Galvani's three frog-muscle contraction experiments* (ca. 1800); the first dealt with stimulation by electrostatic induction, the second with stimulation by electrode potentials and the third truly demonstrated bioelectricity. It was the research carried out by Volta, because of Galvani's explanation of his second experiment, that resulted in the discovery of current electricity produced by what is now known as the galvanic or voltaic cell, consisting of two dissimilar metal electrodes in an electrolyte. Physical scientists devoted their attention to the design of electrodes to obtain the maximum amount of electrical energy from such cells; electrophysiologists concerned themselves with developing electrode pairs that produced no potential difference so that the voltage measured between electrodes encompassing living cells was biological in origin.

In the measurement of a bioelectric event (except when insulated electrodes are used), the two electrodes are always in contact with electrolytes; for this reason it is necessary to know the nature and properties of an electrode-electrolyte junction in order to interpret correctly, with an instrument of known properties, the potential difference measured between the electrode terminals. Therefore, considerable attention is devoted to an accurate description of this junction as it participates in the measurement of bioelectric signals.

Despite the many different configurations and names applied to electrodes used to measure bioelectric events, there are basically only two functional types, extracellular and intracellular. Extracellular electrodes, which take the form of plates or screens, are placed on the integument or on tissues; in the form of probes they are inserted below the integument to establish better contact or to bring the electrode closer to the source of the biopotential, the ionic gradient that exists across all cell membranes. Extracellular electrodes are large and usually distant with respect to the dimensions of the cells on which

* See Geddes and Hoff 1971.

the measurements are made. Intracellular electrodes are much smaller than the cells under study and hence can be advanced through cell membranes without damage. In practice, intracellular electrodes (very slender pointed needles of metal or glass micropipets) usually have diameters in the range of 0.5 to 1.5 μ (microns). In the following chapters, the various electrodes used for the measurement of bioelectric events are described and insofar as possible, their electrical properties are presented. From these data the reader should be able to estimate the appropriateness of an electrode type and a measuring instrument for the measurement of a bioelectric event.

1

The Electrode-Electrolyte Interface

When a metallic electrode is placed in an aqueous solution, there is a tendency for the metal ions to enter into solution; there is also a tendency for ions in the solution to combine with the electrode. Although details of the reaction may be complex in a given situation, the net result is a charge distribution at the electrode-electrolyte interface and its spatial arrangement depends on the way in which the electrode metal reacts with the electrolyte. Several forms of charge distribution are possible; the simplest was conceived by Helmholtz (1879), who described the electrical double layer, later called the Helmholtz layer. He proposed the arrangement (Fig. 1-1a) which represents a layer of ions tightly bound to the surface of the electrode and an adjacent layer of oppositely charged ions in the solution; with such a simple arrangement the potential distribution would be as shown. Because of the thermal motion of ions, it was believed that the simple Helmholtz electrical double layer was not adequate to describe the environment of an electrode, Gouy (1910) suggested another charge distribution in which the fixed (Helmholtz) layer of negative charge was not enough to balance the positive charge of the electrode. To satisfy this requirement, he proposed the existence of a diffuse charge distribution adjacent to the Helmholtz layer. Gouy's arrangement of a fixed and diffuse layer and the accompanying potential distribution appear in Fig. 1-1b. Stern (1924) believed that the fixed layer could contain more negative charges than are required to balance the positive charges on the electrode. This situation, in combination with the diffuse Gouy layer, is presented in Fig. 1-1c, along with the expected potential distribution. Either combination of a fixed and diffuse layer is called a Stern layer. It is also possible for the charge distribution to consist entirely of a diffuse Gouy layer and to exhibit the potential distribution in Fig. 1-1d. As stated previously, the particular charge distribution that exists depends on the species of metal used for the electrode and the type of electrolyte. It is the ionic distribution that endows an electrode with its properties.

3

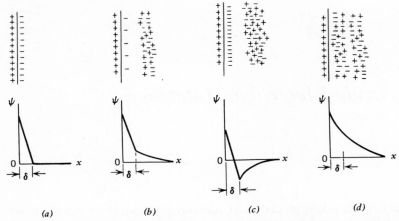

Fig. 1-1. Various configurations of charge and potential distribution at an electrode-electrolyte interface: (*a*) Helmholtz (1879); (*b*) Gouy (1910); (*c*) Stern (1924); (*d*) pure Gouy.

Parsons (1964) described electrodes in terms of the reactions at the double layer. Electrodes in which no net transfer of charge occurs across the metal-electrolyte interface were designated by him as perfectly polarized. Those in which unhindered exchange of charge is possible are called perfectly non-polarizable. Real electrodes have properties that fall between these idealized limits. It should be apparent by this definition that a truly polarized electrode has all the characteristics of a capacitor. MacInnes (1961) stated that the term "electrode polarization" is used in two ways: as previously, and to refer to the condition when an electrode-electrolyte potential is altered by the passage of a current.

Electrode Potential

As a result of the particular charge distribution that occurs when an electrode contacts an electrolyte, the electrode acquires a potential. Because it is not possible to measure the potential of a single electrode with respect to a solution, and because it is impractical to tabulate the various potential values attained between electrodes of different metals in a variety of electrolytes, electrode potentials are measured with respect to a standard electrode that is easily re-produced in the laboratory; such an electrode is the standard hydrogen elec-trode (SHE). Although choice of a particular reference electrode fixes elec-trode potentials with respect to it, it does not effect the difference in potential measured between two electrodes in an electrolyte; in fact it permits calcula-tion of the potential difference by subtraction of the individual potentials measured with respect to the SHE.

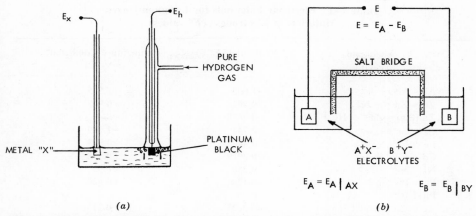

Fig. 1-2. Electrode potential: (*a*) measurement of electrode potential with the standard hydrogen electrode; (*b*) the Galvanic (Voltaic) cell.

The standard hydrogen electrode (Fig. 1-2*a*)* consists of a platinum-black electrode in contact with a solution (usually HCl) containing hydrogen ions of unit activity and dissolved molecular hydrogen; the activity of the latter is specified by requiring it to be in equilibrium with hydrogen at 1 atmosphere in the gas phase. The platinum black has a large capacity for adsorbing hydrogen and probably acts as a catalyst to convert the hydrogen to its ionic form. The hydrogen gas is bubbled over the electrode and the cell is operated at 1 atmosphere. The potential of such an electrode is defined as zero at all temperatures, a fact that is contrary to physical reality but can be circumvented by identifying the temperature at which the measurement of electrode potential is made.

If an electrode of element X is placed in the electrolyte (of unit activity) with the SHE, a galvanic or voltaic cell is created and a potential difference can be measured; the potential will, of course, be relative to that of the hydrogen electrode. Because the potential of the latter is defined as zero, potential measured is called the electrode potential (E_X^0) for the metal X; Table 1-1 lists the electrode potentials for a variety of metals used as electrodes.

If a second electrode X′ of the same metal (X) is placed in the electrolyte, the same potential will be measured between the hydrogen electrode; that is $E_{X'}^0 = E_X^0$. If the potential is measured between electrodes X and X′, the value obtained will be zero.

When two electrodes of different metals X, Y are placed into any electrolyte,

---

* The first practical hydrogen electrode was introduced by Hilderbrand, who described its properties in 1913; it later became the standard hydrogen electrode.

**Table 1–1   Electrode Potentials for Commonly used
Materials in Electrodes[a] (E^0 values)**

Metal and Reaction	Potential ($E^0_{25°C}$) (V)	Temperature Coefficient (mV/°C)
$Al = Al^{3+} + 3e^-$	-1.662	$+1.375$
$Zn = Zn^{2+} + 2e^-$	-0.7628	$+0.962$
$Zn\ (Hg) = Zn^{2+} + Hg + 2e^-$	-0.7627	—
$Cr = Cr^{3+} + 3e^-$	-0.744	$+1.339$
$Fe = Fe^{2+} + 2e^-$	-0.4402	$+0.923$
$Cd = Cd^{2+} + 2e^-$	-0.4029	$+0.778$
$Ni = Ni^{2+} + 2e^-$	-0.250	$+0.93$
$Pb = Pb^{2+} + 2e^-$	-0.126	$+0.420$
$Pt(H_2)H^+$	0	—
$Ag + Cl^- = AgCl + e^-$	$+0.2225^b$	$+0.213$
$Cu = Cu^{2+} + 2e^-$	$+0.337$	$+0.879$
$Cu = Cu^+ + e$	$+0.521$	$+0.813$
$2\ Hg = Hg_2^{2+} + 2e^-$	$+0.788$	—
$Ag = Ag^+ + e^-$	$+0.7991$	-0.129
$Pt = Pt^{2+} + 2e^-$	$+1.2$ approx.	—
$Au = Au^{3+} + 3e^-$	$+1.498$	—
$Au = Au^+ + e^-$	$+1.691$	—

[a] From A. J. de Bethune, in *Handbook of Electrochemistry*, C. A. Hampel, Ed., New York: Reinhold, 1964.
[b] MacInnes (1939).

a galvanic cell is created whose potential depends on four factors: the difference in the potentials of the individual electrodes (E_X^0, E_Y^0 in Table 1-1), the temperature, the concentration of the electrolyte, and the manner in which the electrode metals react with the electrolyte. Although the potential of a galvanic cell is sometimes quite difficult to calculate, the underlying principles can be illustrated by a simple model. Consider a galvanic cell consisting of two metal electrodes A and B, each dipping into an electrolyte which is a salt of each metal (Fig. 1-2*b*); connection between the two electrolytes is established by a salt bridge (see p. 18) to avoid the creation of a liquid-junction or diffusion potential. Thus the galvanic cell consists of the two following half-cells:

$$E_{1/2A} = E^0_{A^{n+}} - \frac{RT}{nF} \ln A_{A^{n+}}$$

$$E_{1/2B} = E^0_{B^{n+}} - \frac{RT}{nF} \ln A_{B^{n+}}$$

In these expressions, the E^0 values are the standard electrode potentials measured with respect to the hydrogen electrode (Table 1-1), T is the absolute temperature, R is the gas constant, n is the valence, A is the activity of the metal ions in the electrolyte, and F is the faraday; the activity is, of course, proportional to the concentration.

The potential difference E_{AB} of such a voltaic cell is the difference between the two half-cell potentials $E_{1/2A}$ and $E_{1/2B}$; therefore

$$E_{AB} = E_{1/2A} - E_{1/2B}$$
$$= E^0_{A^{n+}} - E^0_{B^{n+}} - \frac{RT}{nF} \ln \frac{A_{A^{n+}}}{A_{B^{n+}}}$$

Note that the potential of the voltaic cell is composed of two parts: one (usually the larger) that is the difference in the half-cell potentials of the metals measured with respect to the hydrogen electrode and a second (usually smaller) that is due to the concentrations (strictly activities) of the metal ions in solution. (It should be recalled that in this simple example, a liquid-junction potential was eliminated by the use of a salt bridge.)

When electrodes are applied to a subject to measure a bioelectric event, a galvanic cell is created because each electrode is in contact with an electrolyte, which may be electrode paste, saline solution, or tissue fluids. Therefore, the actual potential measured between the electrode terminals will have several origins, and the magnitude of each may be difficult to establish accurately in a particular situation. Nonetheless it should be remembered that there exist two electrode potentials between the electrode terminals, and these potentials may not be equal. Obviously if electrodes of the same metal contact the same electrolyte, the net electrode potential will be zero. There will, however, be a net electrode potential if the same metals are in contact with solutions of differing concentrations or if different metals are in contact with the same electrolyte. If different metals are in contact with different electrolytes, there may be a liquid-junction potential in addition to the electrode potentials.

The best way of minimizing the existence of an electrochemical potential difference between the electrodes is to use electrodes made from the same metal, arranging to have them in contact with the same electrolyte. Even when this precaution is taken, there is often a small residual potential difference which is not always stable. The residual potential is probably due to differences in surface contamination of the two electrodes or to slight differences within the electrode metal. Usually a residual potential difference can be tolerated and canceled with an opposing voltage. When capacitively coupled recording systems are used, a residual potential is not coupled into the measuring instrument. However, residual potentials are not always stable and they fluctuate randomly, adding artifact to the record of the bioelectric event. It should be emphasized that in most instances the biopotentials measured, especially with extracellular electrodes, are of the order of a few millivolts. This requires that random variations in electrode potentials be small with respect to the magnitude of the biopotential.

Table 1–2   Fluctuations in Potential Between Electrodes in Electrolytes

Electrode Metal	Electrolyte Type	Potential Difference Between Electrodes	Investigator and Year
PbHg	$PbCl_2$ in chamois on human skin	0–600 μV (basal) 1.3–6.8 μV (fluctuations)	Ferris (1934) (av. of 10 electrodes)
Calomel	Saline	1–20 μV	Greenwald (1936)
Zn–$ZnSO_4$	Saline	180 μV	Greenwald (1936)
Zn	Saline	450 μV	Greenwald (1936)
Stainless steel	Saline	10 mV	Lykken (1959)
Zn	Saline	100 mV	Lykken (1959)
ZnHg	Saline	82 mV	Lykken (1959)
Ag	Saline	94 mV	Lykken (1959)
AgHg	Saline	90 mV	Lykken (1959)
Ag-AgCl	Saline	2.5 mV	Lykken (1959)
Pb	Saline	1 mV	Lykken (1959)
PbHg	Saline	1 mV	Lykken (1959)
Pt	Saline	320 mV	Lykken (1959)
Ag, AgCl sponge	ECG paste	0.2 mV 0.07 mV drift in 1 hour	O'Connell et al. (1960)
Ag, AgCl (11-mm disk)	ECG paste	0.47 mV 1.88 mV drift in 1 hour	O'Connell et al. (1960)
Pb (11-mm disk)	ECG paste	4.9 mV 3.70 mV drift in 1 hour	O'Connell et al. (1960)
Zn, $ZnCl_2$ (11-mm disk)	ECG paste	15.3 mV 11.25 mV drift in 1 hour	O'Connell et al. (1960)

A number of investigators have measured the residual potential difference between similar electrodes in contact with the same electrolyte. Table 1-2, which presents the data reported to date, reveals that there often exists an unstable residual potential; the table also shows that some metals readily attain equilibrium with physiological electrolytes. Several interesting techniques have been used by electrophysiologists to minimize undesirable fluctuations in electrode potential. For example, a pair of electrodes made from the same piece of metal when placed in 0.9% saline and connected to a high input impedance graphic recorder frequently produce a fluctuating voltage. Often it is possible to eliminate this fluctuating voltage by connecting the electrodes together and allowing enough time to pass to permit attainment of a stable equilibrium with the electrolyte. This technique has long been employed by electrophysiologists to "quieten" their electrodes. A related observation has been that newly prepared electrodes are often electrically unstable; with the passage of time they become stable if left in the electrolyte. Another method of quietening electrodes is to deposit from a large electrode a uniform film of material covering both recording electrodes. In this way the tiny differences in electrode material are virtually eliminated and a pair of electrodes having a very small and stable potential difference can be made.

Another useful piece of information derives from a consideration of the double layer; namely, the effect of its mechanical disturbance. Experience has revealed that the electrodes that are relatively free of movement artifacts are those in which the electrode-electrolyte interface is removed from direct contact with the subject. Because the double layer is a region of charge gradient (i.e., a source of potential), disturbance of it gives rise to a change in potential that is small electrochemically but nevertheless may be large with respect to the size of many bioelectric events. Because movement artifacts produced by disturbance of the electrical double layer are in the frequency spectrum of many of the bioelectric events electrical filtering techniques can seldom be employed without a loss of important components of the bioelectric signal. Therefore, the electrical stability of an electrode can be enhanced by stabilization of the electrode-electrolyte interface.

The preceding discussion of electrode potentials was presented to alert the reader to recognize the possibility of the presence of voltages of nonphysiologic origin. In order to be able to place confidence in the magnitude of the voltage appearing between the electrode terminals, electrodes should be routinely checked for voltage without the bioelectric event interposed.

Often relatively little attention is given to the large unstable potentials developed when the electrode wires come in contact with electrolytes. Special precautions should be taken after a carefully prepared electrode is joined to the wire connected to the recording apparatus. In the early days of electrocardiography Pardee (1917) recommended that the connecting wire be riveted to the electrode and the use of solder avoided. Henry (1938) noted that if the solder connection joining the electrode to the interconnecting wire became wet with an electrolytic solution, there developed a multimetal electrolytic cell that acquired unstable voltages and caused eventual corrosion and breakage of the connection. The simple practice of covering the solder connection to the electrode with a waterproof coating not only produces a more stable electrode but a longer-lasting one.

Liquid-Junction Potential

A liquid-junction potential is developed at the interface between two solutions having different concentrations and ionic mobilities; the magnitude of the potential difference E is given by the following variant of the Nernst equation:

$$E = \left(\frac{u^+ - v^-}{u^+ + v^-}\right)\frac{RT}{nF} \ln\frac{C_1}{C_2}$$

In this expression u and v are the mobilities of the cations and anions, respectively, R is the gas constant, T is the absolute temperature, F is the faraday and n is the number of charges carried by the ions.

To illustrate the order of magnitude of a typical liquid-junction potential, consider the junction between two solutions of sodium chloride, the concentration of one being ten times that of the other. To calculate the liquid-junction potential, it is necessary to know the mobilities of the Na^+ (i.e., u) and Cl^- (i.e., v) ions; these values are, respectively, 5.19×10^{-4} and 7.91×10^{-4} cm sec^{-1}/V cm^{-1}. Entering these values into the foregoing expression gives a junction potential of 12.2 mV at 25°C, the more dilute solution being negative. If instead of sodium chloride, two solutions of potassium chloride having a 10/1 concentration ratio, are considered, the junction potential is considerably less because the mobility of the potassium ion (7.62×10^{-4}) is almost the same as that for the chloride ion. In this situation the junction potential amounts to only 1.1 mV at 25°C.

From these examples and from the equation, it is obvious that the use of solutions with equal ionic mobilities favors minimization of junction potentials. The nearly equal ionic mobilities of K^+ and Cl^- recommend the use of potassium chloride as the substance of choice for the construction of salt bridges. The low resistivity of potassium chloride, along with the nearly equal mobilities of the K^+ and Cl^- ions, make potassium chloride the electrolyte of choice for filling micropipets (see p. 60).

Although liquid-junction potentials are relatively easy to calculate when a single electrolyte is involved, they are much more difficult to determine when two or more electrolytes containing differing ionic species form the junction. A discussion of some of the assumptions that are made and the difficulties encountered in calculating liquid-junction potentials in such circumstances was presented by MacInnes (1961). Electrochemists, by and large, tend to adopt experimental procedures in which liquid-junction potentials are minimized by the use of appropriate salt bridges.

Chlorided Silver Electrodes

Although Jahn (1900) is usually credited with the introduction of chlorided silver electrodes in electrochemistry, they were in use at an earlier time for recording bioelectric events. For example, in his classic paper on the mammalian electrocardiogram, Waller (1889) reported

> The methods followed have been in the main those described in the paper already referred to [1887], with certain modifications of detail, such as the use of d'Arsonval's chloride of silver electrodes (which proved to be convenient and excellent for the purpose in view), and with this difference, that in order to examine the as far as possible intact and uninjured organ the [cat] heart was examined *in situ*, the thorax being laid open and its walls fixed to a board immediately after the decapitation of the animal.

d'Arsonval's (1886) original description of chlorided silver electrodes was merely a one-page note which did not mention their application beyond the statement that they were "homogeneous and perfectly nonpolarizable."

Since the beginning of the twentieth century the silver-silver chloride electrode has shown itself to be the most stable of the easily constructed electrodes. With care, it is possible to make a pair of electrodes having a difference in potential amounting to a few microvolts and to attain a stability of about the same magnitude; in the measurement of bioelectric events, this is a highly desirable feature. To illustrate this point the author and a colleague (Baker, 1967) constructed and tested a series of cortical electrodes similar to those in Fig. 2-31a. Each electrode consisted of a silver ball 1.35 mm diam., made by heating the tip of a piece of silver wire 0.6 mm diam. in the flame of a bunsen burner. After fabrication, the electrodes were polished with emery paper until the metal became bright and free of all visible surface contaminants; then the wire supporting the ball was coated with insulating varnish. Each silver ball was then chlorided by making it positive with respect to a large silver plate in a beaker of 0.9% saline solution. The chloride deposit amounted to about 500 mA-seconds/cm² of electrode surface.

After the chloriding process, the electrochemical noise level was measured by recording the potentials between four such electrodes in a beaker of 0.9% saline solution and connected to three recording channels in the conventional recording manner (Fig. 1-3a). The illustration, obtained at a slow chart speed, shows that the random variations in electrode potential ranged from a few μV to about 10 μV. The type of randomly fluctuating potentials obtained (Fig. 1-3b) when the chloride deposit was scraped off each electrode demonstrates the importance of chloriding. The electrodes were chlorided again in the same beaker of saline; Fig. 1-3c shows the noise-level record obtained.

Thus one of the chief advantages of chloriding a silver electrode is stabilization of the half-cell potential. Initial studies by the author and his colleagues (1967, 1969) have also demonstrated that optimum chloriding reduces the electrode-electrolyte impedance. The subject of optimum chloriding is discussed further in the section of this chapter dealing with electrode impedance.

An ingenious method of stabilizing chlorided silver electrodes while they are in storage was described by Cooper (1956). The method, which serves to maintain the electrodes connected to each other and maintains them in a chlorided condition, consists of mounting the electrodes with their silver-silver chloride surface immersed in a dish of saline. A carbon rod that projects out of the solution is mounted in the dish, and to the top of the carbon rod Cooper affixed a stainless steel plate and connected all the electrode terminals to it. Thus all electrodes were joined together. In addition, because carbon is slightly electronegative with respect to silver-silver chloride, a small chloriding current was maintained; during storage the electrodes were maintained at the same potential and continuously chlorided. Cooper reported that

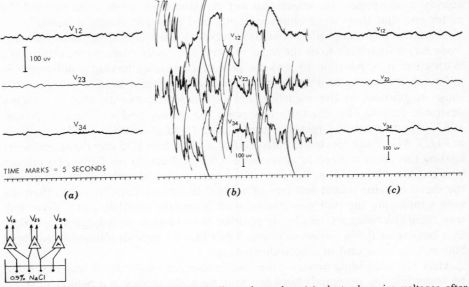

Fig. 1-3. Electrical stability of chlorided silver electrodes: (*a*) electrode noise voltages after chloriding; (*b*) electrode noise voltages after removing silver chloride coating with emery paper; (*c*) electrode noise voltages after rechloriding. [From L. A. Geddes and L. E. Baker, *Med. Res. Engng.* 1967, **6**(3):33–34. By permission.]

electrodes treated in this way were adequately stable for use with high-gain direct-coupled amplifiers.

The Potential of a Silver-Silver Chloride Electrode   It is to be noted that the silver-silver chloride electrode is a chloride -ion electrode with a potential that reflects the activity (A_{Cl^-}) of the environmental chloride ions; the activity is in turn dependent on the molal concentration m and the activity coefficient γ. The potential is given by the following expression:

$$E = E_T{}^0 - \frac{2.303RT}{F} \log_{10} A_{Cl^-}$$

In this expression, $E_T{}^0$ is the potential of the silver-silver chloride electrode (Table 1-3) at the temperature ($T°K$), R is the gas constant (8.317 J/cal), F is the faraday (96,500 coul). The activity A_{Cl} is equal to γm. The foregoing expression can be evaluated to give the potentials under specified conditions.

$$E = 0.2214 - 0.0582 \ \log_{10} \gamma m \quad \text{at} \quad 20° \text{ C}$$
$$E = 0.2225 - 0.05915 \log_{10} \gamma m \quad \text{at} \quad 25° \text{ C}$$
$$E = 0.2251 - 0.06153 \log_{10} \gamma m \quad \text{at} \quad 37° \text{ C}$$

Table 1–3   The Potential of Silver-Silver Chloride Electrodes[a]

Electrode/Electrolyte (at 25° C)	Potential (V)[a]	Temperature Coefficient (mV/°C)
Ag-AgCl/0.001M KCl	+0.401	+0.77
Ag-AgCl/0.01M KCl	+0.343	+0.62
Ag-AgCl/0.1M KCl	+0.288	+0.43
Ag-AgCl/1.0M KCl	+0.235	+0.25
Ag-AgCl/($A = 1$) KCl	+0.222	+0.213 (calc.)

[a] Data obtained from A. J. de Bethune, in *Handbook of Electrochemistry*, C. A. Hampel, Ed., New York: Reinhold, 1964. By permission.

Very frequently the silver-silver chloride electrode is used in conjunction with potassium chloride solutions having various concentrations expressed in molar concentration M. De Bethune (1964) has presented the data covering this situation in Table 1-3; the data are plotted in Fig. 1-4. Sometimes the silver-silver chloride electrode is used with a saturated potassium chloride

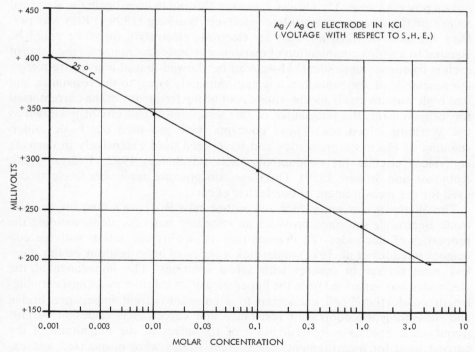

Fig. 1-4. The potential of a silver-silver chloride electrode in contact with potassium chloride solutions.

solution, which at 25°C is equivalent to about $4.7M$; for this situation Castellan (1964) reported an electrode potential of $+0.197$ V.

ELECTRODE IMPEDANCE

Equivalent Circuits

The presence of a charge distribution at an electrode-electrolyte interface not only provides an electrode with a potential, it endows it with capacitance, since two planes of charge of opposite sign separated by a distance constitute a charged capacitor. Because the distance between the layers of charge is molecular in dimension, the capacitance is large. In fact, this property permits electrolytic capacitors to exhibit large capacitance in relatively small packages.

Knowledge of the nature of an electrode-electrolyte interface came from studies by those who desired to measure the resistivity of electrolytes. It was known that when alternating current flowed through such an interface, a bothersome potential drop occurred. Kohlrausch (1897, 1898) minimized this potential drop in the case of platinum electrodes when he introduced the platinization process (see p. 32). Others, however, desired to investigate the electrical nature of the electrode-electrolyte interface. Warburg (1899, 1901) was perhaps the first demonstrate that an electrode-electrolyte interface could be equated to a series combination of resistance R and capacitance C; the value of each is frequency dependent. Moreover, he showed that at a given frequency the reactance of the capacitance is approximately equal to the resistance and that both vary inversely as the square root of the frequency of the current used for measurement; the impedance of this series equivalent circuit is known as the Warburg impedance. These concepts have provided our basic understanding of electrode properties and have been used extensively in deriving equivalent circuits for conductivity cells (Grahame, 1952; Feates, 1956; Robinson and Stokes, 1959). The same concepts are applicable to electrodes used for the measurement of bioelectric events.

The Warburg concept (which the author calls Warburg's law) for an electrode-electrolyte interface provides an excellent basis for understanding the properties of electrodes. To demonstrate its validity the author with his colleagues (Geddes et al, 1971) undertook a series of investigations on the stainless steel surface in contact with saline solutions. The impedance of the electrodes was separated from the impedance of its solution by using a variable-length conductivity cell connected to a constant-current impedance bridge and measuring the equivalent resistance and capacitance values at various frequencies. Studies were also made of the effect of the magnitude of the current used for measurement; the current values were normalized and expressed in terms of milliamperes per square centimeter of electrode area.

The series-equivalent resistance and capacitance values obtained for the different concentrations of saline (0.6–5%) were slightly different, but the same general trend was encountered. Figure 1-5a shows how the resistance and capacitance of a stainless steel electrode in contact with 0.9% saline varies with frequency using a current density of 0.025 mA/cm². It is interesting to note that both R and C vary inversely almost as the square root of frequency and it is perhaps more interesting to compare the resistance with the reactance X represented by the capacitance C (i.e., $X = 1/2 \pi f C$); this comparison has been made in Fig. 1-5b. The data presented in Fig. 1-5 agree with Warburg's law; namely, an electrode-electrolyte interface can be equated to a series resistance and capacitance circuit, and the values of each vary inversely approximately with the square root of frequency. More over, if the resistance is compared with the reactance (Fig. 1-5b) it is seen that they are approximately equal and both vary almost inversely with the square root of frequency. The same general relationship was reported by Gesteland et al. (1956) for platinum and silver-silver chloride electrodes.

The data in Fig. 1-5 are typical of those reported for the series-equivalent capacitance and resistance for an electrode-electrolyte interface measured with low current densities. The information contained in a variety of papers reporting values for the series-equivalent capacitance for various electrode-electrolyte interfaces has been normalized to obtain values for the capacitance per square centimeter of electrode area and to test the accuracy of the relation $C = K f^{-\alpha}$, where C is the series-equivalent capacity ($\mu F/cm^2$) K is a constant that depends on the type of the metal-electrolyte junction, and α is constant describing the rate at which the capacitance decreases with increasing frequency; ideally in the Warburg case $\alpha = 0.5$. Figure 1-6 and Table 1-4 summarize these data, which show that bare metals (and polished metals in particular), have the lowest electrode-electrolyte series-equivalent capacitances and that the value for α is in the vicinity of 0.5. Because the value for K is so dependent on surface condition, it is difficult to specify an accurate value for a given type of metal-electrolyte surface. (The most accurate values for K are those for mercury-electrolyte interfaces). It is clearly evident in Fig. 1-6 that platinization, which increases the effective area, dramatically increases the electrode capacitance and hence reduces the electrode-electrolyte impedance. The data and literature cited show that the value of capacitance is slightly dependent on both the concentration of the electrolyte and the temperature. Roughened surfaces, an increase in concentration of the electrolyte, and an increase in temperature all increase the value of the series-equivalent electrode-electrolyte capacitance.

After reviewing the published literature, Fricke (1932) pointed out that the series-equivalent capacitance C of an electrode-electrolyte interface does not always behave according to Warburg's concept; Fricke noted in particular

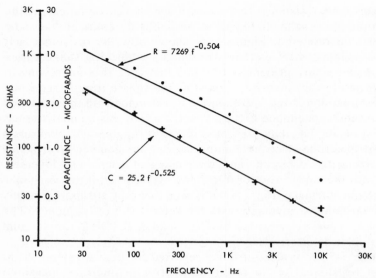

Fig. 1-5a. Series-equivalent resistance R and capacitance C of a stainless steel electrode (0.157 cm²) in 0.9% saline measured with a current density of 0.025 mA/cm² (From Geddes et al., *Med. Biol. Engng.* 1971. In press. By permission).

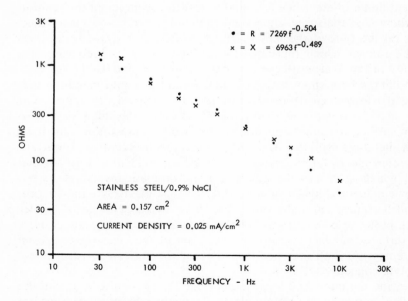

Fig. 1-5b. Series-equivalent resistance R and reactance X of a stainless steel electrode in 0.9% saline. (From Geddes et al., *Med. Biol. Engng.* 1971. In press. By permission.)

16

that C varied as $f^{-\alpha}$, and Table 1-4 verifies this observation. He further stated that although α was not always 0.5 (the Warburg case) the phase shift, ϕ, (the tangent of which is the ratio of reactance to resistance) was relatively constant with frequency. Schwan (1963) verified this fact for platinum electrodes and reported that the phase shift varied only by a factor of 2 over a four-decade range of frequency.

Often it is desired to know the impedance of an electrode-electrolyte interface, which of course requires knowledge of its resistive and capacitance components. It is usually easier to measure the series-equivalent capacitance with more accuracy; knowledge of the phase angle would permit calculation of both the resistance and the impedance. Schwan called attention to the fact that Fricke had derived such a relationship which Schwan called "Fricke's Law". It states that the phase angle ϕ is dependent on α, the exponent describing the manner by which capacitance varies with frequency: Fricke's law states

$$\phi = \frac{\pi}{2} (1 - \alpha)$$

To a first approximation, $\alpha = 0.5$, and under this condition the phase angle, the tangent of which is the ratio of a reactance to resistance, is 45°. This means that the reactance and the resistance are equal; Fig. 1-5b shows that this is approximately true, at least for the case cited. Therefore, the equivalent series resistance R and reactance X values can be easily transformed into parallel equivalents R_p, X_p by the following circuit theorems:

$$R_p = R + X^2/R$$
$$X_p = X + R^2/X$$

Substitution of $X = R$ into the preceding equations and recalling that $X = 1/2\pi fC$, the following is obtained:

$$R_p = 2R \qquad \text{and} \qquad C_p = 0.5C$$

Thus, insofar as $\tan \alpha = 1$ (i.e., $\phi = 45°$), or $X = R$, this simple conversion of series-to-parallel equivalent can be obtained. Although both equivalents are only fair models for the electrode-electrolyte interface, the author prefers the parallel configuration, which symbolizes a finite value for resistance. It must be remembered, however, that neither equivalent can actually be extended to zero frequency (i.e., dc), since a new class of phenomena exists in which electrolytic changes are produced because of the transfer of electrons at the electrodes.

The equivalent values for the resistive and capacitive components of an electrode-electrolyte interface also depend on the current density used to make the measurement. This has been known for some time by electrochemists (cf. Varley 1871; Wien, 1896), who found that the equivalent series capacity in-

Table 1-4 Electrode Polarization Capacitance[a]

Type of Electrode-Electrolyte Interface	Frequency Range, $f(Hz)$	Temperature (°C)	$C(\mu f/cm^2) = K/f^a$ K	α	Curve (Fig. 1-6)	Investigator and Year
Pt Black (heavy)/0.9% NaCl	Calculated from data at 20 Hz	Room?	149,700	0.366	A	Schwan (1964)
Pt black/0.9% saline	10–10K	Room?	11,759	0.221	B	Gesteland (1956)
Pt Ir black/0.9% saline	10–1K	Room?	8,619	0.299	C	Ray (1965)
Pt Black (medium)/0.9% saline	10–10K	Room?	4,950	0.366	D	Schwan (1964)
Ag/0.9% NaCl	10–10K	20	359	0.328	E	Geddes (unpublished)
Pt/1.46N H$_2$SO$_4$	645–3872	18	2,642	0.580	F	Zimmerman (1930)
Ag/0.1N AgNO$_3$	500–4K	25	634	0.434	G	Jones and Christian (1935)
Pt/1% H$_2$SO$_4$	700–10K	0	171	0.307	H	Wolff (1926)
PtIr/0.9% saline	150–1K	Room?	2,696	0.79	I	Ray (1965)
Stainless steel/0.9% saline	20–10K	20	160.8	0.525	O	Geddes et al. (1971)
Stainless steel/canine muscle	20–10K	37	179	0.387	J	Geddes (unpublished)
Au/1.46N H$_2$SO$_4$	645–2581	18	319	0.476	K	Zimmerman (1930)
Pt/0.025N HCL	700–4.35K	25	322	0.495	L	Jones and Bollinger (1935)
Au/0.01N KBr	570–3500	Room?	86	0.327	M	Miller (1923)
Au/1% H$_2$SO$_4$	200–10K	0	24	0.242	N	Wolff (1926)
Au/0.1NKBr	2000–3500	Room?	2,868	0.623	—	Miller (1923)
Au/1.1% H$_2$SO$_4$	0.1×10^6–0.4×10^6	Room?	2,873	0.541	—	Jolliffe (1923)
Pt/1.1% H$_2$SO$_4$	0.1×10^6–0.7×10^6	Room?	349	0.337	—	Jolliffe (1923)
Pt/0.25N H$_2$SO$_4$	0.05–3500	0	13.6	0.11	—	Murdock et al. (1936)
Pt/H$_2$SO$_4$	1810–10^6	Room?	289	0.472	—	Merritt (1921)

[a] Values for K and α calculated from papers cited.

creased as the current density in an electrolytic cell was increased. To determine the true equivalent capacity, measurements were made with lower and lower currents. It was found that there was no further decrease in capacity below a certain current; this value was called the "initial capacity".

To illustrate how the series-equivalent resistance and capacitance values depend on the current density used for measurement, the author (Geddes et al. 1971) measured these quantities using stainless steel electrodes in contact with 0.9% saline solution. Figure 1-7 presents such data and shows that for a given frequency, the equivalent resistance decreases and the capacitance increases as current density is increased. Note that the largest change occurs in the low-frequency region. This type of relationship is in agreement with the studies by Schwan (1965, 1968) and Jaron (1969), who used platinum electrodes and reported that the 20-and 500-Hz capacitance increased and the resistance decreased with increasing current density. They defined the "limit of linearity" at a specified frequency as that current density for which the value of capacitance increased or the resistance decreased by 10% from the value found with very low current density.

Schwan (1968) elaborated on the nature of an electrode-electrolyte interface when operated under high current density conditions, namely, that (with his definition) "the limit of linearity shrinks to zero as the frequency decreases to zero." This relationship, which is indicated by the dashed lines in Fig. 1-7, raises some questions regarding the suitability of small-area metal microelectrodes for the measurement of transmembrane potentials. However, since few measurements have been made of electrode properties below 10 Hz, one can only speculate about the equivalent circuit for 0 Hz (i.e., the situation that obtains when direct current is passed through an electrode-electrolyte interface). It is well known that direct current can be passed through an electrode-electrolyte interface, and for this reason the simple series-equivalent circuit cannot be considered truly representative.

Because it is easier to understand the behavior of an electrical circuit by using reactance and resistance values, it is of interest to compare the resistance and reactance at different current densities; Fig. 1-8 makes this comparison and shows that at low current densities the Warburg equivalent ($R = X$ and both vary approximately as $1/\sqrt{f}$) is reasonably accurate. However, as current density is increased, both resistance and reactance decrease and the manner in which each varies with frequency does not permit a power function approximation (i.e., one that is linear on a log-log plot).

The foregoing indicates that at a given frequency the series impedance $\sqrt{R^2 + X^2}$ of an electrode-electrolyte interface decreases with increasing current density. In addition, because increasing current density preferentially alters the low-frequency values for resistance and capacitance, it would be expected that the low-frequency impedance would be reduced markedly by

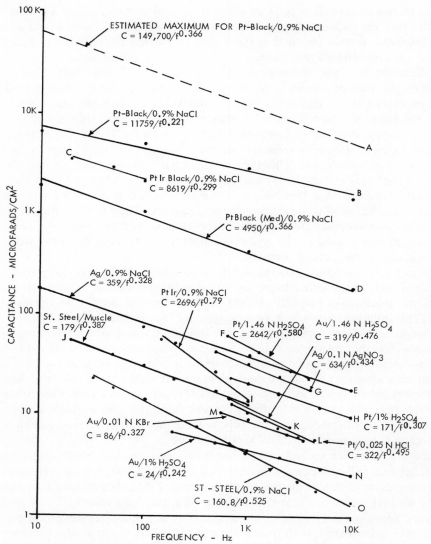

Fig. 1-6. Equivalent series capacitance-frequency data for various electrode-electrolyte surfaces calculated from data published by the following authors: curves *A* and *D*, Schwan (1964); curve *B*, Gesteland et al. (1956); curves *C* and *I*, Ray (1965); curves *E* and *J*, Geddes (unpublished); curves *F* and *K*, Zimmerman (1930); curve *G*, Jones and Christian (1935); curves *H* and *N*, Wolff (1926); curve *L*, Jones and Bollinger (1935); curve *M*, Miller (1923); curve *O*, Geddes et al. (1971).

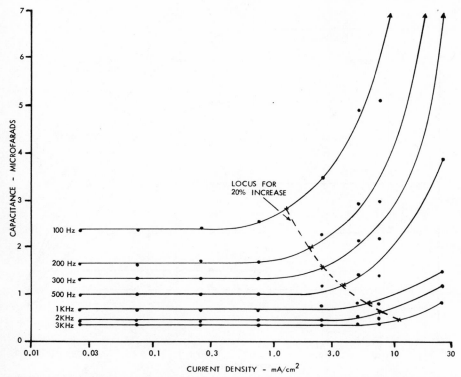

Fig. 1-7a. Dependence of the series-equivalent capacitance on current density and frequency; stainless steel electrode (0.157 cm²) in contact with 0.9% saline. (From Geddes et al., *Med. & Biol. Engng.* 1971. In press. By permission.)

an increase in current density. To illustrate this point, the impedance-frequency characteristic of a silver ball electrode (1.35 mm diam.) immersed in 0.9% saline and paired with a large circular silver disk (3 cm diam.) was measured in the frequency range extending from 20 to 10 KHz using current densities in the range from 0.1 to 40 mA/cm² for the ball electrode. Figure 1-9, which illustrates the results, shows that the electrode impedance diminishes with increasing current density and that the greatest reduction occurs in the low-frequency region.

It should be apparent now that an electrode-electrolyte interface represents a complex electrical circuit and that its impedance depends on the frequency and current density used for measurement. In coupling a bioelectric event to an amplifier via electrodes, consideration should be given to the nature of the impedance of the electrodes. Every precaution should be taken to pass as

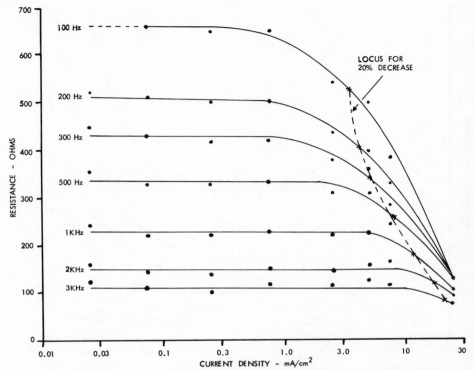

Fig. 1-7b. Dependence of the series-equivalent resistance on current density and frequency; stainless steel electrode (0.157 cm²) in contact with 0.9% saline. From Geddes et al., *Med. Biol. Engng.* 1971. In press. By permission.)

little current as possible through the electrode-electrolyte interface so that a minimum of voltage loss and waveform distortion will be introduced by the electrode impedance. The practical method employed to avoid these distortions is to use an amplifier with an adequately high input impedance. This subject, discussed elsewhere in this chapter, verifies the concern reported by Schwan (1968):

Much work in biological research (for example, studies of membranes with transients), and clinical practice (for example, in cardiac pacemaker design), is conducted with microelectrodes where current densities are applied far in excess of the [above reported] limit of current linearity values [a few mA/cm²]. It appears, therefore, that electrodes often introduce nonlinear characteristics which are erroneously ascribed to the biological system under study.

To close this section it is perhaps useful to summarize the concepts discussed in order to present models for the circuit appearing between the ter-

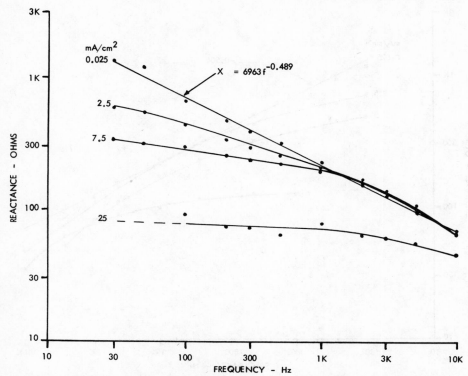

Fig. 1-8a. Dependence of the series-equivalent reactance X on frequency and current density; stainless steel electrode (0.157 cm²) in contact with 0.9% saline. (From Geddes et al., *Med. Biol. Engng.* 1971. In press. By permission.)

minals of a pair of electrodes placed in electrolytic contact with a subject and used to measure a bioelectric event. It will be recalled that an electrode-electrolyte interface possesses a half-cell potential and can also be equated to a series capacitance and resistance. When operating under low current density conditions, the equivalent resistance and capacitance vary approximately inversely as the square root of frequency; moreover the reactance (X) of the capacitance is approximately equal to the resistance (R). Figure 1-10a illustrates two such electrodes (1, 2) placed on a subject to measure a bioelectric event, and Fig. 1-10b presents an approximate equivalent circuit. "Looking" into the terminals (1, 2), three potentials are "seen"; one is due to the bioelectric event (E) and the other two are nonphysiologic and represent the half-cell potentials (E_1, E_2) of the electrodes. Recalling that $X = R$ and that both vary inversely with frequency, the high-frequency impedance measured between terminals 1 and 2 will be that of the subject (R_b). As the frequency is decreased, the

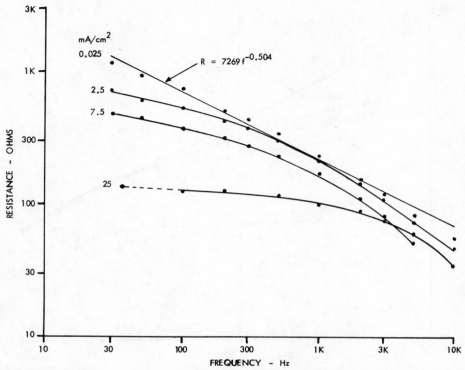

Fig. 1-8b. Dependence of the series-equivalent resistance R on frequency and current density; stainless steel electrode (0.157 cm²) in contact with 0.9% saline. (From Geddes et al., *Med. Biol. Engng.* 1971. In press. By permission.)

reactances of C_1, and C_2 and the resistances of R_1 and R_2 increase, and hence the impedance becomes higher and higher. At zero frequency (i.e., dc) this equivalent circuit would exhibit an infinite impedance. Because it is possible to pass direct current through an electrode-electrolyte interface, the simple series-equivalent circuit cannot be considered valid at very low frequencies. It is possible to convert the series equivalent to a parallel one (see p. 17), however, and therefore the circuit can be rearranged as in Fig. 1-10c. Although conceptually more pleasing, this circuit is no more or less accurate than that in Fig. 1-10b; however, it permits assigning a finite low-frequency limit on the magnitude of the resistances (rather than an infinite value for the series capacitances) and thereby allows for the condition of a finite dc resistance; supportive evidence for this choice is found in Experiment 4. In order to indicate a finite dc resistance for an electrode-electrolyte interface, electrochemists place a resistance called the faradic leakage (which accounts for electrolytic processes) across the double-layer capacitance (see Grahame, 1952). Under these conditions, the impedance measured between terminals 1 and 2 would be high in the low-

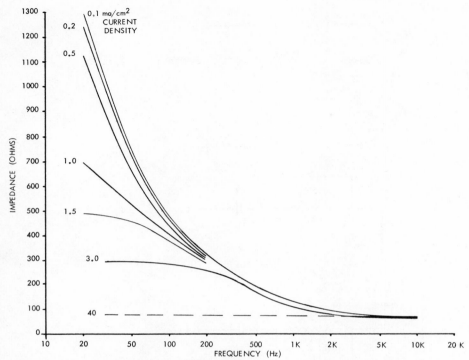

Fig. 1-9. The impedance-frequency characteristics measured between a silver ball (1.35 mm diam.) and a silver disk (3 cm diam.) in 0.9% saline using different current densities. The resistance of the intervening saline is 67 Ω.

frequency region; it would decrease with increasing frequency (because of the manner in which R and X decrease). It must be recognized that aggregates of cells also possess capacitive and resistive properties, as pointed out by K. S. Cole in the first Symposium on Quantitative Biology in 1933. Therefore, because of the combined nature of the electrode — electrolyte interface and tissue properties, the impedance — frequency characteristics of electrode systems exhibit a decrease in impedance with increasing frequency, becoming asymptotic to a value that represents the bulk conducting properties of all of the intervening electrolytes.

The Impedance of Chlorided Silver Electrodes

In a previous section in this chapter it was pointed out that one of the advantages of chloriding a silver electrode is stabilization of the electrode potential. The application of a deposit of silver chloride alters another parameter, the impedance between the electrode and the electrolyte it contacts. The magnitude of the change in impedance depends on the thickness of the chloride deposit. There is, however, substantial variation in the amount of

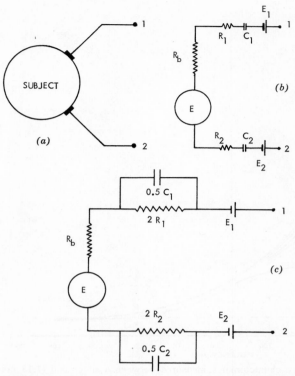

Fig. 1-10. Approximate equivalent circuits for a pair of electrodes arranged to measure a bio-electric event.

deposit considered adequate. This situation probably arises because there are two distinct groups of users of silver-silver chloride electrodes. The electrochemists use them as a reference electrode and require a stable electrode potential; the other group—electro-physiologists—demand a stable electrode potential and a low electrode-electrolyte impedance. It is doubtful that both requirements are fulfilled by the same chloride deposit.

Faraday's law of electrolysis specifies the weight of a material that will be deposited on an electrode in contact with an electrolyte; the important factors are current and time. In fact, the unit of current, the ampere, was first quantified by Faraday's law; the ampere was defined as that steady current which, when passed through a solution of silver nitrate (under conditions in which only silver is deposited), deposits silver at the rate of 1.11800 mg/sec. Thus the unit of current was established by measurement of weight and time.

The unit quantity of electricity is the coulomb (coul), which is one ampere-second (A-sec). From careful measurement of electrodepositions in various cells, it has been found (cf. MacInnes' 1961 review) that a certain number of coulombs of electricity is required to deposit one chemical equivalent (the gram atomic weight divided by the valence) of a substance at a metal-electrolyte boundary. The quantity of electricity is 96,497 coul (i.e., ampere-seconds); in practice the figure used is 96,500 coul, which is defined as one faraday (F). This means that one faraday will deposit the gram atomic weight of a monovalent element; half that weight will be deposited if the element is divalent. It is to be noted that although Faraday's law specifies the amount of deposit, it says nothing about which ions in an electrolyte will take part in an electrochemical reaction in which a current enters or leaves an electrode.

From Faraday's law it can be seen that it is appropriate to describe the amount of chloride deposit on a silver electrode by specifying the magnitude of the current and the time it flows. With this criterion in mind it is informative to examine the amount of chloride deposited by various investigators who reported the production of satisfactory chlorided silver electrodes. In order to make the data comparable, the deposit (mA-sec), has been converted to the deposit per square centimeter of electrode area. Table 1-5 presents these data along with the density of current (mA/cm² of electrode area) that was used in the chloriding process.

From Table 1-5 it can be seen that chloride deposits ranging from 48 to 168,000 mA-sec/cm² have been employed. In order to discover the effect of depositing chloride on silver, the author [with Baker (1967)] compared the impedance-frequency characteristic of bare silver electrodes in saline with that obtained after electrolytic deposition of silver chloride (about 500 mA-sec/cm²). Figure 1-11*a* illustrates the impedance-frequency characteristic of the bare electrodes; Fig. 1-11*b* shows the same relationship for the chlorided electrodes. Clearly evident is a remarkable decrease in the low-frequency impedance and a slight increase in the high-frequency impedance. The relative constancy with increasing frequency of the impedance-frequency characteristic of the chlorided electrodes indicates that the impedance of the electrode-electrolyte interface has become lower and the resistance of the electrolyte now dominates the circuit.

The change in electrode-electrolyte impedance accompanying chloriding silver electrodes was studied quantitatively by the author with Baker (1969). The impedance-frequency characteristics of square and circular silver electrodes of differing areas (225, 100, 78.5, 49, 38.5, 25, 19.6, and 3.14 mm²) were measured bare and with different amounts of chloriding. Each electrode was mounted on a slide that could be fitted into a slot in a rectangular plastic box filled with 0.9% saline solution. A large, heavily chlorided silver electrode (25 × 25 mm) located at the other end of the box was used as the reference

Table 1–5 Parameters Employed for Chloriding Silver Electrodes

Chloride Deposit (mA-sec/cm²)	Chloriding Density (mA/cm²)	Investigator and Year
500–1400	1.7–2.3	Mac Innes and Parker (1915)
4000–5600	3.3–4.7	Mac Innes and Beattie (1920)
3600	1.0	Carmody (1929)
18,000–21,600	0.2	Afanasiev (1930)
8500	2.4	Carmody (1932)
4800	2.7	Brown (1934)
28,800–43,200	8–12	Smith (1938)
18,000–168,000	5.3–31	O'Connell (1960)
432–43,200	0.01–1.0	Bures (1962)
7800	8.7	Goldstein (1962)
48–108	0.4–0.6	Grayson (1962)
300–500	2.5	Cooper (1963)
3600	3	Day and Lippitt (1964)
600	1.0	Skov (1965)
1000–2000	—	Cole (1967)
—	0.4–10	Janz (1968)

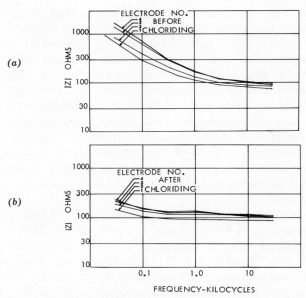

Fig. 1-11. Impedance-frequency curves for bare and chlorided silver electrodes. (From L. A. Geddes and L. E. Baker, *Med. Biol. Engng.* 1969. **7**:49–56. By permission)

electrode for each impedance-frequency measurement. The measurements were carried out over the frequency range of 10 Hz to 10 KHz using the same current density (0.1 mA/cm²) for each electrode.

The procedure for investigating the electrode characteristics consisted of first measuring the impedance-frequency characteristic of the bare silver electrode, which had been polished with emery cloth and washed clean of the abrasive material. Then a known amount of chloride was deposited, and the impedance-frequency curve was measured. An additional amount of chloride was then deposited and the impedance-frequency curve was measured. The procedure was repeated until a heavy coat of chloride had been deposited. Chloriding was carried out in room light by using a constant-current circuit consisting of a 50-V dc power supply in series with a resistance box and milliammeter. With this arrangement, the chloriding current could be preset to the desired value and then connected to the chloriding bath, which consisted of a rectangular plastic box containing 0.9% saline and a large (25 × 25 mm) bare silver cathode. Fresh saline solution was used to chloride each electrode.

Figure 1-12 presents a typical family of curves obtained from a single electrode. The impedance-frequency curve of the bare silver electrode (curve *A*), exhibited capacitive reactance (i.e., the impedance decreased markedly with an increase in frequency). Even with a small chloride deposit, there was a reduction in low-frequency impedance (curve *B*). As chloriding was increased, there was a further reduction in the low-frequency impedance (curves *C*, *D*) and the behavior of the electrode-saline-electrode circuit became much more resistive as shown by the impedance-frequency curve, which became nearly parallel to the frequency axis. With continued chloriding, both the low- and high-frequency impedances increased (curve *E*). With additional chloriding, the impedance values at all frequencies were increased and the impedance-frequency curves still indicated a circuit that was mainly resistive. A continuation of chloriding further raised the impedance at all frequencies, and equal increments of chloriding resulted in parallel, equally spaced curves (curves *F–J*).

It is of some interest to speculate on the reason for the initial decrease, followed by an increase in the electrode-electrolyte impedance resulting from the continued deposition of silver chloride. Although there is still no accurate explanation, it is conceivable that the initial decrease in impedance results from an increase in area because of the manner in which electrolytic deposits are laid down. Although silver chloride is a relatively good insulator, having a resistivity in the range of 0.5×10^5 to 6×10^5 Ω-cm (according to Jaenicke, 1955), a very thin deposit may increase the electrode area more than it decreases its capacitance. With an increase in deposition, the area increase may be offset by the effect of an increase in thickness, which in turn would raise resistance

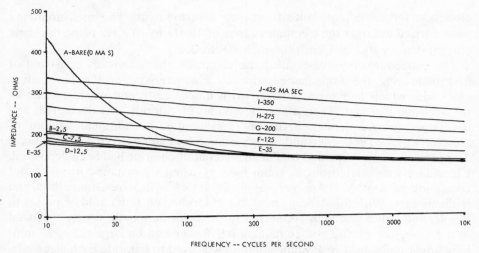

Fig. 1-12. Impedance-frequency curves with different amounts of chloride deposit (mA-sec); electrode area 0.25 cm². (From L. A. Geddes and L. E. Baker, *Med. Biol. Engng.* 1969, 7:49–56. By permission.)

and decrease capacitance; both situations would lead to an increase in impedance. That the deposition of silver chloride causes a considerable increase in area was demonstrated by Marmont (1949), who microscopically examined silver chloride deposits. He also showed that by converting the silver chloride deposit to silver, using photographic developer, the impedance of a silver chloride electrode was dramatically reduced. (see p. 35).

Figure 1-13 was composed to demonstrate that, irrespective of area, there is an optimum amount of chloriding for electrodes to obtain the lowest impedance. In this illustration the 10-Hz (Fig. 1-13*a*) and 10-KHz (Fig. 1-13*b*) impedances are plotted versus the normalized amount of chloriding (mA-sec/cm² electrode area), calculated by multiplying the actual chloriding milliampere-second value used for a particular electrode by 1/electrode area (cm²). In this manner, the changes produced by chloriding each electrode are more readily seen and compared. Figure 1-13 illustrates that, in general, chloriding to between 100 and 500 mA-sec/cm² provided the lowest values of electrode impedance. Prolongation of chloriding beyond 500 mA-sec/cm² increased the impedance of all electrodes at all frequencies. The figure is somewhat lower than that advocated by Cole (1962), who stated that the chloride deposit for minimum impedance is 2000 mA-sec/cm².

Another series of tests was conducted to identify the limits of interchangeability of current and time to achieve a desired chloride deposit. Using the reduction in low-frequency impedance as an indicator of the attainment of

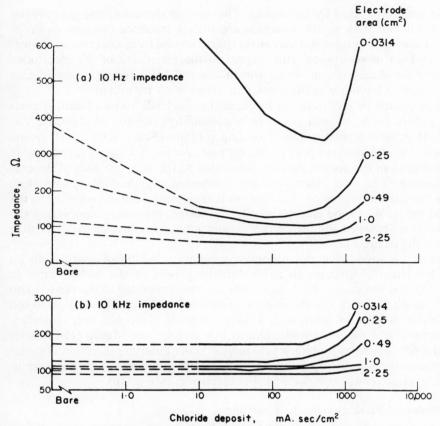

Fig. 1-13. The dependence of the impedance of a silver electrode on the chloride deposit. (*a*) Note that the lowest impedance was obtained with a deposit of about 500 mA-sec/cm². (From L. A. Geddes and L. E. Baker, *Med. Biol. Engng.* 1969, 7:49–56. By permission.)

the desired chloride deposit, it was found that if chloriding was carried out with a current density of 5 or more mA/cm², milliamperes and seconds could be manipulated for convenience to attain the desired chloride deposit. This value is in fairly good agreement with the current densities used by many of the investigators listed in Table 1-5.

The study reported by the author and that reported by Cole indicate that a chloride deposit in the range of 100 to 2000 mA/cm² on a silver electrode achieves the lowest electrode-electrolyte impedance. To attain this deposit, by convenient choice of current and time, a chloriding current density in excess of 5 mA/cm² should be used. Because the impedance of an electrode-electrolyte is altered by passage of current, it is recommended that a constant-current source be used for chloriding. Thus during the chloriding process the current will not vary as the electrode-electrolyte interface changes its properties. A constant-current source can be approximated by connecting a resistor, which is high in resistance with respect to the resistance of the electrodes and the chloriding bath, in series with a controllable voltage source (e.g., a power supply) having a milliammeter in series with its output.

The first step in chloriding is to clean the electrode to be chlorided carefully; failure to do so will result in a nonuniform deposit of chloride. The electrode is then placed in the chloriding solution (NaCl, KCl, or HCl) and connected to the positive pole of the current source. A large silver electrode is also placed in the solution and is connected to the negative pole of the current source. The two electrodes are connected together (short circuited) before the current is applied. The current is then turned on and adjusted to the desired value. When the chloriding is to be started, the connection across the electrodes is removed and the current is allowed to pass through the chloriding bath for the desired time.

It must be added that the literature contains a wealth of useful data on chlorided silver electrodes. In particular, the papers by the authors cited in Table 1-5 and the review by Janz (1961) are recommended to the reader who desires to delve more deeply into this important subject. In addition to the electrolytic method of fabricating a silver chloride electrode, it is important to recognize two other methods. One is due to Burr and Mauro (1948), who coated a silver wire by dipping it into melted silver chloride; the other consists of compressing a mixture of silver and silver chloride under high pressure to form a pure silver-silver chloride pellet electrode, (see p. 86).

The Platinized Platinum (Black) Electrode

In electrochemistry, and to some degree in electrophysiology, the platinized platinum or platinum-black electrode is employed. This electrode was introduced by Kohlrausch in the late 1890s for the measurement of the conductivity of electrolytic solutions; his work is well summarized in his book (1898). Since Kohlrausch's time, the platinum-black electrode has been used with gaseous hydrogen as the standard reference electrode; it is still the most practical and accurate electrode for electrolytic conductivity cells because the electrode-electrolyte impedance is reduced to a very low value by the platinization process.

Although it was realized at the time that Kohlrausch's method of electrolytically depositing a finely divided deposit of platinum on a bare platinum electrode dramatically reduced the electrode-electrolyte impedance, probably by virtue of the considerable area of electrode in intimate contact with the electrolyte, it was more than a quarter-century later that Jones and Bollinger (1935) conducted studies to quantify the effect of depositing different amounts of platinum-black on platinum electrodes. After each deposition they measured the impedance-frequency characteristics in a frequency range extending from 500 to 3070 Hz. They found that a bare (bright) platinum electrode exhibited a decreasing impedance with increasing frequency, which is characteristic of the capacitance of the electrode-electrolyte interface. Even after a thin deposit of platinum-black (424 mA-sec/cm^2) the electrode-electrolyte impedance dropped 67-fold, although the deposit was barely visible. Additional deposition caused a steady but much smaller decrease in impedance; with a deposit of 5950 mA-sec/cm^2, the impedance had dropped 856-fold. Further deposition did not reduce the impedance proportionally. In fact, Jones and Bollinger noted a slight increase in impedance when the deposit was doubled.

From these data it would appear that, in round numbers, 500 mA-sec/cm^2 constitutes a light platinum-black deposit and 6000 mA-sec/cm^2 is perhaps a more desirable value. In an earlier paper, Jones and Bollinger (1931) reported that a light deposit was 750 mA-sec/cm^2 and heavy deposit was 73,500 mA-sec/cm^2. Cole (1962) found that a deposit of 10,000 mA-sec/cm^2 provided an electrode impedance of 1 Ω (at 1 kHz) for a 1-cm^2 electrode. Schwan (1968) found that the optimum deposit was in the range of 30,000 mA-sec/cm^2; his studies are discussed subsequently.

The value of platinizing platinum-iridium electrodes to reduce their resistance and reactance for depth recording with conventional amplifiers was described by Ray et al. (1965). Figure 3-5 illustrates the remarkable reduction in resistance and reactance obtained with this process.

The method advocated by Kohlrausch is still used to produce a platinum-black deposit on a platinum electrode. The electrode is first cleaned thoroughly and placed in a solution of 0.025N hydrochloric acid containing 3% platinum chloride and 0.025% lead acetate. Such solutions are available from many laboratory supply houses. Failure to include the lead acetate yields a flaky gray deposit; inclusion of the lead acetate results in a stronger deposit having a velvety black appearance. A direct current is passed through the solution via a large-area platinum anode, and the electrode to be blackened is made the cathode. There is no unanimity on the current density to be employed; Kohlrausch employed 30 mA/cm^2. Jones and Bollinger used 6 mA/cm^2 for a light deposit and 27 mA/cm^2 for a heavy deposit. Schwan (1963) carried out important studies which indicated that there is an optimum current density to obtain the lowest electrode-electrolyte impedance. He studied the change in the series-equivalent electrode-electrolyte capacitance with different platinizing current densities and different amounts of platinum deposit, which he expressed in coulombs (i.e., ampere-seconds) per square centimeter of electrode surface area. His data, which appear in Fig. 1-14, illustrate that sandblasting is the best method of cleaning a platinum electrode and that the maximum capacitance, which is synonymous with the lowest electrode-electrolyte

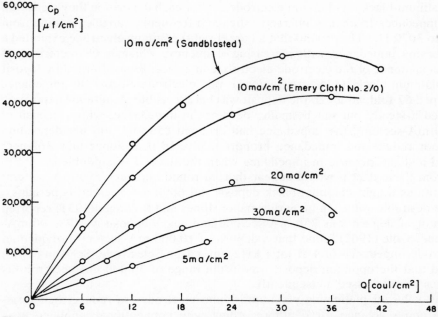

Fig. 1-14. The increase in the capacitive component of the impedance of a platinum electrode with platinization, (coulombs/cm² of electrode area). The effect of platinizing on different surfaces with different current densities is shown. From these data, measured at 20 Hz, the maximum capacitance (and hence the lowest impedance) was obtained by using a platinizing current of 10 mA/cm² and obtaining a deposit corresponding to about 33 A-sec. (From H. P. Schwan, in *Physical Techniques in Biological Research*, Vol. VI B, W. L. Nastuk, Ed., New York: Academic Press, 1964. By permission © Academic Press.)

impedance, occurs when a current density of 10 mA/cm² is used to platinize the electrode and the optimum deposit is 33 A-sec/cm². Jones and Bollinger (1935) obtained a maximum capacitance with 64.3 A-sec/cm² platinum black deposit. Schwan pointed out that his parameters apply for electrodes with an area greater than 1 mm²; smaller electrodes require the use of a higher current density for optimum platinizing. In summary, with the above-mentioned restriction reported by Schwan, it would appear that the optimum parameters for platinizing a platinum electrode are a current density of 10 mA/cm² of electrode area and a deposit of about 30 to 60 A-sec/cm² of electrode area.

The platinum-black electrode is now used frequently in electrophysiology For example, its low impedance makes it ideal for use in voltage-clamp studies (Cole, 1949; Moore and Cole, 1963). In addition, an interesting hybrid platinum-black electrode was employed by Cole and Kishimoto (1962). They stated that the platinized platinum electrode had a poorly defined potential and their goal was to obtain a low-impedance electrode with a stable electrode potential. Therefore they started with a silver electrode, which was given a heavy chloride deposit of 10,000 mA-sec/cm² (using a current density of 2 mA/cm²); the electrode resistance obtained was 300 Ω for a 1-cm² electrode. The chlorided silver electrode was then platinized with 1000 to 5000 mA-sec/cm² using a current density of 2mA/cm². The electrode was then chlorided lightly. This procedure resulted in an electrode with an equilibrium potential within a few millivolts of the original chlorided silver electrode and exhibited a 20-Hz impedance of a few ohms for one square centimeter of area. The electrode thus approached the best properties of both the chlorided silver electrode and the platinized platinum electrode. No doubt the properties of this electrode will be put to good use in the future.

The Silver-Black Electrode

To reduce the impedance of a silver electrode, a principle similar to that for platinizing was described by Marmont (1949). Since the surface of a chlorided silver wire, observed under a microscope, looks like an agglomerate of very fine particles with a surface area much in excess of the wire, Marmont decided to use photographic developer to remove the chloride from the silver. He reasoned that the area of the silver surface would be increased considerably and hence the electrode-electrolyte impedance would be reduced. To test the theory, he chlorided a pair of silver electrodes (2.7–7.5 A-sec/cm² at 1 mA/cm²) and measured their impedance-frequency characteristic in seawater. The silver chloride coatings were then converted to molecular silver by placing the electrodes in photographic developer for 3 minutes, and impedance-frequency measurements were repeated. The results (Fig. 1-15) clearly indicate the dramatic reduction in impedance at all frequencies, amounting as it does to between one-twentieth to one-sixtieth of the undeveloped (i.e., chlorided silver) value; the impedance measured was essentially that of the resistance of the fluid column in the measuring cell.

It is interesting to speculate on the possible reasons for the significant reduction in impedance. Marmont proposed an increase in the surface area by the deposition of finely divided silver; however, another possibility exists. Silver chloride has a high resistivity, amounting to between 0.5×10^5 and 6×10^5 Ω-cm (depending on the current density used for deposition as reported by Jaenicke et al., 1955); and removal of even a small amount by conversion to silver, which has a resistivity of 2×10^{-6} Ω-cm, would also favor a reduction in impedance. It is not known which of these two factors predominates. Despite the remarkable reduction in impedance produced by the developing method, no use has been made of the phenomenon; nor have attempts been made to optimize it. There is, however, no doubt that further investigation will be carried out.

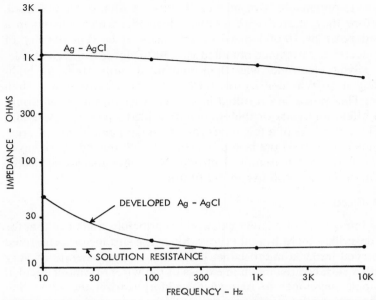

Fig. 1-15. Impedance-frequency characteristics of a pair of silver-silver chloride electrodes in seawater before and after treatment with photographic developer. (From data presented by G. Marmont, *J. Cell. Comp. Physiol.* 1949, **34:**351–382.)

The Calomel Electrode

Although the standard hydrogen electrode (SHE) is an easily constructed electrode, its use is not without difficulties. For example, the platinum-black deposit can be poisoned by impurities in the hydrogen gas; this hinders the establishment of equilibrium at the electrode-electrolyte interface. Sulfides, disulfides, and proteins in solution also alter the platinum-black surface and lead to error in potential measurement. In addition, the potential of the hydrogen electrode is dependent on pressure, with respect to both the depth of immersion of the platinum-black electrode and the barometric pressure. For these reasons it is often simpler to employ another type of reference electrode, and the most popular of these consists of a pool of mercury covered by a paste of mercurous chloride ($HgCl$; i.e., calomel) and mercury in contact with a solution of potassium chloride. Three concentrations of potassium chloride solution are in use: saturated, normal ($1.0N$), and decinormal ($0.1N$). Naturally the potential of the calomel electrode (with respect to the SHE) depends on the concentration of the potassium chloride and temperature; the values for the three types referred to are presented in Table 1-6.

Table 1–6   Standard Half-cell Potentials of Calomel Electrodes[a]

Temperature (°C)	Saturated[b]	1N (at 25°C)	0.1N (at 25°C)
0	0.2602	0.2854	0.3338
10	0.2541	0.2839	0.3343
15	0.2509	—	—
20	0.2477	0.2815	0.3340
25	0.2444	0.2801	0.3337
30	0.2411	0.2786	0.3332

[a] Data from G. J. Hills, and D. J. G. Ives, in *Reference Electrodes*, D. J. G. Ives, and G. J. Janz, New York: Academic Press, 1961. By permission © Academic Press.
[b] With respect to the SHE.

Three popular types of KCl calomel electrode are presented in Fig. 1-16. The sidearm type (Fig. 1-16a) is used in electrochemical determinations—potassium chloride solution completely fills the reservoir above the calomel and including the sidearm, which is the electrical connection to the solution from which measurements are to be made; the terminal T of the calomel electrode is a platinum wire dipping into mercury that is in electrical contact with the pool of mercury in the electrode. When potential measurements are to be made, the junction J is placed in the solution to be measured and the stopcock S on the sidearm is opened. When used with solutions having concentrations different from those used to fill the calomel electrode, diffusion of the filling solution will occur and if the mobilities of the ions are different, a junction potential will develop. To eliminate the error that would be caused by such a junction potential, the calomel electrode is placed in contact with a solution—the same solution that is used to fill the electrode. Electrical contact between this solution and the solution from which measurement is to be made is established by the use of a salt bridge (to be described), which consists of a saturated solution of potassium chloride in agar placed in a U-tube. Potassium chloride is usually chosen because the mobilities of the K^+ and Cl^- ions are nearly the same, hence both will diffuse at about the same rate, minimizing the junction potential (see p. 9).

Frequently the standard electrochemist's KCl-calomel electrode (Fig. 1-16a) is used in electrophysiology. Sometimes a wick soaked in an appropriate electrolyte and placed in the measuring junction is led to the preparation under measurement. Occasionally the measuring junction is coupled directly to a micropipet via a fluid-filled glass or plastic tube.

The KCl calomel electrodes in Fig. 1-16b are usually employed as reference electrodes in pH meters. The KCl solution in these electrodes communicates

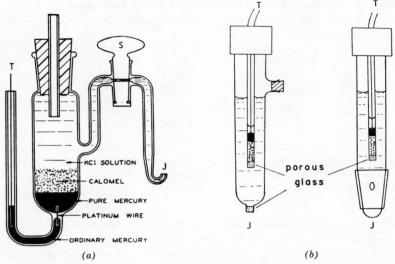

Fig. 1-16. Types of mercury-calomel-KCl reference electrodes: (*a*) sidearm type used in electrochemistry; (*b*) reference electrodes used with pH meters (J = measuring junction. T = electrode terminal. [(*a*) From *Instrumental Methods of Analysis*, H. H. Willard, L. L. Merritt, and J. A. Dean, Princeton, N. J.: D. Van Nostrand, Company, 1956. By permission)]. [(*b*) From *Reference Electrodes*, D. J. G. Ives and G. J. Janz, New York. Academic Press, 1961. By permission)].

with the solution to be measured through a porous glass membrane which serves as a salt bridge.

Although mercury-calomel-KCl electrodes are available commercially in a wide variety of forms, they are not difficult to construct. However, if a high degree of reproducibility is desired, special precautions must be taken regarding the purity of all the component substances and the enveloping glass container. Excellent reviews of the fabricating procedures were presented by Spiegler and Wyllie (1956) and Hills and Ives (1961). The latter investigators reported that a reproducibility to within 0.1 mV can be obtained with carefully prepared materials. They also stated that, although there have been many careful studies of the potential of the calomel electrode, its absolute potential "cannot be fixed beyond debate," and it is therefore not yet appropriate to base fundamental thermodynamic calculations on the published values for potential. However, Hills and Ives believe that values listed in Table 1-6 are the most accurate to date.

The Salt Bridge

A method for making a salt bridge was presented by Spiegler and Wyllie (1956); their procedure consists of mixing 35 g of potassium chloride and 3 g of agar in 100 ml of water, which is heated in a vessel placed in a water bath. Heating is continued until the agar and potassium chloride are completely dissolved. While this mixture is still hot, it is poured into the U-tube that is to be used as the salt bridge. After time has elapsed for cooling, one arm of the U-tube is placed in the potassium chloride solution of the calomel electrode and the other is placed in the solution to be measured.

Amplifier Input Impedance Considerations

When a bioelectric event is measured with surface, subintegumental, or intracellular electrodes, the quantity to be measured is potential. This means that the measuring device, which is calibrated in volts, must not draw current from the source of the biopotential. It is therefore necessary to consider the conditions that must be fulfilled to attain this goal. Because both intracellular and extracellular measurement techniques are employed, the conditions that apply to each situation must be discussed separately. The following paragraphs deal with the conditions for accurate measurement of bioelectric events with extracellular electrodes; the special conditions pertaining to the intracellular electrode technique are covered in Chapter 4.

When a pair of electrodes is placed to encompass a group of irritable cells, a potential difference can be measured between them when the cells become active and recover. As stated previously, the potential between the electrode terminals will be representative of the potential presented to the electrodes only if the potential-measuring device draws no current; this means that its input impedance must be infinite. In practice, a small amount of current is drawn from the source of the biopotential. In order to draw the least amount of current, the input impedance of the voltage-measuring instrument must be many times higher than that appearing between the electrode terminals. The area of the electrodes and the nature of the tissues and fluids comprising the bioelectric generator determine the impedance measured between the electrode terminals; therefore, it is worthwhile to examine the equivalent circuit in order to discover how to specify input characteristics for the potential-measuring device.

Consider a pair of electrodes placed on tissue containing an aggregate of irritable cells. When the cells become active and recover, they will send current through the volume conductor that constitutes their environment. The electrodes are always placed to measure as much of the voltage drop produced by the environmental current as possible (Chapter 6). However, because extracellular electrodes are both large and distant (in terms of cell dimensions), only a fraction of the available potential is sensed; this situation is schematically depicted in Fig. 1-17a. In this circuit, R and r are resistors representing the equivalent resistance of the bioelectric generator and the tissue surrounding it. Because r is small with respect to R only a fraction of the available potential

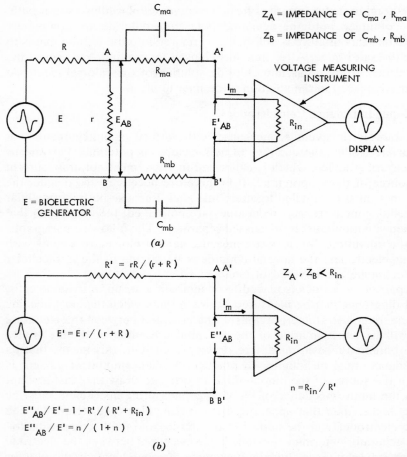

Fig. 1-17. Approximate equivalent circuits: (*a*) electrodes on a subject; (*b*) simplified equivalent circuit.

E is detected by distant electrodes. Thus, neglecting electrode potentials, the potential E_{AB} presented to the electrodes C_{ma}, R_{ma} and C_{mb}, R_{mb} is the voltage that appears across r. Because the voltage-measuring instrument has a finite input resistance R_{in}, a small current I_m will flow through it. Therefore, the potential (E'_{AB}) appearing across the electrode terminals A′B′ will be less than that presented to the electrodes in the absence of the measuring instrument. The voltage E'_{AB} seen by the measuring instrument is

$$E'_{AB} = E_{AB} - I_m \ (Z_A + Z_B)$$

where Z_A and Z_B are the impedances of electrodes A and B.

It can readily be seen that the voltage across the measuring instrument E'_{AB} will be equal to the voltage presented to the electrodes E_{AB} if I_m is zero. This situation can be approached if R_{in} the input resistance—strictly speaking, impedance—of the measuring device is high with respect to $r + Z_A + Z_B$; the goal is therefore to make the ratio $R_{in}/(r + Z_A + Z_B)$ as high as possible. To shed light on how high this ratio should be, the circuit of Fig. 1-17a can be simplified to a series-equivalent circuit consisting of an equivalent voltage and resistance. This transformation can be carried out by the use of Thévenin's theorem. To apply it, it is only necessary to consider E, R, r to be a generator with terminals A, B which sends current into a load consisting of $Z_A + R_{in} + Z_B$. Thevenin's theorem states that the voltage of the equivalent generator is found by removing the load (in this case $Z_A + R_{in} + Z_B$) and measuring the voltage across the terminals AB; this voltage is $Er/(R + r)$. To find the impedance of the equivalent generator, the voltage sources (in this case, E) are short-circuited and the impedance between the terminal AB is measured; this impedance is $rR/(r + R)$. Thus the circuit of Fig. 1-17a can be represented by the series equivalent in Fig. 1-17b. The voltage to be measured E_{AB} is separated from the measuring points AB by a resistance R'. If the impedance of the electrodes Z_A, Z_B are considered to be low* with respect to the resistance R_{in} of the voltage-measuring instrument, the voltage indicated by it E''_{AB} (as connected in Fig. 1-17b) is easily calculated; $E''_{AB} = E' - I_m R'$. The current I_m that flows is the voltage E' divided by the resistance of the circuit $R' + R_{in}$; therefore, $I_m = E'/(R' + R_{in})$. Substituting for I_m in the equation for voltage E''_{AB}, the following relation is obtained:

$$E''_{AB} = E'\left(1 - \frac{R'}{R' + R_{in}}\right)$$

Of central interest is the ratio of E''_{AB} to E' (i.e., the ratio of what is measured to what is available). Dividing the foregoing expression by E' gives

$$\frac{E''_{AB}}{E'} = 1 - \frac{R'}{R' + R_{in}}$$

Ideally, the ratio E''_{AB}/E' should be as close to 1.0 as possible. The actual value depends on the relation between R_{in} and R', the ratio of the resistance of the measuring device R_{in} to the resistance of the generator R'. If this ratio is given the symbol η (i.e., $\eta = R_{in}/R'$) and substituted in the preceding expression, the following equation is obtained,

$$\frac{E''_{AB}}{E'} = \frac{\eta}{\eta + 1}$$

It is illuminating to plot the ratio of the measured to available voltage (i.e., E''_{AB}/E') versus η, the ratio of the resistance of the measuring instrument to

* The consequences of not fulfilling this requirement are discussed in Chapter 3, p. 107.

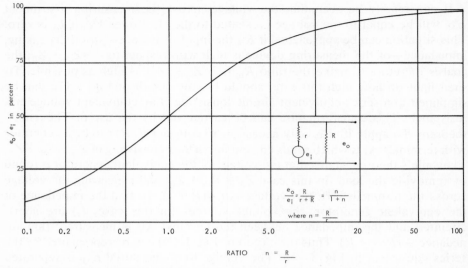

Fig. 1-18. The effect of increasing the ratio n of the input impedance of the measuring instrument to that of the generator on the measured voltage.

that of the generator. Figure 1-18 presents these data and shows that when the ratio is 100/1, the measured voltage is 0.99 (i.e., 99% of the available voltage). When the ratio is 1000/1, the measured voltage is 99.9% of that available.

From this perhaps oversimplified analysis, two important conclusions can be drawn. First, in order to measure the maximum amount of the potential available, the resistance of the voltage-measuring device should be as high as possible with respect to the resistances of the source of the voltage; (this statement can be further generalized by substituting the word impedance for resistance). Second, if the resistance of the measuring device is lowered, the voltage indicated will decrease. Connecting successively lower values of resistance across a generator is designated "loading the generator," and a plot of voltage versus load resistance constitutes a load curve. It is important to note that when a load is placed across the generator terminals so that the open-circuit (i.e., unloaded or $R_{in} = \infty$) voltage drops to one-half, the resistance of the load is equal to the resistance of the generator. This technique of loading a generator to reveal its loss of output voltage is frequently performed to determine its output resistance and impedance; the technique also permits calculation of the open-circuit voltage if it cannot be measured.

To establish the important facts just presented it is necessary to recall that the following simplifying assumptions were made: the bioelectric generator was surrounded by a homogeneous resistive volume conductor, and the impedance of the electrodes was small with respect to the resistance of the generator and the voltage-measuring instrument. These assumptions are reasonably valid for the argument presented. However, when it is necessary to establish the relation between electrode impedance and the input impedance of the voltage-measuring instrument, it should be obvious that causing appreciable current to flow through the electrode impedances (by the use of a voltage-measuring instrument that does not have a high input impedance) will result in appreciable voltage drops across the electrode impedances. These voltage drops will not only affect the various sinusoidal frequency components of the bioelectric event differently, they will also introduce time displacements between them (i.e., phase distortion).

It should therefore be apparent that, in the measurement of a bioelectric event, it is essential to avoid creating potential drops across the electrode impedances. This condition is achieved by making the input impedance as high as possible with respect to the sum of the impedance of the bioelectric generator (R' in Fig. 1-17) plus those of the two electrodes Z_A, Z_B. Because electrode impedance is primarily related to electrode area in an inverse manner, the range of magnitudes encountered are presented when the various electrode types are discussed. In many instances loading tests have been performed to indicate the order of magnitude of electrode-bioelectric generator impedance and thereby enable specification of amplifier input impedance.

The total electrode-subject impedance dictates the magnitude of the amplifier input impedance required to detect bioelectric events, and the previous discussion is focused on this rule. Although it is relatively easy to obtain amplifiers having high input impedances, there is a practical advantage to be gained by making the electrode-subject impedance as low as possible. In nearly all recording situations with ground-referred (i.e., power-line-operated) recording instruments, there is a considerable amount of environmental power-line interference available which can enter the amplifying system along with the desired signal. The amount appearing at the output of the amplifying system depends on the amount of interference present and the type of amplifier (e.g., single-sided or differential), but the attainment of a low electrode-subject impedance (consistent with the constraints imposed by the particular measurement situation) will favor rejection of interference when bioelectric signals are measured.

2

Surface Electrodes

When an electrode is placed on the integument to measure a bioelectric event, it is desired to detect the potential variations that arise due to the current that flows through the body tissue and electrolytes as a result of the potential changes that are produced by the bioelectric generator (i.e., the excursions in the potential of its membranes). Because most bioelectric recording instruments are voltage recorders with a finite (rather than infinite) input impedance, it becomes necessary to establish adequate ohmic contact between them and the conducting tissues and fluids of the body through which the bioelectric currents flow. For this reason it is of value to know the impedance of the electrodes employed. Because the impedance (but not the electrode potential) is inversely related to the area of the electrode-electrolyte-subject junction, a practical means for comparing electrodes is on the basis of geometric area*— and therefore an important characteristic of a pair of electrodes is the impedance measured between the terminals. When large-area electrodes are used, it is customary to measure the dc resistance between the electrode terminals. However, such a resistance value does not by itself describe the electrical circuit constituted by the electrodes and the biological material. To describe the circuit adequately, it becomes necessary to know the resistive and reactive components at all frequencies.

The order in which electrodes are described in this book is based on area: the electrodes with the largest areas (i.e., skin-surface and subintegumental electrodes) are described first, intracellular electrodes last.

The Skin

To better understand the nature of the circuit when an electrode is placed on the surface of the skin, it is informative to examine its anatomical structure. The following brief description is presented as an aid to understanding how

---

* The area referred to here is calculated from the physical dimensions of the electrodes and not the effective area, which is usually much larger because of surface roughness.

electrodes are in electrical communication with the tissues and fluids that surround bioelectric generators.

The skin, or integument, is the important boundary between the body and its environment. It protects the underlying parts from invasion and injury and at the same time plays an important role in sensation. It also permits the outward selective passage of substances, and in man and some animals the skin plays a major part in temperature regulation. Although the skin of different regions varies, there are basically two types of human skin, hairless and hairy. The former is found, for example, on the palms and soles; the latter is found over the rest of the body. There are only minor structural differences between the two.

Histologically the skin is divided into three layers; the outermost is the epidermis, and below it is the dermis or corium, which rests on subcutaneous tissue. The epidermis and the dermis are in turn characterized by cells, tissues, and organs that endow the skin with its properties. Figure 2-1 illustrates a section of skin of the face.

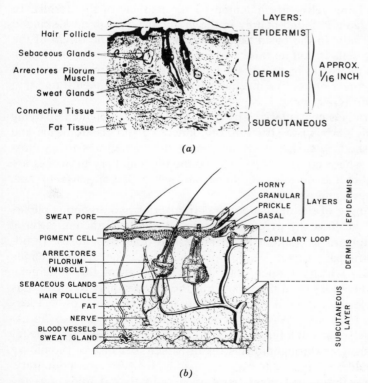

Fig. 2-1. Structure of facial skin. (From G. L. Sauer, *Teen Skin*. Springfield, Ill.: Courtesy of Charles C. Thomas, Publisher, 1965. 57 pp. By permission.)

About three-quarters of the outermost layer of the epidermis, the stratum corneum, consists of a sheet of cells; the outer horny layers are dead and are continually cast off. Allen (1965) reported that the average person sheds about 6 to 14 g of dead cells each day. These cells are, of course, replaced from below. The thickness of the stratum corneum varies considerably over the surface of the body. On the forehead it varies between 0.02 and 0.04 mm in thickness; on the palm and sole it is 0.4 to 0.7 mm. There are regional permeability differences in the stratum corneum to water.

Below the stratum corneum, in the skin of the palm and sole is the stratum lucidum, a clear zone consisting of several rows of flattened nonnucleated cells amounting to about 10 μ in thickness. The time taken for the cells of the stratum corneum and the stratum lucidum to be shed is about two days. Immediately below the stratum lucidum is a layer composed of granular, pickle, and basal (germinative) cells. From two to five layers of spindle-shaped cells arranged with their long axes parallel to the surface of the skin constitute the granular layer. The cytoplasm of these cells contains pigment (melanin) granules. The prickle and basal cell layers—the latter believed to be the main source of the overlying cells—are indented by the papillae of the dermis. In the white race, melanin is found mainly in the granular layer; in dark-skinned races, most of the cells in the basal layer contain melanin. The origin of the melanin-producing cells is still an unsettled matter according to Allen (1965). Exposure to ultraviolet light increases the amount of melanin in the basal cell layer; this response is a protective mechanism because it has been found that melanin absorbs ultraviolet light.

While the epidermis is characterized by dead cells on top of pigmented dying and living cells, the dermis is a structure of connective and elastic tissue and living cells, nourished by a rich vascular supply that penetrates to its most superficial layers, where numerous papillae containing capillary networks are located. Also within the dermis are numerous nerve fibers, sense organs, smooth muscle fibers, and glands.

The vascular supply of the dermis consists of arteries and a venous plexus that is located in the lowest part of the dermis. A richly branching arterial network springs upward to nourish the papillae and other structures. In addition to this return path to the venous plexus, there is an arterial-venous shunt (anastomosis) which, if open, can divert blood from the superficial layers and reduce heat loss from the skin. Thus the blood supply to the dermis serves a double role—nutrition of the cells within it and regulation of heat loss from the surface of the skin. Associated with the vascular system is an extensive capillary network and a lymphatic drainage system. There is a transfer of water and electrolytes throughout this region which, because of the nature of these fluids, would be expected to be a region of high electrical conductivity.

In hairy skin, the basal cell layer (stratum germinativum) folds into the dermis to form the hair follicles. At the bottom of the hair follicle is a bulbous cellular matrix from which the hair grows; the bulb is nourished by a vascular papilla. Near the root of the hair, the shaft is continuous with the follicle wall. The hair shaft is composed of keratinized cells. Opening into the hair follicle is a duct leading to a sebaceous gland composed of a mass of cells whose central axes are filled with vacuoles. The secretion of a sebaceous gland is accompanied by breaking down of the central cells, and the contents of these cells is poured into the follicle and passes up the hair on to the surface of the skin as an oily protective deposit. When the hair is made to stand more erect by contraction of the piloerector muscles, it is believed that a small quantity of sebum is expelled; both processes are believed to be heat conserving.

Located deep within the dermis are the eccrine sweat glands, which deliver their secretion (mainly a dilute saline solution) via a duct to a pore on the surface of the skin. Sweat glands are distributed over most of the surface of the human body; they are simple, convoluted tubules composed of cells containing secretory granules. The ducts are lined with a double layer of cells; one layer which is longitudinal, is believed to be muscular and functions to expel the sweat.

Sweat is the most dilute of all animal fluids. It is about 99% water, but the remaining 1% contains a rich variety of other substances. The composition of human sweat was described by Whitehouse (1935) and his data are presented in Table 2-1. For the purpose of establishing electrical contact with the skin, sweat can be considered as a weak saline solution with a concentration varying between 0.1 and 0.6%, averaging 0.3%.

Table 2-1   Composition of Human Sweat[a]

Substance		Amount (g%)
Water		99.742–99.221
Solids		0.258– 0.779
Organic solids		0.030– 0.290
Ash		0.144– 0.566
Chlorine (usually over 0.15 per cent)		0.059– 0.346
Lactic acid	about	0.070
Sulfate	"	0.004
Sodium	"	0.150
Potassium	"	0.017
Urea	"	0.030
Sugar	"	0.004

[a] From Starling's *Human Physiology*, 13th ed., Dawson and Eggleton, eds., Philadelphia: Lea and Febiger, 1962, 1579 pp. By permission.

In the axilla, groin, genital, and anal areas, in close relation to the hair follicles, are apocrine glands which secrete a viscid, turbid sweat. Apocrine sweat is odorless, but when it reaches the surface of the skin where it is attacked by flora, it develops a characteristic odor. Some believe that herd and race recognition can be made on the basis of this odor.

Thus the surface of the skin, on which an electrode is placed, is lubricated by an oily secretion from the sebaceous glands and a weak saline solution from the eccrine sweat glands. These secretions rest on a layer of dead cells, which give unprepared skin its high resistance. It is the task of the electrolytic compound applied to an electrode to provide an ionic path through the epidermis to the electrolytic layers in the dermis where the bioelectric currents establish potential fields.

Because the outer layer of the skin, the stratum corneum, is fairly dry and is shed at regular intervals, it hinders the development of colonies of microorganisms. However, it should not be concluded that the skin is devoid of such colonies; on the contrary even so-called clean skin is well inhabited by a rich variety of living organisms, and some of these are potential pathogens. Furthermore, the secretions that exist on the skin—forming a dilute solution of saline, lipid exudates, carbohydrates, and nitrogenous material—are all warm and constitute a good culture medium if accumulation is allowed. In relation to the application of electrodes covering certain areas of the skin, it is easy to see that although short-term application may not affect the local ecology, long-term application may, and serious thought should be given to this problem under such circumstances. In addition, if the skin is damaged, potentially pathogenic organisms can gain entry to the body and produce their own type of response.

Figure 2-2 illustrates some of the types of microorganisms found on human skin. Basically, the inhabitants consist of fungi, yeasts, bacteria, and follicle mites. Of all these, bacteria (both gram positive and negative) are the most numerous. It is difficult to present a typical census of skin inhabitants because of large regional variations; nonetheless inventories have been made. Marples (1969) reported that the adult male axilla has the highest population density, with 2.41 million aerobic bacteria per square centimeter of skin surface. Only slightly behind is the scalp with 1.46 million/cm²; the forehead has 0.2 million/cm². The back and the forearm have the lowest population density with 314 and 105 to 4500/cm², respectively; in regions of sebaceous glands, the density of anaerobic microorganisms may be ten times as great. Many of the organisms are harmless, but a few are potential pathogens, particularly if the skin is damaged.

From this survey of skin inhabitants, it should be apparent that there is a ready supply of invaders; most are innocuous, but some are dangerous. The application of a surface electrode closes off an area of skin and, if the time of

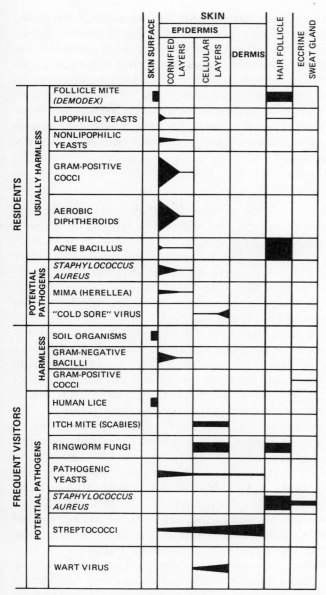

Fig. 2-2. Inhabitants of the human skin. (From M. J. Marples, Life on the human skin. *Sci. Amer.* 1969, **220**:108–115. By permission.) © Copyright 1969 by Scientific American, Inc. All rights reserved.

closure is long, favorable conditions can develop for the proliferation of the residents. Whether an important skin response occurs will depend on the situation created. Experience has shown that the short-term application of electrodes for routine clinical measurement of bioelectric events is a safe procedure. Not enough experience has been accumulated to indicate the optimum situation for the safe, prolonged application of electrodes. Conceivably, electrode-electrolytes for such a situation may require the incorporation of agents that simultaneously create an unfavorable environment for the proliferation of microorganisms and permit establishment of an adequate electrical contact with the subject.

Electrode Electrolytes

When metallic electrodes are placed on the skin, an electrolytic compound is used to establish an ohmic communication with the underlying body fluids. The electrolyte is often contained in a paste, jelly, or some other vehicle. Such electrolytic compounds were developed in the early string-galvanometer days of clinical electrocardiography, when there was a need to establish a low-resistance communication between the subject and the string galvanometer. This communication was provided by immersion electrodes, which required that the subject be seated, with both hands and both feet in large saline-filled containers. Experience with these electrodes indicated that the electrocardiograph (ECG) was distorted if the electrode resistance was high. Under these conditions, the string tension had to be reduced in order to obtain adequate sensitivity, and, as a consequence, the response time of the string was prolonged. Thus obtaining a satisfactory ECG necessitated a tight string, which meant that the electrode resistance could not exceed a certain value.

To eliminate the inconveniences of immersion electrodes, investigators began to study the behavior of electrodes consisting of sheets of metal wrapped in saline-soaked bandages and applied to the skin. James and Williams (1910) were the first to introduce electrodes made of German silver* plates that were wrapped in saline-soaked gauze and applied to the subject. Cohn (1920) described a more practical electrode of soft lead (22 × 7 cm) backed by a rubber sheet. The electrode was applied to the skin, which had been rubbed with a saline solution. The ECGs taken with German silver and lead electrodes were essentially the same as those taken with immersion electrodes because the investigators had succeeded in obtaining a low electrode-subject resistance.

In 1935 Russell, in the United Kingdom, reported that entirely satisfactory ECGs could be obtained with the string galvanometer using small-area electrodes applied to skin that had been prepared by having a new type (Cam-

---

* German silver, or nickel silver, is an alloy of nickel, copper, and zinc. It contains no silver; its origin is oriental.

bridge) of electrode jelly rubbed into it. The electrode-subject resistance was very comparable with that obtained with immersion electrodes. In a similar study in the United States, Jenks and Graybiel (1935) presented their own recipe for electrode paste which was made available by the Sanborn Company. The use of such electrode preparations increased, and the reason for their success was given in a paper by Bell et al. (1939). Using lead electrodes (14 × 5 cm) on human subjects, these investigators measured the dc resistance and 300-Hz impedance with the following substances under the electrodes: (*a*) 1% saline; (*b*) a paste of saline, glycerine, water, and pumice; (*c*) soft green soap; and (*d*) the recently introduced electrode jelly, which contained crushed quartz. They found that the dc resistance was highest (3080 Ω) when the electrodes were wrapped in gauze soaked in 1% saline, and applied to the subjects. With the remaining three preparations in direct contact with the electrodes and skin, the resistances were 2010, 2040, and 1100 Ω, respectively. Analysis of the results quickly revealed that the presence of an abrasive reduced the resistance considerably. Bell and his colleagues were able to show that the resistance with green soap was divided by three when crushed quartz was added and the mixture rubbed into the skin. They also found that lightly rubbing the dry skin with glass paper (fine sandpaper) "so that it lost its sheen and white color" (i.e., the stratum corneum had been removed) and then applying the electrolyte, yielded very low and extremely stable dc resistance and impedance values. This early observation demonstrated the need for abrasives in electrode pastes and jellies.

As shown, in a preceding section, the requirement for a low electrode-subject resistance was dictated by the relatively low impedance (3–5 KΩ) of the string galvanometer. Thus electrode area and the resistivity of the electrolytic preparation were important considerations. To obtain a low-resistance communication with the subject required the use of large-area electrodes and a low-resistance electrolytic preparation containing an abrasive. When the string galvanometer was supplanted by the vacuum-tube electrocardiograph, which had an input impedance in the megohm range, the need for a low-resistance contact with the subject virtually disappeared; however, this situation was not immediately recognized.

Littmann (1951) placed a solution of 2% electrolyte (presumably sodium chloride) in 50% propernol (with a small quantity of glycerin to retard evaporation) under conventional electrodes, and obtained with a vaccum-tube electrocardiograph ECGs that were indistinguishable from those obtained with conventional electrode jelly under the electrodes. In an effort to speed up the process of taking routine ECGs with vacuum-tube instruments, Krasno and Graybiel (1955) developed a plaster-of-paris pad for a placement between the electrode and skin. To activate the pad it was merely necessary to moisten it was a few drops of water.

Kniskern (1961) obtained clinically satisfactory ECGs by using a mixture of 1 part liquid detergent in 100 parts water as the electrode electrolyte. In a similar study, Lewes (1965) reported that, in a series of 4000 ECGs obtained with a remarkable variety of weak electrolytes, there were no clinically significant differences in the records when compared with those obtained with conventional electrode jelly. The substances he used were lubricating compounds (K-Y jelly, Lubrifax), culinary compounds (mayonnaise, marrons glacées, French mustard, tomato paste), and toilet preparations (hand cream and tooth paste).

To further emphasize his point, Lewes employed dry-polished electrodes (15 cm²) on dry skin and in 6 minutes obtained entirely satisfactory ECGs indistinguishable from those taken with standard jelly. Examination of the skin under each electrode revealed the presence of a small amount of sweat. An analysis of the sweat showed approximately 6 mg of sodium chloride. Additional evidence that strong electrolytes were unnecessary was provided when entirely satisfactory ECGs were obtained in 15 sec after a single drop of distilled water has been placed under each electrode.

Another ingenious method of avoiding the use of conventional electrode preparation in electrocardiography was described by Fischmann et al. (1962). To accomplish this goal they vacuum impregnated thin pads of balsawood the size of the electrodes with a variety of deliquescent salts. In practice, a plate electrode was applied to the skin and an impregnated balsawood pad was slipped under it. Soon the pad became moistened by sweat that formed an electrolytic solution and established electrical contact between the skin and the electrode. After examining the properties of many electrolytic compounds. these investigators chose lithium chloride; and when they compared the ECGs obtained with such electrodes with those from conventional electrodes, they found no clinically significant differences.

The chief advantage of this form of lithium chloride electrode is its ease of application; its disadvantages are the higher resistance and time required for equilibration. Fischmann et al. reported that the resistance of the impregnated balsawood electrodes (50 × 35 × 10 mm) was only slightly higher than that obtained with conventional electrodes and electrode preparations.

The observations of these investigators prove that when high input impedance instruments are employed to record the ECG, a low electrode-subject resistance is not required. However, these findings relative to the relation between amplifier input impedance and high-impedance electrodes resulting from the use of electrolytic solutions of low ionic content have not yet received wide acceptance. For single- or multiple-channel recording with bipolar electrodes, the observations are in accord with theory. However, King (1964) pointed out that when monopolar recording techniques are used in which several electrodes are connected through resistors joined to a common point, a high electrode-subject resistance is incompatible with the low value of averaging resistors presently in use (5–20 KΩ). Others have since made the same observation, and King further illustrated his point by making recordings with low- and high-resistance electrodes. Such a condition can exist when high-resistance electrodes are used, but provision could easily be made (e.g., by the use of buffer amplifiers) to electronically add the voltage from each electrode, providing a high input impedance for each electrode and thereby removing the need for a low electrode-subject resistance in procedures employing averaging resistors.

The important point to be derived from these studies is that the tolerable magnitude of electrode-subject impedance is dictated by the input impedance of the bioelectric recorder. To avoid loss of amplitude and distortion of the components of the waveform of a bioelectric event, the input impedance of the bioelectric recorder must be 100 to 1000 times that of the impedance of the electrode-subject circuit. The magnitudes of these impedances are identified for each electrode type described in this chapter.

As a result of the studies reported previously and in response to the requirements encountered in recording a variety of bioelectric events, a remarkably wide variety of electrode preparations is available commercially. In addition, various investigators have proposed their own recipes; among these are Jenks and Graybiel (1935), Bell et al. (1939), Marchant (1940), Thompson (1958), Shackel (1958), Lykken (1959), Edelberg (1962), Asa et al. (1964), and Fascenelli (1966). Table 2-2 summarizes the data that illustrate the properties of many of the commercially available preparations; in making the table, the author surveyed the published literature and measured the resistivities of various preparations in his laboratory.

Because there are so many electrode types and electrolytes for use with each, it is usually possible to find the combination that is suitable for a particular task. However, unconventional circumstances often arise; for example, if the electrodes are to be left in place for extended periods, evaporation of the electrolytic solution usually occurs. In some instances it may be of value to locate the electrodes in body cavities and use the fluids in these regions as the electrolytic conductor; at other times, the cavity can serve as a container for the electrolyte. Although not all the body cavities can be employed in unanesthetized subjects, consideration should be given to use of the nose, ear, mouth, axilla, navel, rectum, vagina, and urethra. It is important to note that electrodes in these sites have a relatively fixed anatomical location and hence, the sites can be designated as standard.

In using an electrode preparation with surface electrodes, it is important to note that the electrode impedance is high immediately after application; with the passage of time, the impedance decreases. This characteristic was reported by Roman (1966) and Almasi and Schmitt (1970), who studied the temporal

Table 2-2   Resistivities of Electrode-Electrolytes

Preparation and Supplier	Resistivity[a] (Ω-cm)
Redux Electrode Paste	9.4
Sanborn Div., Hewlett Packard; Waltham, Mass.	
Electrode Cream EC-2	30.0
Grass Instrument Co.; Quincy, Mass.	
Cambridge Electrode Jelly	10.4
Cambridge Instrument Co., Inc.	
Ossining, New York	
Beckman-Offner Paste	5.9
Offner Division, Beckman Instruments Inc.	
Chicago, Ill. U.S.A.	
EKG-Sol	200.0
Burton, Parsons & Co., Washington, D.C.	
Burdick Electrode Jelly	10.0
Burdick Co., Milton, Wis.	
Cardiopan	120.0[b]
Leichti; Berne, Switzerland	
Cardette Electrode Jelly	313.0[b]
Newmark Instrument Co.; Croydon, Surrey, England	
Electrode Jelly	118.0[b]
Smith and Nephew Res. Ltd., Harlow, Essex, England	
Cardioluxe Electrode Jelly	84.0[b]
Philips Electrical Ltd., Balham, London, England	
Electrode Jelly	196.0[b]
Data Display, Ltd., Liverpool, England	
NASA Flight Paste	13.0
National Aeronautics and Space Administration (NASA); Houston, Tex.	
Electrode Cream	82.0[b]
National Aeronautics and Space Administration NASA; Houston, Tex.	
K–Y Lubricating Jelly	323.0[b]
Johnson & Johnson; Slough, Buckinghamshire, England	
0.9% (physiological) saline solution	70

[a] At room temperature.
[b] From Hill and Khandpur, *World Med. Instr.* 1969, **7:**12–22. By permission.

decrease in 10-Hz impedance of standard (3 × 5 cm) plate, suction-cup (15 mm), and recessed chlorided silver electrodes applied to various sites on human subjects. A typical 10-Hz impedance-time record for a suction-cup electrode applied to the inner forearm with electrode paste (Redux) lightly rubbed into the skin appears in Fig. 2-3. Almasi and Schmitt reported that the decrease in impedance was exponential in nature, with a time constant (time to fall to 37% of the initial value) for 11 subjects of 6.9 minutes (1 standard

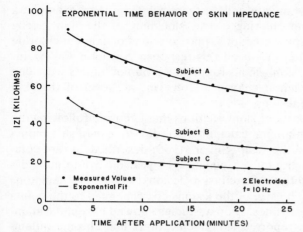

Fig. 2-3. Exponential decrease in electrode-subject impedance with time. Data are for a 15-mm suction-cup electrode applied to the inner surface of the forearm, which was rubbed lightly with electrode paste. (From J. J. Almasi and O. H. Schmitt, *Ann. N.Y. Acad. Sci.* 1970, **170**:508–519. By permission.) © The New York Academy of Sciences.

deviation (SD) = 2.7). In about 30 minutes the impedances had settled to their final values; the mean ratio of initial to final value of impedance for 11 subjects was 3.2 (1 SD = 1.6).

The investigation by Almasi and Schmitt also uncovered differences in electrode-skin impedances for different body sites. The impedance values with electrodes on the outer forearm were highest and those found with electrodes on the forehead were lowest; the average ratio was 15. Differences between males and females were also noted; the values for women were higher, typically by as much as 50%. The investigators also remarked that dark-skinned people appeared to exhibit a higher electrode-skin impedance. In commenting on these findings, Almasi and Schmitt pointed out that the impedance values they found followed a log-normal distribution and that, for a typical leg electrode, 1 out of 100 individuals could be expected to exhibit an impedance in excess of 0.36 mΩ. They recommended that the input impedance of the bioelectric recorder be designed for the electrode-subject impedance encountered just after electrode application, since bioelectric recordings are usually made at this time.

Skin Response to Electrolytes When a bioelectric event is measured with surface electrodes, it is often necessary to pay attention to the type of metal and the electrolytic preparation used with it, because each may produce its own physiological response. The constituents of certain electrolytes can produce, in some subjects, allergic reactions, erythema, or discoloration of the skin. Some species of ions stimulate cells; others are toxic. For example, a high concentration of calcium chloride causes sloughing of the skin. Seelig (1925) showed, by subcutaneous injections, that solutions of calcium chloride stronger than 1% concentration produced this response. Sneddon and Archibald (1958) reported that the skin lesions found in a group of miners were due to water containing 3.5% calcium chloride. However, abrasion of the skin preceded development of the lesions.

There have been a few reports of skin reactions that occurred following the use of bentonite paste (containing calcium chloride) to establish contact between metal electrodes and the scalp. Silver (1950) described "a rare complication of electroencephalography" in which a 14-year-old white male developed small, annular, slightly elevated skin lesions having erythematous centers and yellow borders. The size of the lesions corresponded to the points of application of the scalp electrodes. Biopsy studies showed basophilic staining of the connective tissue (reported as "calcification" and chronic inflammatory reaction). Skin-patch tests revealed that the patient was sensitive, to bentonite powder but not to Caclz [sic—$CaCl_2$?] or glycerin, which were contained in the paste.

Clendenning and Auerbach (1964) reported two cases of skin lesions following prolonged exposure (6–8 hours) of the skin to an electroencephalographic (EEG) paste composed of bentonite, calcium chloride, and glycerin. In both patients the electrodes were on the scalp, face, and ear lobes. A test was made by applying electrodes with the same paste to the arms of five normal subjects; none developed irritation. In one subject, histologic examination of the skin was carried out at the electrode site and a deposit of calcium was found in the dermis. A small linear white plaque appeared later. In commenting on their findings, Clendenning and Auerbach pointed out that the electrodes and paste were in contact for a much longer time than that required for routine electroencephalography. They stated that it was calcium that caused the lesions, that calcium is found in bentonite, and that a saturated solution of calcium choride is an ingredient in many EEG pastes. Most important, they implicated skin abrasion as a precipitating factor and pointed out that the abrasion could occur during preparation of the skin at the electrode site or during the long period of contact. In the bibliography the authors cited personal communications indicating that the occurrence of lesions in routine electroencephalography, although quite rare, is not unknown. For example, a tissue response in a 13-month-old infant following the taking of a routine EEG with bentonite-calcium chloride-glycerin paste was reported by Giffin and Susskind (1967). An investigation of the incident caused the authors to implicate calcium and silver ions from the electrodes. Many of the newer electrode preparations used in electroencephalography contain no calcium. In some laboratories mild detergents or other weak electrolytes are employed.

Faced with the difficult problem of telemetering the human fetal ECG, Asa et al. (1964) developed an electrode paste that produced a surprisingly low contact resistance. By mixing equal parts of a commercial electrode paste with silver oxide, resistance values of 1500 Ω were observed. Without the silver oxide in the paste, the resistance was 4000 to 15,000 Ω. The method was evaluated by the author, who confirmed the reduction in resistance; with some subjects, however, the paste produced a reddening of the skin. Whether this response is a necessary accompaniment to obtaining a low resistance cannot be confirmed without further study.

When recording the galvanic skin response (GSR), the ionic composition of electrolytes merits special consideration. For example, Edelberg (1962) conducted a series of ingenious experiments in which the responses at test and control sites were compared. He found that solutions of molar ($1.0M$) calcium chloride, ammonium chloride, and potassium sulfate potentiated the GSR by 100 to 300%. Aluminum chloride potentiated the GSR by 1000%, and zinc chloride ($0.5M$) approximately doubled the response. Very dilute acids, alkalis, and detergents decreased the response. A solution of $0.05M$ sodium chloride had negligible effect on the GSR and he recommended its use for this purpose. Thus in the routine recording of a bioelectric event from skin surfaces containing sweat glands, what may appear as an artifact may be an enhanced GSR. On the other hand, if one is attempting to record the GSR, the electrolyte may enhance or diminish the response. Scarification of the region under the electrode can also produce unwanted voltages. Edelberg (1962) reported that although cuts or skin punctures reduce skin resistance, they also reduce the GSR.

The accumulation of considerable experience has proven that most of the currently available electrolytic preparations for use with surface electrodes are safe in routine clinical situations. However, it should not be concluded that all are safe for long-term use. Recovery-room, intensive-care, and prolonged spaceflight monitoring will undoubtedly call attention to the need for electrolytic preparations with special properties; in fact, new preparations are beginning to appear for these applications.

At present the principal constituent of electrode preparations is sodium chloride. Some preparations contain a little potassium chloride, but calcium chloride has been eliminated almost entirely. Glycerin is usually included to retard evaporation. Beyond this point the various preparations differ substantially; some include pumice, gum tragacanth, carbolic acid, and preservative. It is not now possible to generalize on the composition of electrode preparations nor to state that one type is superior to another.

The flight paste described by Day and Lippitt (1964) for long-term monitoring of the ECG and impedance respiration during spaceflight underwent considerable development and careful testing. It can be considered to be almost

Table 2-3   NASA Flight Paste[a]

Components	Amount
Methyl-*p*-hydroxybenzoate	1 g ⎫ preservatives
Propyl-*p*-hydroxybenzoate	1 g ⎭
Hydroxyethylcellulose[b]	80 g
Polyvinylpyrrolidone[c]	35 g
Sodium chloride	90 g
Potassium chloride	3.1 g
Calcium chloride	3.3 g
Water (deionized)	1 liter

The benzoates are added to 1 liter of deionized water and blended by using a mixer with plastic coated beaters and a glass or plastic container. (An electric beater-type mixer is highly satisfactory for this purpose). After the salts, the polyvinylpyrrolidone and then the hydroxymethylcellulose are slowly blended in. Food coloring may be added to facilitate visual observation. The pH is adjusted with $6N$ NaOH or HCl to 7.0 ± 0.1 using a pH meter. Plastic tubes or other containers are used for the storage of this paste to avoid reaction with metal, since the salt concentration makes the paste extremely reactive and it will quickly become contaminated if in contact with metal.

[a] J. L. Day, and M. Lippitt, *Psychophysiology*, 1964, **1**:174–182. By permission. © 1964, The Williams & Wilkins Co., Baltimore, Md. 21202 U.S.A.
[b] Natrosol 250 GR Hercules Powder Co. More can be added for thickening.
[c] K-90, General Aniline Co.

nonirritating, having been used for six years as the conducting material between chlorided silver electrodes and the thoraces of astronauts and left *in situ* for weeks. The recipe for this paste, containing no abrasive, is given in Table 2-3.

Montes et al. (1967) performed extensive dermatologic studies on the electrode preparation described by Day and Lippitt. They applied six 1-cm² electrodes to the heads and four 2-cm² electrodes to the chests of ten normal subjects and left them in place with the Day–Lippitt preparation for periods up to 14 days. On the scalp, the hair in the electrode site was clipped and cleaned with acetone. The stratum corneum was then removed by the skin-drilling technique described by Shackel (p. 59). At three electrode sites a depilatory was used (Surgex, Crookes Barnes Laboratories) for 10 minutes; the scalp under the other three electrodes received no additional preparation.

The electrode containing the electrolytic compound was then applied and secured, using Eastman 910 adhesive (Eastman Kodak Co., Rochester, N.Y.). The skin under the thoracic electrodes was cleaned but not drilled before application of the electrodes, which were secured with Stomaseal (Minnesota Mining and Manufacturing Co., St. Paul, Minn.). During the 14-day test period at least one electrode was removed each day and a biopsy was taken from the site.

Montes et al. (1967) reported that there was excellent tolerance of the skin by the electrolytic preparation used with the electrodes. They reported that

Histological comparison between specimens from control biopsies and others obtained throughout the two-week period showed no significant differences. However, specimens from under several EEG electrodes showed that proliferation of bacteria and fungi may take place under the electrode housing.

This latter finding is not surprising since even so-called clean skin is known to harbor numerous microorganisms.

Other electrode preparations have been advocated for measuring different bioelectric events or for recording under special circumstances. For example, those who record emotionally produced electrodermal phenomena (i.e., a change in resistance or a change in voltage) have strong preference for certain electrolytes. The papers by Thompson (1958), Shackel (1958), Lykken (1959), and Edelberg (1962) present useful data, as well as the authors' justifications for their choices. As stated previously, most investigators use a weak sodium chloride solution because it is the main constituent of extracellular fluid. It is therefore of value to know how well it conducts an electric current. To provide information on this point, Fig. 2-4 presents the resistivity of saline solutions at different temperatures, along with the same data for potassium chloride (which is used in electrolytically filled micropipets). Additional data can be obtained from a review by Deware and Hamer (1969).

Skin-Drilling Technique

A novel method for obtaining a low-resistance communication between an electrode and a subject was described by Shackel (1959). The method, which he called the skin-drilling technique, is painless when properly applied. The area of skin where the electrode is to be placed is first cleaned with an antiseptic solution. The site is then abraded with a dental burr in a hand tool. Only the upper layers of the epidermis are eroded, and no blood is drawn. The amount of abrasion required depends on the type of skin. Kado (1965) reported that deeply pigmented skin requires more abrasion. In a few seconds, tissue fluid can be seen seeping into the drilled depression. The site is then cleaned with alcohol or acetone; if the skin has been drilled to the proper depth, the subject should feel a slight tingling sensation when the region is

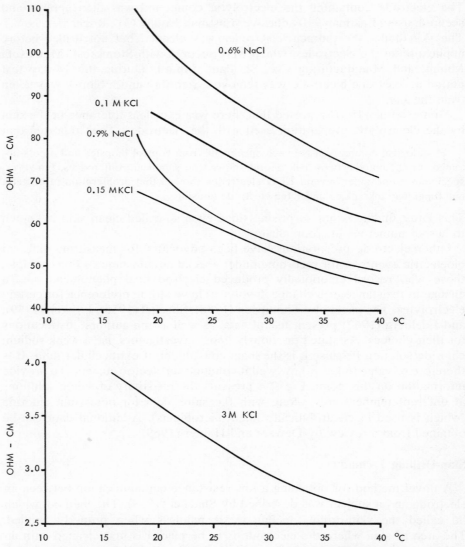

Fig. 2-4. Resistivity-temperature data for solutions of sodium chloride and potassium chloride.

cleaned with either of these solutions. The electrode electrolyte is then applied and the electrode secured.

To test the value of the technique, Shackel compared the resistance values obtained with and without drilling; the drilled sites consistently exhibited values one-fifth to one-tenth of those of undrilled sites. Upon removal of the electrodes, the drilled sites are again cleaned with an antiseptic solution. Lanolin cream is then rubbed in and the sites soon become invisible.

The author has seen some remarkably artifact-free ECGs recorded with miniature recessed electrodes (see p. 78) placed over drilled sites at each end of the sternum of human subjects. Neither tapping the electrodes nor vigorous exercise caused movement of the recorded baseline. There is no doubt that this combination will be of value for some measurement situations which involve acceleration forces.

Immersion Electrodes

Just prior to the start of the twentieth century, the first mammalian ECGs were obtained using the capillary electrometer (see Geddes and Hoff, 1961), connected to two types of "nonpolarizable" electrodes. One type consisted of a metal plate wrapped in chamois soaked with saline; the other was a saline-filled container into which an extremity was placed. In his first studies on the human ECG, Waller (1889) applied the metal plate type to the chest and back; but he sometimes used saline-filled containers and induced his pet bulldog Jimmie to stand in them (Fig. 2-5a). This type of electrode, adopted by Einthoven (1903) when he introduced his string galvanometer, gave birth to clinical electrocardiography; the type of electrode and method of use are shown in Fig. 2-5b.

Because the string galvanometer is a direct-coupled recorder capable of reproducing steady as well as changing potentials, it was necessary to employ what were designated as "nonpolarizable electrodes" (i.e., electrode pairs which, when connected to the subject, exhibited no steady potential); this of course requires the same half-cell potential for each electrode. To attain this condition, Einthoven employed the electrodes appearing in Fig. 2-5b; each consisted of an amalgamated zinc cylinder filled with saline solution into which the subject placed his hand or foot. The zinc cylinder was placed in a vessel filled with zinc-sulfate solution into which was placed an amalgamated zinc electrode, which was in turn connected to the string galvanometer. The potential of such an electrode was constituted by the following half-cells: subject/saline/HgZn/$ZnSO_4$/HgZn. Evidently, in practice, the sum of the potentials of two such electrodes (with the subject between) produced only a small residual potential. However, in some instances it became necessary to add a compensating (bucking) voltage, derived from a potentiometer, to reduce

Fig. 2-5a. Waller's use of immersion electrodes to obtain an ECG on his pet bulldog Jimmie. The lead employed (left forelimb to left hindlimb) was later designated as lead III by Einthoven. (From A. D. Waller, *Physiology, The Servant of Medicine* (Hitchcock Lectures), London: University of London Press, 1910.)

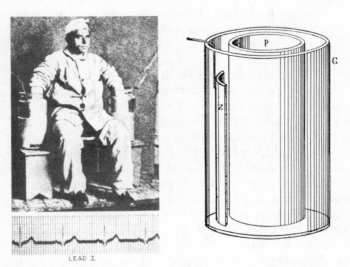

Fig. 2-5b. Immersion electrode system used by Einthoven. (Photograph from *Arch. Int. Physiol.* 1906–1907, **4:**132–164; cross section from *Arch. Ges. Physiol.* 1908, **122:**517. By permission.)

the sum of the electrode potentials to zero; in this application no standing potential is presented to the string galvanometer, thus allowing the string to operate centrally in the magnetic field of the galvanometer.

The method employed by Waller to obtain ECGs on his pet bulldog is ideally applicable to larger and smaller animals when it is not permissible to remove the hair or abrade or penetrate the skin with needles, or when the animal attempts to dislodge the electrode (as it would a biting insect, by shaking the skin vigorously), thereby producing large movement artifacts. The author and his colleagues (Geddes et al., 1971) have enjoyed considerable success in using immersion electrodes to obtain ECGs from standing horses, ponies, cows, calves, goats, and dogs. Each of the electrodes employed (Fig. 2-6a) consists of a metal screen (hardware cloth) placed in the bottom of a plastic tray. Over the metal screen is a felt pad on which the animal stands. Electrical contact is established with the animal by half-filling the tray with 0.9% saline. Figure 2-6b shows a cow standing in four immersion electrodes and the ECG obtained from lead aVF.

It is noteworthy that high quality ECGs can be obtained with immersion electrodes; this is so probably because they provide a low-resistance contact with the subject. In fact, the resistance measured between a pair of Einth-

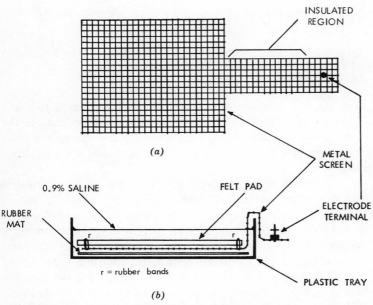

Fig. 2-6a. Tray-type immersion electrodes (*a*) metallic portion of electrode; (*b*) assembled electrode. (From Geddes et al. *Vet. Med.* 1970, **65**:1163–1168; VM/SAC. Dec. 1970. By permission.)

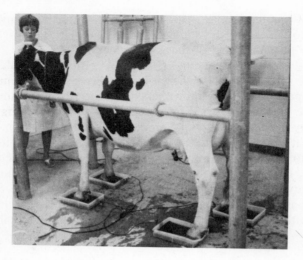

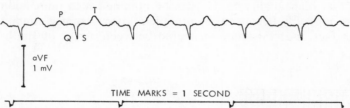

Fig. 2-6b. Use of the tray-type immersion electrode to record the bovine ECG. (From Geddes et al. *Vet. Med.* 1970, **65**:1163–1168; VM/SAC, Dec. 1970. By permission.)

oven's electrodes, as used in human electrocardiography, amounted to a few thousand ohms. Such a low-resistance contact was necessary because of the low resistance of the string galvanometer. The need for low-resistance electrodes for electrocardiography disappeared when vacuum-tube electrocardiographs were introduced.

To illustrate the resistance range of the tray-type immersion electrodes, the author conducted a series of loading tests (see p. 42) while recording ECGs on horses, dogs, and cows. The amplitudes of the P, R, and $\hat{T}$ waves were measured for each resistance-value place across the electrode terminals; the amplitudes obtained were expressed as percentages of those obtained without any connection across the electrode terminals. For each species, the percentages for the P, R, and T waves were averaged. Figure 2-6c presents the results, demonstrating that a remarkably good contact was established with the animals studied; the resistance of these electrodes is in the low kilohm range (3000–5000 Ω) and indicates that they can be used with any conventional electrocardiograph.

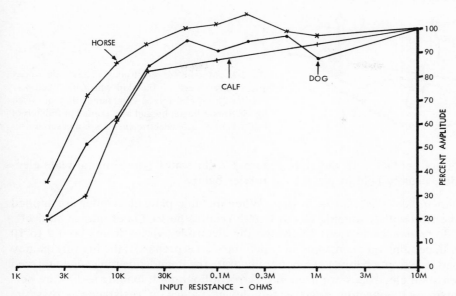

Fig. 2-6c. Characteristics of tray-type immersion electrodes as revealed by the reduction in amplitude of the *P*, *R*, and *T* waves of the ECG resulting from placing known resistance values across the electrode terminals. (From Geddes et al. VM/SAC, Dec. 1970. By permission.)

Although abandoned long ago for human electrocardiography, immersion electrodes should not be totally discarded, for they offer some highly attractive characteristics for obtaining ECGs on animals and even for rapid screening of humans. For example, Maness (1970) demonstrated excellent lead-I ECGs obtained from people who merely dipped the index finger of each hand into a 50-ml beaker of saline containing a silver-silver chloride electrode that was connected to an electrocardiograph.

Metal-Plate Electrodes

The most popular electrodes used for recording bioelectric events are the metallic plate electrodes employed in electrocardiography. They usually consist of rectangular (3.5 × 5 cm) or circular (5 cm) plates of German silver (nickel silver) or chrome or nickel-plated steel. Around 1910 such plate electrodes replaced the more cumbersome immersion electrodes that had been introduced by Einthoven, along with the string galvanometer which gave birth to clinical electrocardiography. When plate electrodes were first employed, they were much larger and were wrapped in cotton gauze, felt, or chamois soaked in saline. Thus the electrodes did not come in direct contact with the skin. It was the introduction of electrode jellies and pastes and vacuum-tube amplifier instruments that permitted placement of small metal electrodes

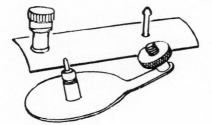

Fig. 2-7. Metal-plate electrodes. These electrodes are usually placed on the skin which has been prepared with an electrolytic compound; they are used for electrocardiography and as a ground or indifferent lead when other bioelectric events are measured.

in direct contact with the skin. Figure 2-7 illustrates typical metal-plate electrodes, usually held in place by a rubber band.

Impedance of Plate Electrodes   When metallic-plate electrodes are applied to the skin with a suitable electrolyte to record the ECGs of human subjects, the dc resistance measured between the electrodes varies from about 2 to 10 KΩ, depending on the manner in which the skin is prepared; the high-frequency impedance is much lower. To better illustrate the impedance range encountered when such electrodes are used in electrocardiography, loading tests were performed by connecting successively lower values of resistance across the

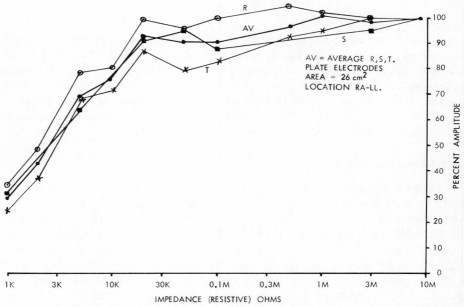

Fig. 2-8.   Loss of amplitude of the *R*, *S*, and *T* waves of the human ECG as input impedance (resistive) is reduced; lead II, electrode area 26 cm².

electrode terminals. The amplitudes obtained with a 9-MΩ resistive input impedance were taken as 100% (controls) and the amplitudes with the other resistive input impedances were expressed as percentages of the control values.

Figure 2-8 shows the loss in amplitude of the R, S, and T waves of the ECG as the input resistance of the amplifier was lowered in steps down to 1000 Ω. Note that when the input resistance was below 20 KΩ there occurred a significant loss in amplitude; at about 3000 Ω the loss was approximately 50% for the three waveforms. The data in Fig. 2-8 indicate that, with an input impedance greater than 2 MΩ, an insignificant loss in amplitude will occur. (The variations in amplitude above the 100% line merely reflect the small respiratory variations in amplitude of the R, S, and T waves that occurred during the period of recording.)

The impedance-frequency characteristics of a typical pair of ECG plate (26 cm²) electrodes applied to a human subject with electrode jelly are shown in Fig. 2-9. A low current density (1.8 μA/cm²) was employed to ensure that the characteristics would be obtained under linear and representative conditions. The impedance values for the frequency spectrum for the electrocardiogram (0.05–100 Hz) range from about 7 to about 3 KΩ, a range that is in general agreement with that found by the loading tests (Fig. 2-8).

Although little difficulty is encountered when large-area metallic-plate

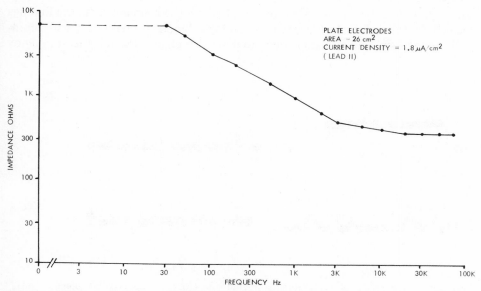

Fig. 2-9. Impedance-frequency characteristic for plate electrodes (26 cm²) obtained with a current density of 1.8 μA/cm² on a human subject (lead II).

electrodes are used to record bioelectric events with high input impedance
recording instruments, it was found in the early days of electrocardiography
that, if such electrodes were employed with the string galvanometer (which
had an input impedance of about 3000–5000 Ω), serious waveform distortion
occurred. Lewis (1914) was one of the first to call attention to this fact by
comparing ECGs obtained with the standard large-area "nonpolarizable"
immersion electrodes and platinum plate electrodes wrapped in gauze soaked
with saline. Although the relation between electrode area and amplifier input
impedance is discussed in this chapter and in chap. 3, Lewis's observations
are included here for historical accuracy and to alert the reader to the potential
ability of electrodes to distort bioelectric events if they are not employed with
a recording instrument having an adequately high input impedance.

Figure 2-10 is Lewis's illustration of the calibration record (*a*) and the ECG
of a normal subject (*b*) obtained with contemporary immersion electrodes.
When the platinum electrodes covered with saline-soaked gauze were employed, he obtained the calibration record and ECG in Figs. 2-10*c* and 2-10*d*,
respectively. Comparisons of (*c*) and (*d*) with (*a*) and (*b*) illustrate that the
type of distortion encountered is consistent with the capacitive nature of an
electrode-electrolyte interface and low input resistance of the string galvanometer. Note that both the calibration wave and the ECG show evidence
of electrical differentiation, which manifests itself as a loss of low-frequency
response. The calibration signal is only reproduced at the "on" and the "off";
the *P* and the *T* waves are diphasic and the *S* wave is enhanced. These findings
clearly call attention to the need to employ a bioelectric recorder with an input impedance much higher than that of the electrodes for the frequency

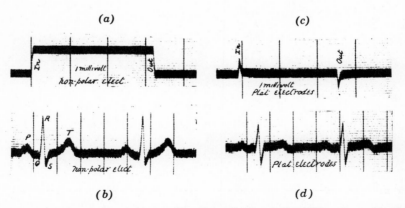

Fig. 2-10. Comparison of (*a*) calibration record and (*b*) ECG taken with "nonpolarizable"
immersion electrodes with recordings of (*c*) and (*d*), the same events obtained with platinum
electrodes wrapped in saline-soaked gauze. (From T. Lewis, *J. Physiol.* 1914–1915, **49**:L–LII.
By permission.)

spectrum of the bioelectric event to be measured. In Lewis's study, the low-frequency impedance of the platinum electrodes was many kilohms, that of the string galvanometer was between 3 and 5 KΩ.

Adhesive-Backed Metal Screens A variant of the plate electrode that permits quick application, is one contained in a strip of adhesive tape. One such electrode (Fig. 2-11; Telemedics, 1961) consists of a light-weight metallic

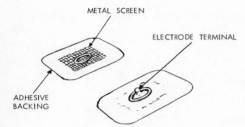

METAL SCREEN

ELECTRODE TERMINAL

ADHESIVE
BACKING

Fig. 2-11. Adhesive-backed metal-screen electrodes (Telemedics, Inc., Vector Division, United Aircraft Corp. Southampton, Pa., U.S. patent 3,085,577.)

screen backed by an adhesive pad for retaining electrolytic paste. Measuring approximately 1.5 in.², the electrode adheres well to hairless skin. The edges of the adhesive backing hold the electrode in place and retard evaporation of the electrolyte. Although there are no published data on these electrodes, because of the slightly smaller area, it can be assumed that their impedance in contact with a suitable electrolyte on the human skin will be slightly higher than the conventional plate electrodes used for electrocardiography.

Suction Electrodes

The use of negative pressure to hold electrodes against living tissue for stimulating or measuring bioelectric events is probably not new. Recently such electrodes have proven to be very practical in many applications; a very useful model (Fig. 2-12*a*) is used routinely to record the precordial ECG. According to Burch and DePasquale (1964) it was introduced by Rudolph

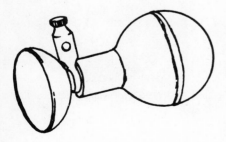

Fig. 2-12*a*. Metal suction-cup electrode.

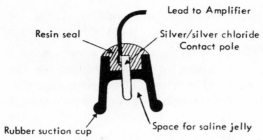

Lead to Amplifier

Resin seal

Silver/silver chloride
Contact pole

Rubber suction cup

Space for saline jelly

Fig. 2.12b. Shackel's rubber suction-cup electrode (Redrawn from B. Shackel, *J. Appl. Physiol.* 1958, **13:**153–158. By permission.)

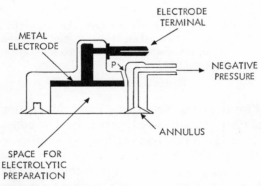

ELECTRODE
TERMINAL

METAL
ELECTRODE

P

NEGATIVE
PRESSURE

ANNULUS

SPACE FOR
ELECTROLYTIC
PREPARATION

Fig. 2-12c. Stangl's recessed suction electrode. (Redrawn from German patent #1,108,820.)

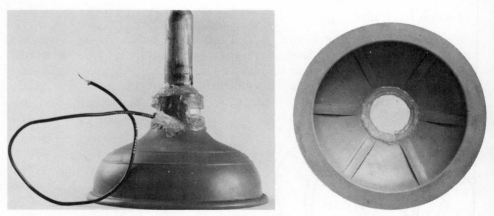

Fig. 2-12d. Rubber suction-cup electrode used by Spencer et al. (1967) for whale electrocardiography. (Personal communication from M. P. Spencer. 1970.)

70

Burger; the type illustrated was employed by Welch (1951). The forerunners of this type of suction-cup electrode were described by Roth (1933–1934) and Ungerleider (1939).

The impedance of a typical suction-cup electrode (3-cm diam.) can be estimated by the loss in amplitude of the *P*, *R*, and *T* waves of the human ECG, recorded between the right arm (large plate electrode) and chest (suction electrode), as the input impedance of the recording apparatus is reduced by placing known values of resistances across the electrode terminals. Figure 2-13 presents the results of such a procedure and illustrates that, for the frequency spectrum of the human ECG, the impedance of the suction-cup electrode is slightly higher than that of the conventional plate electrode. With an input impedance of about 80 KΩ there was an average of 10% loss in amplitude: 50% loss in amplitude occurred with an input impedance of 6 KΩ. These data indicate that, because of the smaller area of the suction-cup electrode, the input impedance of an amplifier for electrocardiography should be chosen on the basis of the impedance of the suction-cup electrode. From these studies it would appear that an input impedance of several megohms would insignificantly load (i.e., reduce the amplitude), of the *P*, *R*, and *T* waves of the human ECG.

In addition to the popular precordial suction-cup electrode used in electro-

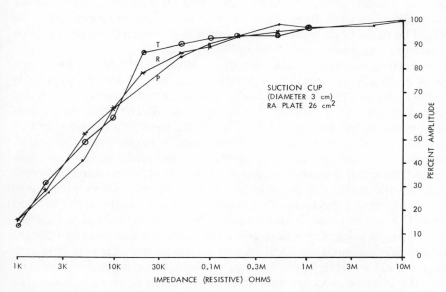

Fig. 2-13. Loss of amplitude of the *P*, *R*, and *T* waves of the human ECG measured with a suction-cup electrode on the chest paired with a 26-cm² electrode on the right arm.

cardiography, a number of different types of suction electrodes with interesting properties have been described. For example, Andrews (1939) reported the successful use of a miniature (1-cm diam) model for recording the EEG and the ECG. The metallic portion of his electrode was cut from a sheet of silver 0.5 mm thick and consisted of a disk 5 mm diam. with a long silver tail that was folded back and pushed through the top of a cup-shaped rubber eraser of the type usually placed on the top of an ordinary pencil. The cup was filled with electrode jelly, the sides were then squeezed together lightly to expel the air; the cup was placed against the skin, and the sides were released. The electrode paste provided an airtight seal and established an electrical connection between the skin and the silver-disk electrode. It is interesting to note that this type of electrode, now designated a recessed electrode (see p. 78), is remarkably free from movement artifacts; this observation is due to Andrews, who used his electrode to record EEGs from subjects undergoing drug withdrawal. He also stated that the silver disk could be chlorided and either type (bare or chlorided) would remain attached to the skin for more than an hour, despite bodily movements. Andrews pointed out that his electrodes should only be used with vacuum-tube amplifiers because of their high resistance (8 KΩ between a pair on a subject).

Shackel (1958) developed another high-stability, recessed general-purpose suction-cup electrode (Fig. 2-12*b*), consisting of a chlorided silver rod mounted centrally in a small rubber cup filled with electrode jelly. He found that the electrode adhered well to the skin and, despite movement, produced almost no artifacts. The range of dc resistance between a pair of such electrodes applied to the forearm was found to be 2000 to 7000 Ω.

An interesting recessed suction electrode, which ought to exhibit a high degree of electrical and mechanical stability, was patented by Stangl (1961) and is illustrated in Fig. 12*c*. The metallic portion of the electrode is mounted in a cup-shaped housing made of insulating material. The space in front of the metal electrode can be filled with an electrolytic paste or a sponge saturated with any desired electrolyte; access to this space is available via the port *P*. In the rim of the cup-shaped housing there is an annular space that communicates with a fitting connected to a source of negative pressure. With the electrode on the surface of the skin, the application of negative pressure to this fitting will cause the electrode housing to adhere snugly to the subject.

Although it offers some unique practical features, Stangl's electrode has not seen much service. His patent gives very little information regarding its properties and use; nonetheless it should be noted that, because it is self-adhering, it can be quickly applied and probably sticks tenaciously to the skin. Because the electrode-electrolyte surface is protected from disturbance, the electrode ought to exhibit a high degree of electrical stability and thus should be useful for recording bioelectric events on exercising subjects. It is to be noted that this electrode, like other suction electrodes, is only suited for short periods of application because of the alteration in capillary pressure gradient imposed by this electrode type.

Probably the largest suction electrode, representing perhaps one of the most intriguing uses for such electrodes, was described by Spencer et al. (1967); in their study they obtained ECGs from submerged whales. The electrodes employed (Fig. 2-12*d*) consisted of a stainless steel disk mounted in the center of a plumber's suction plunger of the type used for unclogging sinks and toilets. The edge of the suction cup was lubricated with silicone grease and, when the electrode was placed in the water and pressed against the whale's precordium, the seawater remaining in the plunger cup established electrolytic contact between the stainless steel disk and the animal. Electrocardiograms were obtained using one suction-cup electrode and an indifferent electrode consisting of a stainless steel rod (6 × 0.5 in.) totally immersed in the water adjacent to the animal.

Hartman and Boettiger (1967) recorded action potentials by applying a slight negative pressure to micropipet electrodes, having tip diameters in the range of 5 to 100 μ to hold axons against the tips of the electrodes. This method of holding active tissue in contact with an electrode has prompted one manufacturer (Cal-Science Co., Riverside, Calif. 92507) to offer a similar suction electrode (Fig. 2-14). Negative pressure $-P$ is applied by a tube on the side that communicates with the tip T of the glass capillary and lumen of the tube, which are both filled with electrolyte; a silver electrode, which is in contact with the electrolyte, is connected to a BNC connector at the end of the electrode assembly. The manufacturer states that these electrodes can be used "to record extracellularly from invertebrate hearts [and] intestinal nerves and stimulate muscles indirectly by sucking the nerve into the end of the suction electrode." No electrical characteristics of these practical electrodes are available as yet.

Suction electrodes lend themselves to quick and easy, but temporary, application to the skin surface or to irritable tissue directly. The term temporary is used because the negative pressure will alter the normal capillary pressure gradient. Prolonged maintenance of an increased capillary pressure

BNC

−P

T

Fig. 2-14. Micropipet suction electrode. (Courtesy Cal-Science Corp., Riverside, Calif.)

gradient may cause local injury. With a large-area cup electrode, such injury may be slight or undetectable. However, an injury potential may be produced when a small electrode is placed directly against irritable tissue and a high negative pressure is maintained for a prolonged period. With these precautions in mind, it is important to recognize that in many instances the suction electrode is the only one that can be used. For example, if it is not possible to secure the electrode with a retaining band or adhesive, and when it is not permissible to insert a pin or needle below the integument, the suction electrode is the electrode of choice. It is particularly well suited for service on wet surfaces, which do not take adhesives. An excellent example of this application is the previously mentioned study by Spencer et al. (1967), in which a large suction-cup electrode was used to record the precordial ECGs of submerged whales weighing between 1000 and 3000 kg.

The electrical stability of suction electrodes depends on their design. Those in which the electrode-electrolyte interface is protected from disturbance can exhibit a high degree of stability. The reasons for the stability of this design are discussed in the section describing recessed electrodes (see p. 78).

Metal-Disk Electrodes

To obtain the EEG from the human scalp, it is convenient to use metal-disk electrodes about 8 mm diam. (Fig. 2-15); usually silver is employed. Some-

Fig. 2-15. Metal-disk electrodes. This type of electrode is frequently applied to the scalp for recording the EEG. The disk (about 8 mm diam.) is usually made slightly concave to hold the electrolytic preparation.

times the disks are slightly cupped to permit retention of the electrolytic compound that establishes ohmic contact with the cleaned scalp. The electrodes are generally secured to the scalp with an adhesive or covered by a small square of gauze saturated with collodion or a low melting-point wax.

The impedance measured between a pair of silver-disk electrodes applied to the human scalp depends on the area of the electrodes, the manner in which the scalp has been prepared and the type of electrolytic compound employed. In clinical electroencephalography, technicians aim to obtain a dc resistance in the range of 5000 to 15,000 Ω. If the scalp is abraded, the resistance can be as low as about 2000 Ω.

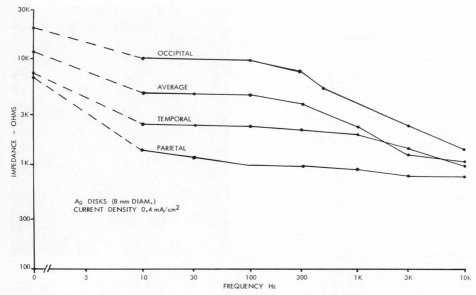

Fig. 2-16. Impedance-frequency characteristics of 8-mm silver disks applied with bentonite paste to the scalp of a human subject.

Typical impedance-frequency curves measured between three pairs of 8-mm silver disks applied to the scalp with bentonite paste are shown in Fig. 2-16. The difference between the curves merely reflects variations in the technique of application. In the frequency spectrum of the human EEG (1–60 Hz), it can be seen that the impedance-frequency characteristic is essentially uniform and that the extremes of the impedance range are 20 to about 2 KΩ. The highest value of impedance (20 KΩ) indicates that an input impedance no lower than 2 MΩ (i.e., 100 times 20 KΩ) would not appreciably reduce the amplitude of the electroencephalographic signals. Currently available EEGs meeting the standards set by the Council on Physical Medicine of the American Medical Association (1947), have an input impedance of about 2 MΩ. However, the Council did not specify a minimum value for input impedance.

Low-Mass Electrodes

It is frequently necessary to record bioelectric events from exercising subjects or from subjects experiencing large vibration or acceleration forces. Under these circumstances it is essential that the electrode be as small and as light in weight as possible. Sullivan and Weltman (1961) developed a low-mass

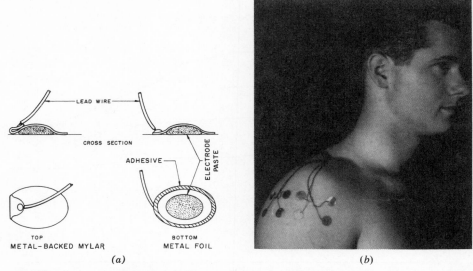

Fig. 2-17. Low-mass electrode. (Sullivan and Weltman, 1961, U.S. Patent #3,151,619; From *J. Appl. Physiol.* 1961, **16:**939–940. By permission.)

electrode (weighing 2 mg) to record electromyograms (EMG) from the deltoid muscles of exercising human subjects. Their electrode (Fig. 2-17) consisted of a strip of 0.001-in. thick mylar ribbon on which was deposited a metal film. A small quantity of electrode paste was placed below the metalized film, which was cemented to the subject using Eastman 910 adhesive. The investigators stated that the film prevents evaporation of the electrode paste and that the adhesive produced no skin reaction. No electrical data were given for the electrodes, but because of their small area it can be concluded that a moderately high input impedance amplifier is required.

Low-mass versions of other types of electrodes have been described. Edelberg's (1961) dry electrode, which consists of an electrodeposition of silver metal into the skin, qualifies as a low-mass electrode. Likewise, the conducting-film electrodes described by Thompson (1956) and Roman (1966) are light in weight and ideal for application to mobile subjects. All these electrodes are described in the section in this chapter that describes metal-film electrodes. Low-mass versions of the recessed electrode (see p. 78) were described by Roman and Lamb (1962), Lucchina (1962), Adey et al. (1963), Kado et al. (1964), and Simons (1965). The electrical stability of the recessed electrode when combined with low mass is perhaps the best type of electrode for obtaining artifact-free recordings of bioelectric events on exercising subjects.

As stated previously, low-mass electrodes are used to study subjects experiencing vibrational forces. In this situation, a low-mass electrode, in which the electrode-electrolyte interface is protected from displacement, does not alter its potential and hence is relatively immune from "movement" artifacts. However, movement of the wires connecting the electrodes to the amplifier causes a change in the capacitance between the wires and to ground. In addition, movement of the wires in the earth's magnetic field causes the induction of a voltage proportional to the rate of displacement of the wires. Although the latter voltage is usually small, the former may not be, and the standard minimization technique employs a shielded wire between each electrode and the amplifier input terminals. However, deformation of ordinary shielded wire results in a change of capacitance between the central conductor and the shield; if there is a potential difference between the shield and inner conductor, a transient change in capacitance will cause a transient current to flow through the input impedance of the amplifier, which will amplify the transient current as well as those due to the bioelectric event. Possible sources of voltage on the shielded cable are the electrode potential or locally induced electrostatic voltages. Adey and his colleagues (1960 1961) and Kado and Adey (1968) pointed out that the use of a special low-noise shielded cable (Mininoise Cable, Microdot Corp., Pasadena, Calif.) eliminated such cable deformation artifacts. This shielded cable, which has a capacitance of 28 pF/ft, has the interesting property of maintaining a relatively constant capacitance between the shield and the central conductor despite bending. This property is achieved by placing a layer of aluminum powder under the outer shield (i.e., on the dielectric surrounding the inner conductor); a second layer of aluminum powder is applied to the insulation in contact with the central conductor. Thus with bending, the distance between the inner and outer surfaces of the dielectric stays almost constant; hence the capacitance is virtually constant. This useful property has permitted successful recording of the EEG on animals and human subjects experiencing large-amplitude vibrational forces.

Kado and Adey (1968) called attention to an important precaution to be observed when using this cable, reporting that the terminal portions of the cable must be carefully denuded of the conducting powder before connections are made. Heeding this advice will permit retention of a high resistance (i.e., low leakage) between the inner conductor and outer shield of the cable.

Silver Cloth Electrodes

The availability of an interesting new material, silver cloth woven of nylon threads coated with silver, prompted Morrison (1958) to employ it to detect skin resistance changes on the feet of subjects exposed to various environmental stresses. The cloth was cut to fit the region of the foot extending from the metatarsal arch to the instep; the area of the electrode was approximately

8 in.^{2}. The cloth was maintained in contact with the sole by a foam backing or by affixing it to a snug-fitting sock with double-sided adhesive tape. The electrical connection was established by sewing a tinned flexible wire to the silver cloth.

Morrison reported that GSR changes could be recorded without difficulty for periods up to 3 days. The basal resistance measured between a pair of such plantar electrodes varied between 5 and 15 KΩ; superimposed on these basal resistance values were the transient resistance changes due to the environmental stimuli.

Although Morrison reported that the electrodes were employed for periods up to 7 days, artifacts were encountered. From his description it would appear that variations in the galvanic potential between the tinned wire and silver cloth in a saline milieu might have been the cause. If so, this source of artifact could be easily eliminated by sewing a silver wire to the silver cloth to establish electrical contact with this potentially useful electrode material.

Edelberg (1967) established an artifact-free communication with a silver cloth electrode by twisting the connecting wire to a tab of silver cloth at the edge of the electrode. The tab and the twisted wire were then insulated with dental impression wax or a flexible cement. The tab was then folded out of the way and cemented down over a sponge backing provided for the electrode. The electrode was in electrical contact with the skin via an electrolytic preparation developed by Edelberg (1967).

Silver cloth (Swift Textile Metalizing Corp., Hartford, Conn.) is inexpensive and is available in three grades (light and heavy rip and marquisette); 3 to 4 ft wide, and in lengths up to 100 yards. The resistivity range for the material is quoted by the manufacturer as "2 to 10 ohms per square yard." The material is gray in color, feels like cloth, and can be sewn with any sewing machine.

Silver cloth presents some interesting possibilities for electrodes. Being soft, it easily conforms to curved surfaces, and being silver, it can be chlorided. Because of its woven texture, it resembles a grid capable of acting as a reservoir for a large amount of low-viscosity electrode jelly; the grid structure provides a large surface area. The only difficulty with this material is that associated with making adequate contact with the silvered filaments. The techniques described previously may solve this problem. However, an essential requirement is avoidance of a bimetal electrode-wire junction (which is exposed to an electrolyte).

The Recessed Electrode

Many investigators have found that much of the movement artifact encountered when a metallic electrode is in direct contact with tissue is reduced by moving the metal a short distance from the tissue and bridging the intervening gap with an electrolytic jelly or paste. In this way, the electrical double layer of charge is protected from mechanical displacement and the electrolytic bridge provides a flexible electrical communication between the skin and the electrode. It will be recalled that the double layer of charge constitutes a potential, and even slight mechanical disturbance will change the electrode potential by an amount that can be large with respect to many bioelectric events. Displacement will also alter the electrode impedance hence the recessed (floating or liquid-junction) type, in which the electrode-electrolyte junction is protected, enjoys considerable popularity among those who record bioelectric events on moving subjects.

Perhaps the first to employ the principle embodied in the recessed electrode were Forbes et al. (1921), who devised funnel electrodes to record ECGs and EMGs on elephants. When standing, elephants sway from side to side, which makes it difficult to obtain artifact-free records with ordinary plate electrodes. The electrodes employed by Forbes consisted of a zinc rod in the neck of a funnel filled with zinc sulfate. The large opening of the funnel, which was placed against the animal, was covered with a permeable membrane soaked in saline. Two rubber-gloved assistants held these electrodes against the inner surfaces of the forelimbs of an animal and the ECG was successfully recorded by a string galvanometer. In a similar manner, EMGs were recorded from the elephant's trunk muscles.

Probably many investigators were aware of the high stability of what is now known as the recessed electrode; however, very few reports have appeared in the literature. One of the first was due to Baudoin et al. (1938), who published a picture (Fig. 2-18a) of their electrode; it consisted of a small rubber cup (6–20-mn. diam.), which was applied to the skin with rubber cement.

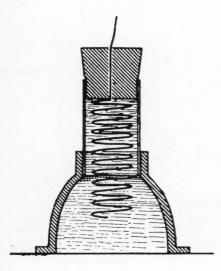

Fig. 2-18a. Baudoin's recessed electrode, consisting of a small rubber cup cemented to the skin and filled with an electrolyte into which was placed a helical silver wire. (From Baudoin et al. *C. R. Soc. Biol.* 1938, **127**:1221–1222. By permission.)

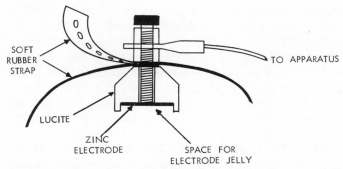

Fig. 2-18b. Haggard's recessed electrode. (Redrawn from E. A. Haggard and R. Gerbrands, *J. Exp. Psychol.* 1947, **37**:92–98.)

The cup was filled with "physiological serum" and a helical silver wire dipped into the electrolyte. Baudoin et al. reported that these electrodes were very convenient for recording skin and muscle action potentials and the EEG; they were also useful for percutaneous stimulation in studies to determine the chronaxie.

Apparently the next to discover the high stability of the recessed electrode

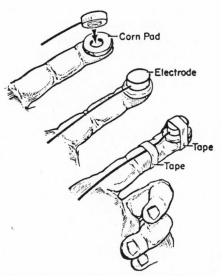

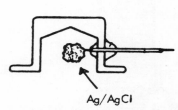

Fig. 2-18c. Lykken's corn-pad recessed electrode. (From D. T. Lykken, *J. Comp. Physiol. Psychol. 1959,* **52**:629–634. By permission.) © 1959 by The American Psychological Association.

Fig. 2-18d. O'Connell's "top-hat" recessed electrode. (From D. N. O'Connell, and B. Tursky, and M. T. Orne, *Arch. Gen. Psychiats.,* 1960, **3**:252–258, By permission.)

were Haggard and Gerbrands (1947), who placed a 7/8-in.-diam. zinc electrode in a lucite cap (Fig. 2-18*b*). The surface of the electrode was in contact with the palm via a film of electrolytic jelly about 1/8 in. thick. The electrode was used for recording changes in palmar skin resistance. An electrode of similar design was used for the same purpose by Clark and Lacey (1950).

In a study of the corneo-retinal potential in human subjects, Kennedy and Travis (1948) found that movement artifacts were not recorded if the electrode was embedded in a small sponge (1/2-in. diam. and 3/32 in. thick). The sponge was soaked with saline and secured to the subject's forehead with surgical tape. The authors reported success in recording the EEG with their electrodes mounted in a head band.

An ingenious form of recessed electrode was described by Lykken (1959). This electrode (Fig. 2-18*c*) which was used for recording the GSR consists of a corn pad applied to the cleaned and lightly sanded volar surface of the index finger. The hole in the pad was filled with electrode paste and then covered by a disk of metal, which was held in place by plastic adhesive tape; another strip of tape was used to secure the wire from the disk to the finger, thus providing strain relief.

Lykken reported that the electrodes took about 2 minutes to prepare and apply, and that they were comfortable and provided stable recordings for periods of several hours. When two such electrodes were applied to the skin, the potential difference measured between a pair of these electrodes was "dependably" less than 1 mV when zinc served as the metal and was rubbed bright with emery paper just prior to use. Although Lykken used zinc and a zinc-sulfate paste, other metals and electrolytes can be used in this most practical, easily constructed electrode.

O'Connell et al. (1960) found that artifact-free records of changes in skin potential, evoked in response to emotional stimuli, could be recorded with electrodes consisting of a silver-silver chloride sponge placed in a small enclosure resembling a top hat (Fig. 2-18*d*). (The use of a silver-silver chloride pellet did not become popular until much later.)

Those who desired to record the ECG on a moving subject soon discovered the advantage of protecting the electrode-electrolyte interface. Atkins (1961) found that the main source of artifacts was contact variations between the electrode metal and skin. When he tried separating the electrodes from the skin by a layer of filter paper or gauze, soaked with an electrolyte, electrode artifacts almost entirely disappeared from ECGs obtained on perspiring miners. Rowley et al. (1961) eliminated movement artifacts when recording the ECG from exercising subjects by employing electrodes made from 5-mm lengths of rubber tubing (9/16 in. o.d. and 5/16 in. i.d.). On each length of tubing was cemented a 2-in.^{2} gauze pad that had been saturated with rubber cement and allowed to dry; the portion of the gauze pad below the lumen of the tube was then cut away. Two tinned-copper wires were pushed through the walls at right angles across two diameters of the rubber tube midway between its top and bottom; the wires were then soldered where they crossed and thus constituted the metallic portion of the electrode. The gauze pad was glued to the skin with surgical latex adhesive. (If sweating was expected, the skin was first coated with an antiperspirant.) After the electrode was applied to the skin, the lumen of the rubber tube was filled with ECG electrode jelly and then the open end of the rubber tube was capped with adhesive tape. The ECGs obtained by Rowley et al. (using a vacuum-tube recording instrument) were indistinguishable from those taken with standard plate electrodes. These workers reported success in recording for periods up to 72 hours and pointed out that, although movement artifacts were not recorded with such electrodes, they possessed no immunity from pickup of muscle action potentials. To eliminate these effects, they placed their electrodes on the chest, thereby obtaining clear R waves from counting heart rate during vigorous exercise.

Rowley et al. reported that the high stability of their electrodes was attributable to the fact that the electrolytic jelly acted as a flexible electrical connection between the metallic electrode and the skin; and therefore, the resistance remained relatively constant during movement. Although this is true, it is only one of the factors in providing stability; the other is that the potential of the electrical double layer at the electrode-electrolyte interface is not disturbed, and hence its potential is stable.

Rowley et al. measured impedance between a pair of electrodes immediately after application and at various times thereafter. Although they did not give the frequency of the current used for measurement, they stated that initially the impedance was in the range of 21 to 110 KΩ. Within 2 hours the impedance decreased to values ranging from 2 to 60 KΩ, where it remained for 24 hours; this is a common finding with skin-surface electrodes (see p. 53).

A low-profile recessed electrode, that has gained considerable popularity in aerospace medicine and in exercise studies was described by Hendler (1961). In this electrode (Fig. 2-18e), a Monel-wire screen is mounted in a flat rubber or plastic washer which is cemented to the skin by special adhesives (Eastman 910, or Stomaseal—double-sided adhesive). The washer holds the electrode away from the skin, and contact is established via a thick film of electrolytic paste. Roman and Lamb (1962) employed similar electrodes (SAM-4) featuring stainless steel screens. When applied to the drilled skin (see p. 59) at either end of the sternum, these electrodes produced some truly remarkable records of the ECG; no artifacts were observed when the electrodes were struck or displaced, or when the subject was jumping or engaged in vigorous activity. These electrodes were employed for monitoring the ECG in pilots flying high-performance aircraft.

Lucchina and Phipps (1962, 1963) also demonstrated that a recessed elec-

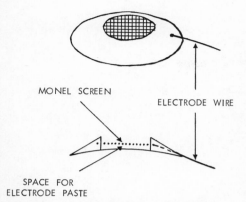

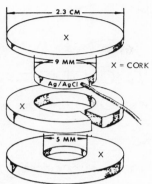

Fig. 2-18e. Hendler's monel-screen electrode. (Redrawn from E. Hendler and L. J. Santa Maria, *Aerosp. Med.* 1961, **32**:126–133.)

Fig. 2-18f. Lucchina's silver-silver chloride pellet electrode. (G. G. Lucchina and C. G. Phipps, *Aerosp. Med.* 1962, **33**:722–729. By permission.)

trode containing a silver-silver chloride pellet (Fig. 2-18*f*) was free from artifacts when pressure was applied to it or when it was displaced. Indeed, they succeeded in obtaining high-quality ECGs from ambulatory subjects. Kahn (1964) developed a similar type of electrode (Fig. 2-18*g*) in which a mixture of silver (40%) and silver chloride (60%) powder was compressed to form a silver-silver chloride pellet. (During compression, a silver wire was included to facilitate making electrical contact with the pellet.) The pellet was mounted in a rigid plastic housing with several holes drilled on the face; a small space for electrode paste was provided between the drilled face plate and the pellet. Recent reports indicate that Kahn's recessed electrodes have been employed successfully to record ECGs on laborers, subjects exercising vigorously, and even swimmers.

The electrodes used by NASA to record ECGs (and impedance pneumo-grams) on the astronauts in projects Mercury, Gemini, and Apollo (Fig. 2-18*h*) were described by Day and Lippitt (1964). The electrode consists of a chlorided silver plate (0.75 in. diam.) which is mounted in a silicone rubber housing. Holes are drilled in the plate to permit overflow of the electrode paste. The region where the connecting wire is soldered to the silver plate is covered with insulating material (epoxy) to eliminate exposing a bimetal junction to the electrode paste. The final step in assembly consists of chloriding the silver plate (3600 mA-sec/cm^2).

A very easily constructed type of recessed electrode was described by Burns and Gollnick (1966), who mounted a 50-mesh brass screen between two neoprene-foam washers 1/16 in. thick with holes 3/4 in. diam. (Fig. 2-18*i*). Even though brass (which is not known for its electrical stability) was used as

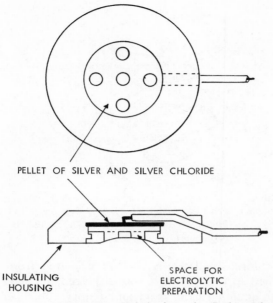

PELLET OF SILVER AND SILVER CHLORIDE

INSULATING
HOUSING

SPACE FOR
ELECTROLYTIC
PREPARATION

Fig. 2-18g. Kahn's electrode. (Redrawn from U.S. Patent #3,295,515.)

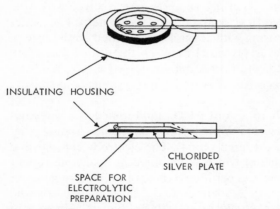

INSULATING HOUSING

CHLORIDED
SILVER PLATE

SPACE FOR
ELECTROLYTIC
PREPARATION

Fig. 2-18h. Day's chlorided-silver disk recessed electrode. (Redrawn from U.S. patent #3,420,223.)

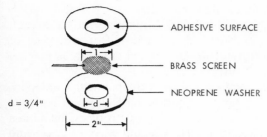

ADHESIVE SURFACE

BRASS SCREEN

NEOPRENE WASHER

d = 3/4"

Fig. 2-18*i.* Burns's brass screen electrode. (Redrawn from D. C. Burns and P. D. Gollnick, *J. Appl. Physiol.* 1966, **21:**1889–1891.)

the electrode metal, virtually artifact-free ECGs were obtained from exercising subjects using the manubrium-xiphoid electrode location. The success of this investigation and the easy construction of the electrode indicate that investigators now have an inexpensive method for examining the properties of the recessed electrode.

To record the EEG on moving subjects Adey et al. (1963) and Kado et al. (1964) constructed an interesting type of recessed electrode in which the metallic conductor was a tin rod centrally mounted in a small porous ceramic chamber of hydrous magnesium silicate, filled with a paste of tin chloride, kaolin, silica, and natrosol. The ceramic chamber was activated by soaking in distilled water for 2 to 3 hours. Contact between the ceramic chamber and the skin was established via a sponge soaked in physiological saline. Other than removing oil from the scalp, no special precautions were required prior to the application of the electrodes. When prepared and applied in the stated manner, these electrodes produced remarkably stable EEG recordings in subjects exposed to vibration forces. Even higher performance was obtained by mounting a transistor preamplifier (emitter follower) in the "top-hat" enclosure used to hold the electrode to the scalp.

Recessed electrodes need not be small in size; the author and his colleagues (Geddes et al., 1967) have used 3-in.-diam. recessed electrodes containing a chlorided silver disk to record the ECGs of large and small elephants. As mentioned previously, since these animals rarely stand still, it is difficult to obtain artifact-free recordings with ordinary plate electrodes.

Choice of the electrode materials and electrolytes for the recessed electrodes depends on the event and circumstances of measurement. The following materials have been employed with considerable success: Monel-wire screens (Hendler, 1961); crossed tinned-copper wires in a segment of rubber tubing (Rowley, 1961); stainless steel screens (Roman, 1962; Mason and Likar, 1966); silver disks (Boter et al., (1966); chlorided silver screens and plates (Shackel, 1958; Day, 1964; Skov, 1965; Geddes, 1967); disks of a compressed mixture of silver and silver chloride (Lucchina and Phipps, 1962; Kahn, 1964); tin (Kado et al., 1964); and brass screen (Burns and Gollnick, 1966). For each of these electrodes, the impedance varies with the method of preparing of the skin, the type of conducting electrolyte, and the area of the metallic portion of the electrode and the column of electrolyte.

Lucchina's and Kahn's electrode studies, which were published independently, are noteworthy because they focus attention on important features relative to the stability of silver-silver chloride pellet electrodes. Lucchina's (1962) electrode (Fig. 2-18*f*) consisted of a disk of equal parts of silver and silver chloride made by first grinding and then compressing the mixture under a pressure of 20,000 lb/in.^{2}. The disk was then mounted in a cork ring that held the electrode away from the skin. Lucchina and Phipps examined the properties of a series of electrodes having different amounts of silver and silver chloride. They measured the voltage difference and resistance between similar electrodes in contact with the electrolyte Graphogel (Tablax Corp., New York, N.Y.) in a test jig and found that decreasing the amount of silver chloride decreased the dc resistance and increased the voltage difference. Although

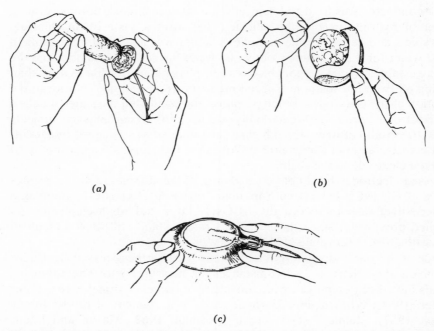

(a)

(b)

(c)

Fig. 2-19. Preparation and application of the recessed electrode: (*a*) filling with electrode paste; (*b*) removal of protective covering on adhesive surface; (*c*) application of electrode to skin.

they noted that the presence of a very tiny amount of silver chloride reduced the potential difference between a pair of electrodes, they recommended that a mixture of 30% silver and 70% silver chloride was the best compromise between voltage difference and resistance. With this version they reported a dc resistance range of 500 to 2000 Ω when the electrodes were applied to abraded skin. In Kahn's electrodes (Beckman Instruments, Spinco Division, Palo Alto, Calif.), the pellet consists of 40% silver and 60% silver chloride.

In many instances recessed electrodes, such as those in Fig. 2-18 are cemented to cleaned bare skin with nonirritating adhesives (e.g., Eastman 910). Such electrodes can be applied in much less time by the use of plastic tape that has adhesive on both sides (e.g., Stomaseal 3-M Co., St. Paul, Minn.). The procedure employed to apply the electrode (Fig. 2-19) consists of first cleaning the skin by any convenient method. Next washers are cut from the adhesive, which comes with a protective covering over both sticky surfaces. The outer diameter and the size of the hole in the washer are made appropriate for the size of the electrode to be used. Then one of the protective coverings on the adhesive is removed and the washer is applied to the portion of the supporting ring of the electrode that is to face the subject. The space in front of the metal electrode is filled with the electrolytic compound employed (Fig. 2-19*a*) and the excess can be wiped off over the protective coating on the adhesive; then the protective covering is removed (Fig. 2-19*b*) and the electrode is applied to the skin (Fig. 2-19*c*). Several manufacturers now supply packaged electrodes in which this method is employed.

Impedance of Recessed Electrodes   In many instances, recessed electrodes are used to record the ECG. The impedance range measured between a pair applied to the human thorax is higher than that found with plate electrodes, and therefore a higher input impedance is required in the amplifier. To illustrate this point the author made a study of the loss in amplitude of the ECG as amplifier input impedance was lowered. The electrodes were placed on the skin at the ends of the sternum (i.e., the manubrium and xiphoid process) in the MX lead configuration used for taking ECGs on exercising subjects (Geddes et al., 1960). Using electrodes made of stainless steel wire screen approximately 20 mm² in area, ECGs were recorded, the resistive input impedance of the input stage was lowered in steps, and the amplitudes of the ventricular waves (R and T) were measured for each input impedance. The amplitudes obtained with a 9-MΩ input impedance were taken as 100%, the amplitudes of the R and T waves obtained with the other input impedances were expressed as a percentage of the amplitudes obtained with the 9-MΩ input impedance. Figure 2-20 presents the data obtained, revealing that with a resistive input impedance of less than 5 MΩ the R and T wave amplitudes are attenuated 2 and 5%, respectively. With an input impedance of approximately

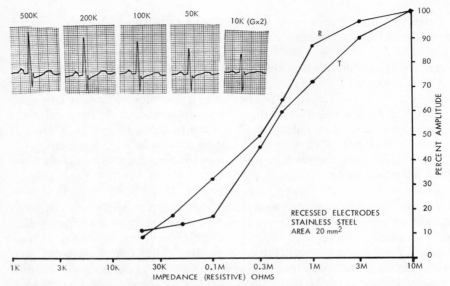

Fig. 2-20. The loss of amplitude of the *R* and *T* waves of the human ECG with recessed electrodes (area = 20 mm²) as amplifier input impedance (resistive) is reduced.

0.3 MΩ, there occurred an amplitude loss of 50%. Also illustrated are typical waveforms obtained with various input impedances. It is quite obvious that the lower the input impedance, the more the *R* and *T* waves of the ECG become distorted because of electrical differentiation by the capacitve nature of the electrode polarization impedance that begins to dominate when a low input impedance is used. Such distortion was originally reported by Lewis (1914–1915; Fig. 2-10) and more recently by Roman and Lamb (1962) and Geddes et al. (1968); this subject is discussed in detail in Chapter 3.

A thorough investigation of the relation between the area of recessed silver-disk electrodes and input impedance was carried out by the author and his colleagues (Geddes et al., 1968). The quantitative effect of connecting different resistances across the electrode terminals was determined by careful measurement of the records. For each resistance value, the amplitudes of three *R*, *S*, and *T* wave were measured. The average *R*, *S*, and *T* amplitudes were than calculated. The amplitudes obtained with the 9-MΩ input impedance were taken as 100% (controls) and the average amplitudes obtained with the various values of input impedance were expressed as percentages of these control values. Finally, the percentage amplitudes for *R*, *S*, and *T* for each input impedance were averaged and the values plotted. Presented in this way, (Fig. 2-21), the data show the average loss in amplitude of the ventricular

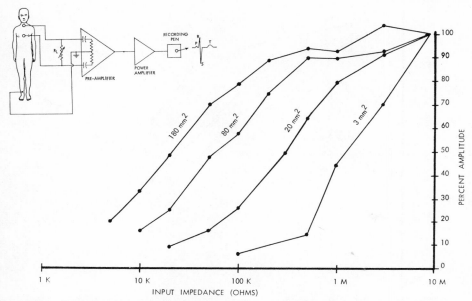

Fig. 2-21. Average loss in amplitude of R, S, and T waves as input impedance (resistive) is reduced using silver-disk electrodes having areas 180, 80, 20 and 3 mm². (From L. A. Geddes et al. *J. Electrocardiol.* 1968, **1**:51–56. By permission.)

components of the ECGs for various resistive input impedances. The figures 3, 20, 80, and 180 mm² on the curves identify the areas for the various electrodes.

When recording the ECG or any other bioelectric event, it is, of course, desirable to avoid a loss in signal—in other words, to record the true amplitude. From the data presented in Fig. 2-21, it is possible to specify the input impedance that will produce a known loss in amplitude with electrodes of various areas. Figure 2-22 gives the relation between input impedance and electrode area for various average percentage reductions in amplitude of the R, S, and T waves as derived from the data obtained. From this illustration it is apparent that if a recessed electrode with a 10-mm-diam. circular silver plate (approximately 80 mm² in area) is used with an ECG amplifier with a 2.7 MΩ input impedance, a 5% loss in average amplitude of the R, S, and T waves will occur. If the input impedance of the amplifier is 0.6 MΩ the average amplitude loss will be 10%. Similarly, the amplitude loss with other areas and input impedances can be predicted.

Although a slight loss of amplitude encountered with an inadequately high input impedance may seem unimportant, the situation is accompanied by the

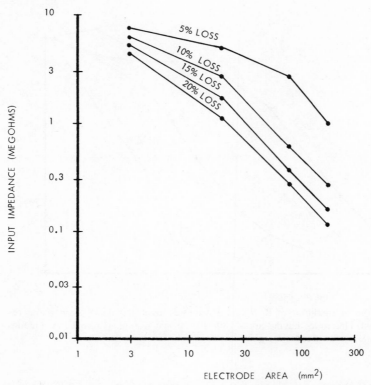

Fig. 2-22.   Average loss in amplitude of *R*, *S*, and *T* waves when various input impedances (resistive) are used with recessed electrodes of various areas. (From L. A. Geddes, et al. *J. Electrocardiol.* 1968, **1**:51–56. By permission.)

possibility of waveform distortion (see inset, Fig. 2-20) and it means that the electrode impedance is beginning to become an appreciable part of the impedance of the input circuit and that, moreover, the relative amplitudes and temporal relationships of the complex of bioelectric waves will be altered. Although slight loss of amplitude coupled with insufficiently high input impedance arises most frequently with needle electrodes, the condition has been observed with small-area recessed electrodes; for example, Roman and Lamb (1962) demonstrated that with their recessed electrodes (0.2 cm² in area, made from stainless steel screens), the use of an amplifier with an inadequately high input impedance resulted in distortion in the ECG of considerable clinical significance. The type of distortion, which is recognizable as electrical differentiation, consists of a prominent displacement in the level of the *ST* segment and the addition of small second phases to the *P* and *T* waves. This phenomenon is discussed in detail in Chapter 3. The value of input impedance where such distortion begins to appear can be determined by performing a loading test as described here.

Following the loading tests described previously, impedance-frequency curves were obtained with each pair of electrodes applied to the chest; a current of a few microamperes was employed in each case. The data obtained (Fig. 2-23) show the inverse relation between impedance and electrode area and the decrease in impedance with increasing frequency, a characteristic of the capacitance of the electrical double layer of charge at each electrode-

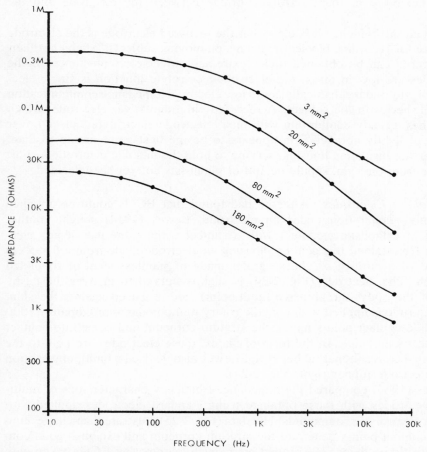

Fig. 2-23. Impedance frequency curves for recessed electrodes of various areas. (From L. A. Geddes et al. *J. Electrocardiol.* 1968, **1**:51–56. By permission.)

electrolyte boundary. The high impedance of small-area recessed electrodes focuses attention on the need to use an amplifier with an adequately high input impedance if undistorted bioelectric events are to be recorded.

Although considerable attention has been devoted to use of the recessed electrode for recording ECGs, it can, and has been used for a variety of bioelectric events because of its high electrical stability, which is primarily the result of protection of the electrode-electrolyte interface (the site of the electrode potential) from mechanical disturbance. The high stability that has been obtained with different metals indicates that protection of the double layer from mechanical movement is the major contributor. Another factor adding to stability is the relative constancy of impedance with movement due to the flexibility of the electrolytic bridge between the electrode and the subject.

From all the evidence it is clear that the recessed electrode is the electrode of choice for recording bioelectric events on moving subjects. Almost artifact-free records can be obtained under extremely adverse circumstances if the electrodes are low in mass, the electrode-electrolyte junction is stable electrochemically and mechanically, and a low electrical impedance communication is established with the subject. A recessed chlorided silver electrode (which establishes a ready equilibrium with most electrode electrolytes) placed over regions of lightly abraded skin appears to be the best combination to date. Whether the insulating housing, serving to hold the metallic electrode a short distance from the skin, should be stiff or pliant has not yet been decided.

Multipoint Electrodes   After he had found that ECGs could be obtained with only perspiration under an electrode, Lewes (1954) sought another method of establishing contact with a subject without the use of any electrolyte. He attained this goal by devising an electrode made from a 6 × 5 cm segment of a nutmeg or cheese grater made of stainless steel or tin-plated soft iron. The electrode (Fig. 2-24) is slightly curved to fit over the fleshy parts of the body; the abrasive (multipoint) side is placed against the skin. The electrode is applied with a gentle rotary motion, and approximately 1000 fine active contact points pierce the stratum corneum and come into contact with the tissue fluids. In the form of plates, these electrodes are held to the skin with a conventional rubber strap; Lewes also devised a multipoint suction chest electrode (upper right, Fig. 2-24).

Lewes (1966) compared the impedance-frequency characteristics of multipoint electrodes with those obtained using standard electrodes applied with various electrolytic compounds. His data (Fig. 2-25) illustrate that a large number of contact points penetrate the stratum corneum and establish good contact with tissue fluids. He proved his point by comparing ECGs taken with multipoint electrodes with those obtained with conventional electrodes, and

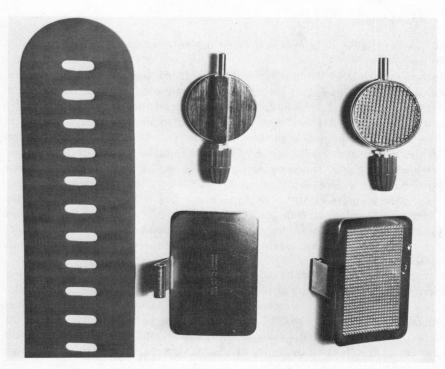

Fig. 2-24. Lewes's multipoint electrodes. (Courtesy D. Lewes, Bedford, England.)

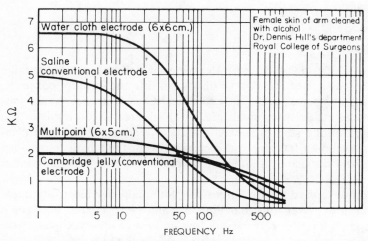

Fig. 2-25. Impedance-frequency curves for multipoint and conventional electrodes. (Courtesy D. Lewes and D. Hill.)

found no detectable differences; a conventional vacuum-tube electrocardiograph was used.

Multipoint electrodes offer several advantages for electrocardiography; speed of application and removal are perhaps the most important. In a demonstration to the author, Lewes applied four limb electrodes, obtained recordings from leads I, II, and III, and removed the electrodes in 80 sec. Another attractive feature is elimination of the need to prepare the skin before and after use of the electrodes. The absence of the requirement for electrode jelly permitted use of the electrodes to obtain ECGs in the freezing, low-pressure environment of the Himalayas at 13,000 ft. They have also been used successfully to obtain ECGs from expensive animals, when preparation of the skin was not permissible. The author has been able to obtain ECGs in the horse using multipoint electrodes applied to the clipped skin. It would therefore appear that the multipoint electrode will see extensive service in a variety of situations.

A word of caution must be added regarding the long-term application of multipoint electrodes that are applied to unclean human skin. It will be recalled that normal skin harbors numerous bacteria and fungi; it is also warm and becomes moistened with sweat if covered for long periods. Such an environment favors the multiplication of microorganisms and, if the population becomes large, there is the possibility of a local skin response to an invasion via the tiny holes made by the multipoints. However, as stated previously, this situation would not occur without prolonged application of the electrodes to a skin rich in microorganisms. There are no reports of such an occurrence with routine use of these electrodes, which are both very practical and ideal for short-term application.

Metal-Film Electrodes

Several types of electrodes have been described which consist of a metallic film placed on the surface of the skin or deposited into the stratum corneum. Thompson (1958) reported success in recording ECGs from exercising subjects with electrodes made from small pieces of silvered nylon applied to a lightly sanded area of the skin (a special conducting adhesive held the electrodes in place). Although the electrodes worked well, their high impedance (150,000 Ω) required the use of a differential amplifier with a very high input impedance.

Edelberg (1963) described a most interesting low-mass dry electrode made by electrodeposition of silver into the skin. In man, the resistance between a pair of silver depositions was remarkably low and no electrolytic paste was required; the silver spots were virtually terminals on the subject.

No further development of Edelberg's interesting electrode has taken place.

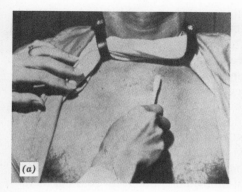

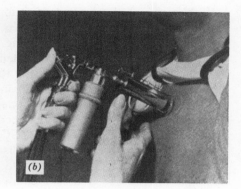

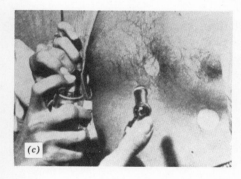

Fig. 2-26. Roman's spray-on electrode made by the application of a conducting glue containing silver powder: (*a*) skin preparation; (*b*) electrode application; (*c*) dry electrode. (From J. Roman, *Aerosp. Med.* 1966, **37**:790–795. By permission.)

At present, the only drawback is the relatively short life of the electrode. As time passes, the stratum corneum is shed; in addition, the silver undergoes chemical changes and the electrode eventually disappears.

A rapidly applied, low-mass "dry" electrode was developed for aerospace monitoring by Roman (1966). The electrode, which consists of a film of conducting adhesive, is applied to the skin by first rubbing electrode jelly into it for 3 sec using a toothbrush; after this the skin is wiped dry. Next a film of conducting adhesive [43 g Duco Household Cement (Dupont S/N 6241), 23 g silver powder (Handy and Harman Silflake 130)], and 125 cc acetone is painted or sprayed on the skin to form a conducting spot about 2 cm in diam. Then an 0.004-in. silver-plated copper wire is placed in the conducting adhesive and is captured as drying proceeds; when the assembly is dry, a coat of insulating cement is applied over the electrode. Figure 2-26 illustrates the process of preparing the skin *a* and applying the metallic film with a spray gun *b*; *c* is the circular film electrode in which the electrode wire is captured.

With a pair of these electrodes applied to the subject's chest, remarkably clean ECGs have been recorded. To date more than 500 hours of flight and 700 hours of ground recording of the ECG have been logged successfully by U.S. Air Force personnel. No incidence of skin irritation has been observed.

From the literature it appears that Roman's electrodes are ideally applicable for recording ECGs on exercising subjects. Frank et al. (1968) reported using these electrodes successfully with one over the manubrium and the other over the apex. Lategola et al. (1969) placed a number of them on the thorax, forehead, and ear lobe to telemeter ECGs from normal subjects and those with known histories of coronary artery disease. In this investigation it was pointed out that serious consideration should be given to selection of electrode locations that provide the greatest amount of information about myocardial metabolism with a minimum of muscle (not electrode) artifacts. This point has been repeatedly brought forth, but no agreement exists for the optimum location for electrodes on exercising subjects.

Roman measured the equivalent parallel resistance and capacitance values of a number of electrodes of differing areas on several subjects; Fig. 2-27*a* gives the data obtained. Although the frequency of measurement was not

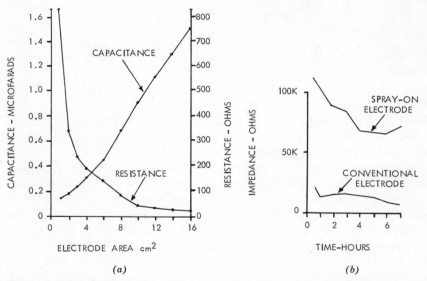

Fig. 2-27. The characteristics of spray-on electrodes: (*a*) parallel-equivalent resistance and capacitance values as a function of area for conductive-film spray-on electrodes applied to a human subject; (*b*) the reduction in electrode impedance with time for a spray-on conductive-film electrode and a conventional recessed electrode on a human subject. (Redrawn from J. Roman, *Aerosp. Med.* 1966, **37**:790–795.)

supplied, the order of magnitude of these components does appear, and it is consistent with the factors present (i.e., increasing the area increases the capacitance and decreases the resistance of the parallel-equivalent circuit).

Roman recognized the high-impedance nature of his electrodes by recommending that an amplifier input impedance of 2 MΩ (or higher) be employed. The 15-Hz impedance of electrodes without skin preparation was measured over a period of 7 hours; Fig. 2-27*b* presents the evidence that the initial impedance was about 150 KΩ and fell to about 60 KΩ in about 4 hours. With skin preparation, the impedance values were reduced by about 30 KΩ. In the same illustration the impedance-time data are given for a 1-cm recessed electrode on prepared skin. This characteristic is discussed further on p. 54).

Conducting Rubber Electrodes

Jenkner (1967) claimed that a low-resistance contact could be established with the skin when carbon-loaded silicone rubber (Silastic S-2086, Dow Corning, Center for aid to Medical Research, Midland, Mich.) was used as an electrode without electrolytic paste or jelly. A disk (1 cm diam. 1.15 mm thick) of this material, which has a resistivity of 15 Ω-cm, was placed between conventional EEG electrodes and the scalp; between pairs of such electrodes Jenkner reported "testing of electrodes prepared in the above manner showed them to have an electrode resistance of 1.5 KΩ when applied over the unprepared dry skin."

Jenkner's observations are difficult to reconcile in view of the high resistance of the superficial layers of the skin. It is not clear whether the 1.5-KΩ resistance represents (a) the resistance measured after the passage of enough time for the accumulation of perspiration, which is known to lower resistance, or (b) the value at 800 Hz, where he made other measurements. (At 800 Hz the impedance for any electrode is much lower than the dc resistance). It is therefore too soon to comment on the merits of these interesting electrodes, which do enjoy the feature of flexibility and ease of application to curved surfaces.

Silicone rubber electrodes have been used successfully by Artz (1970) for recording the ECG; the electrodes were applied to skin that had been prepared by abrading with pumice added to chlorhexidine digluconate in ethanol. By this process it was possible to obtain a resistance of 5000 Ω measured between two electrodes 8 cm^2 in area. Artz also reported that such electrodes have been used for periods up to one week by injecting electrode jelly under the electrode with a syringe and needle. The future will certainly see further investigation of the value of such pliant conducting rubber electrodes, without and with electrode pastes.

As stated previously, the types of dry skin-surface electrodes described in this section are relatively high in impedance. As Lewes (1965) showed, even without an electrode paste, if the electrodes are over a region containing sweat glands, perspiration accumulates and reduces the impedance in a relatively short time. However, any amplifier that is to be used with dry electrodes should have an input impedance that is high with respect to the initial impedance measured in the frequency spectrum of the bioelectric event to be recorded.

Dry Electrodes

One of the major desires among nearly all who measure bioelectric events is to be able to place dry electrodes on unprepared skin and proceed with the measurement. This time-saving feature would be accompanied by the ability to monitor a bioelectric event for prolonged periods without skin irritation or deterioration of the electrode-subject coupling. This desire is about to be fulfilled insofar as application of electrodes to human skin is concerned. In addition to Lewes's multipoint electrode, there are two types of electrodes that do not require skin preparation or paste—one is the dry electrode, in which a dry metallic plate is placed on the skin; the other is the insulated electrode, which is described in the next section.

Bergey et al. (1971) investigated the impedance-frequency characteristics of dry, 2.5-cm disks of anodized aluminum, stainless steel, gold, and silver applied to the unprepared inner surface of the forearm with Micropore adhesive tape. The current employed was 2 μA (peak-to-peak) and the frequency

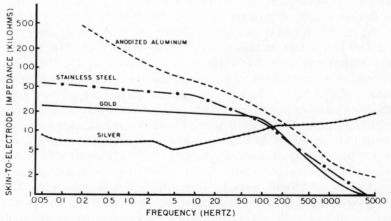

Fig. 2-28a. Impedance-frequency characteristics of single disks (2.5 cm diam.) of different metals placed on the inner aspect of the forearm; no skin preparation or electrode paste were employed. Measurements were made within 10 minutes of application; a peak-to-peak current of 2 μA was employed.

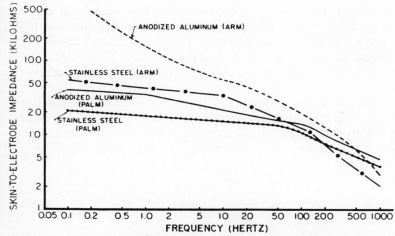

Fig. 2-28b. The effect of perspiration on the impedance-frequency characteristics of metal disks (2.5 cm diam.). The data for dry skin were obtained by placing the electrodes on the forearm; those for moist skin were obtained by placing the electrodes on the palm. (From F. G. Bergey et al. *IEEE Trans. Bio-Med. Eng.* 1971, **BME-18**:200–211. By permission.)

range extended from 0.05 to 5 KHz; measurements were made within 10 minutes of application of the electrodes. The data they obtained appear in Fig. 2-28*a*.

Bergey et al. pointed out that the presence of perspiration lowered the impedance of the electrodes (Fig. 2-28*b*). They also noted that, although silver exhibited the lowest impedance, they had chosen stainless steel for their electrode, because the least amount of movement artifact was obtained with this metal in a series of electrocardiographic tests with the various electrodes mounted directly on source-follower amplifiers (as described by Richardson in the next section).

The success demonstrated by Bergey et al. should open the door for further study of such electrodes placed on thick and thin, dry and moist skin, with measurements carried out over prolonged periods. In particular, it would be of value to know the range of dc resistance to establish the requirements for the measurement of steady (dc) voltages using these electrodes.

Insulated Electrodes   A truly dry electrode of considerable promise was described by Richardson et al. (1967, 1968) and Lopez et al. (1968) of the U.S. Air Force. It consists of an aluminum plate (2.5 × 2.5 cm) that is anodized on the surface placed in contact with the skin. On the back of the electrode is mounted a field-effect transistor (FET) with the gate terminal connected to the electrode. Surrounding the anodized electrode is an insulating block of potting compound contained by a circular metal ring that acts as an electrostatic shield (Fig. 2-29*a*); the FET is connected as a source follower (Fig. 2-29*b*). To protect the FET from acquiring a high electrostatic voltage, a high-resistance leakage path is provided by using two diodes (1N3600) placed in series opposition. The diodes can be omitted if a higher input impedance is desired.

Richardson reported that the resistance of the anodized electrode, measured between the gate of the transistor and the subject, is between 1000 and 30,000 MΩ; thus the electrode is capacitively coupled to the subject. According to the various investigators who have used them, the high input impedance of

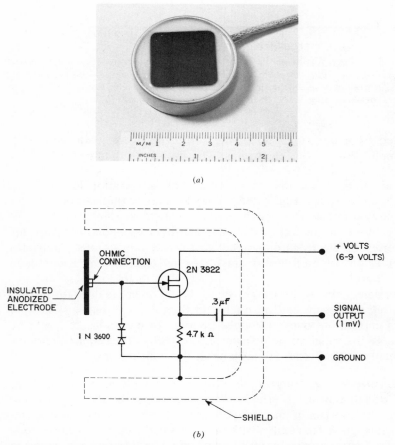

(*a*)

(*b*)

Fig. 2-29. Insulated electrode: (*a*) Richardson's insulated electrodes; (*b*) source follower circuit for use with the insulated electrode. (Courtesy of P. Richardson, 1969.)

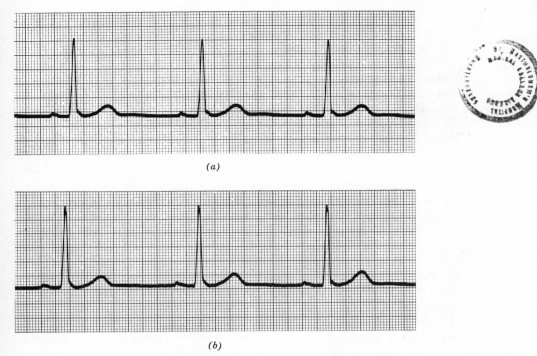

(a)

(b)

Fig. 2-30. Simultaneously recorded ECG: obtained with conventional electrodes (*a*) and insulated electrodes (*b*). (Courtesy P. C. Richardson. Personal communication, 1969).

the FET provides a time constant long enough for reproduction of the human ECG. In use, two electrodes are applied to the subject with no skin preparation. The outputs of the source followers can be connected to a conventional electrocardiograph via an adequately large coupling condenser. Figure 2-30 is an example of records taken with insulated electrodes and conventional plate electrodes to which electrode jelly has been applied.

Following Richardson's lead, Wolfson and Neuman (1969) developed a small insulated electrode consisting of a chip of *N*-type silicon (6 × 6 mm, 0.23 mm thick) on which was thermally deposited a circular layer of silicon dioxide (0.2 μ thick, 4.5-mm diam.). The region outside of the circle and on the edges of the chip was insulated with a layer of silicon dioxide 1.5 μ thick. The electrode was connected and mounted to an ultra-high input impedance amplifier that employed a metal-oxide semiconductor field-effect transistor (MOSFET) as the input stage arranged in a source-follower configuration. A second MOSFET was used as an electronic switch to protect the input transistor from damage due to stray electrostatic voltages. The input impedance attained was

10^{10} Ω and the sinusoidal frequency response measured from the subject to the output of the source follower extended to 0.005 Hz, which indicates a capacitive coupling to the subject of 3200 pf and provides an input constant of 32 secs.

Wolfson and Neuman reported that their electrode was designed for electrocardiography; however, they stated that it could be used for electroencephalography and electromyography. The overall time constant obtained is more than tenfold longer than that required for electrocardiography and even greater than that required for electroencephalography and electromyography. Its use in the two latter applications has not been reported.

Although it is too soon to be able to describe any extensive experience with the insulated electrode, certain of its interesting characteristics can be identified. Like other dry electrodes, it can be applied quickly without skin preparation, eliminating all danger of tissue response which accompanies the use of some conventional electrolytes. There is no doubt that the amount of capacitive coupling and input impedance attainable is adequate for the measurement of a variety of bioelectric events. By calculation from the data given by the various authors and by Rylander (1970), it appears that the capacitances attained range from about 5000 to 20,000 pF/cm² of electrode area. In using such capacitively coupled electrodes with FETs, it is necessary to take precautions to prevent electrostatic puncture of the gate insulation. When completely isolated instruments (e.g., battery-operated telemeters) are used, this requirement is less severe.

With the insulated electrode, although there is no electrode-electrolyte interface, which is a source of artifacts with conventional electrodes, another type of movement artifact can be encountered. Displacement on the skin will change its capacitive coupling and hence alter the charge distribution and potential. Whether the magnitude of the voltage so developed is comparable to bioelectric events remains to be seen.

At present there are attempts to provide ultrathin films of insulating materials having high dielectric constants and strengths so that a high electrode-to-subject capacitance will be attained and the insulation will not be punctured by the voltages that such insulated electrodes may acquire. Heroic efforts in this direction will not be amply rewarded because it must be remembered that the other "plate" of the electrode is in reality constituted by the tissue fluids below the outer horny layer of the skin, which varies in thickness from 20 to 500 μ. Thus the true thickness of the dielectric is the thickness of the horny layer plus that of the insulating film on the electrode.

Recently Lagow et al. (1971) called attention to the fact that the insulation deposited on some insulated electrodes is permeable to perspiration (a dilute NaCl solution). In particular, films of aluminum oxide are corroded by exposure to saline; this results in a dramatic reduction in resistance, thereby invalidating use of the term "insulated" to describe the electrode. To eliminate this problem, they first demonstrated that oxidized tantalum is saline resistant and then anodized this material to fabricate insulated electrodes with a stability much in excess of that encountered with aluminum oxide.

Electrodes for Exposed Tissues   In many situations it is desirable to surgically expose groups of cells for application of the electrodes. One example of this technique is recording of the electrical activity of the human cortex during neurosurgical procedures. In this application it is convenient to use small spherical metal electrodes, bare or covered, with cotton soaked in saline (Fig. 2-31a). In other circumstances it is desirable to use the wick or "nonpolarizable" electrode (Fig. 2-31b).

The construction and use of the silver-ball electrodes was described by Geddes (1948, 1949) after he had learned the technique in Jasper's laboratory in the Montreal Neurological Institute. The ball is formed by holding the end of a clean silver wire (about 0.025 in. diam.) in the hottest part of a bunsen burner flame (i.e., in the tip of the internal cone). Just after the silver becomes red, it suddenly melts and the surface tension causes the molten silver to form a ball, the size of which can be controlled by the time the ball is permitted to remain in the hottest part of the flame. If a large ball is desired, the wire must be advanced slightly and rotated so that the ball will be symmetrically located at the end of the silver wire. When the ball attains the desired size, the wire is quickly withdrawn from the flame and allowed to cool in the air. The ball is then thoroughly cleaned and polished with emery paper to remove all the sur-

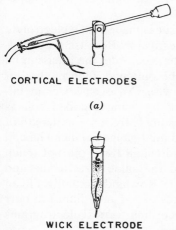

CORTICAL ELECTRODES

(a)

WICK ELECTRODE

(b)

Fig. 2-31. Electrodes for exposed tissues: (a) cortical electrodes; (b) wick electrode.

face contaminants produced by the flame. The shaft of the wire is mounted to whatever holder will be used to manipulate the electrode, and the wire is given an insulating covering. The ball is then chlorided with a deposit of about 500 mA-sec/cm^2 with a current density of at least 5 mA/cm^2. If cotton-wool pads are to be used, they should be applied before the chloriding process.

The impedance of silver-ball electrodes depends on their size and the nature of the surface (i.e., whether it is chlorided). Figure 1-11 presents the impedance-frequency curves of 1.3 mm diam. bare and chlorided electrodes measured versus a large indifferent chlorided electrode in a beaker of 0.9% saline. The magnitude of the impedance of a pair of silver-ball electrodes (1.7-mm diam.) placed on the auditory cortex of a cat was determined by noting the diminution in the amplitude of the evoked cortical potential as the input impedance of the associated amplifier was reduced. With the bare electrode, reducing the input impedance (resistive) to 50,000 Ω reduced the amplitude of the evoked potential by 10% when compared with that obtained with 9.4 MΩ input impedance. Using a chlorided pair of electrodes, a loss in amplitude of 10% was encountered with an input impedance of 30,000 Ω. These data indicate that an input impedance of 3 to 5 MΩ would be satisfactory for silver-ball electrodes serving in this application.

The wick or nonpolarizable electrode (Fig. 2-31*b*) obviously derives its origin from the field of electrochemistry. As used in electrophysiology, it is usually made from a glass medicine dropper. Through the small drop-forming end is passed a wick, which is often a thread or string. The shaft of the dropper is filled with a solution that is compatible with the fluid bathing the cells from which recordings are to be made. Into the solution is placed a metal electrode (usually a chlorided-silver wire supported by a cork, which makes the dropper airtight and prevents leakage of the fluid.

In using the wick electrode, contact with the tissue is made via an electrolyte of composition and dimensions chosen by the investigator. Because the metal-electrolyte junction is contained in the shaft of the electrode, the area can be made large and a low interface impedance can be obtained. One method of providing a large area is to coil the end of the wire. An even lower metal-electrolyte interface impedance can be obtained by using a coiled coil as described by Rowland (1961). His coiled coil was made from a 5-in. length of 30-gauge silver wire that had been wound around a 20-gauge needle. The coil thus formed was stretched slightly so that the individual turns did not touch each other. This coil was then wrapped around the same size needle and inserted into the shaft of the medicine dropper. Rowland actually affixed several such electrodes with tapered capillary ends (0.5–1 mm diam.) to burr holes in the skull of an animal for the purpose of recording cortical electrograms over a three-month period.

Burr (1944, 1950) reported that a pair of nonpolarizable wick electrodes could be made to have a potential difference of 10 μV or less. With Rowland's electrode, the potential difference between a typical pair was 1 mV or less; the random noise fluctuations amounted to 2 to 3 μV. Kahn (1965) employed a compressed mixture of silver and silver chloride as the material in contact with the electrolyte (which was saline and, on occasion, plasma). Between pairs of such electrodes the potential difference reported was 5 to 10 μV. These observations clearly demonstrate that an extremely high stability can be realized with these electrodes.

The impedance of wick electrodes depends on the area and type of the metal-electrolyte interface, the resistivity of the electrolyte employed, the area and length of the tapered tip, and the type of wick used. If the metal in contact with the electrolyte is silver, it can be chlorided to obtain a low impedance and stable potential. The order of impedance encountered can be estimated using Rowland's data. With 0.5 to 1 mm tips (of unspecified length) filled with saline and without wicks, Rowland reported resistances in the range of 30,000 to 50,000 Ω. Larger-diameter tips would be expected to have a lower resistance.

The Thermode When it is desired to measure a bioelectric event in a region that will be heated or cooled, it is advantageous to employ the thermal electrode or "thermode," an electrode incorporating a chamber through which hot or cold water can be circulated. One version of this type of electrode (Fig. 2-32) has been used by Hoff with the author (1970) to record the action

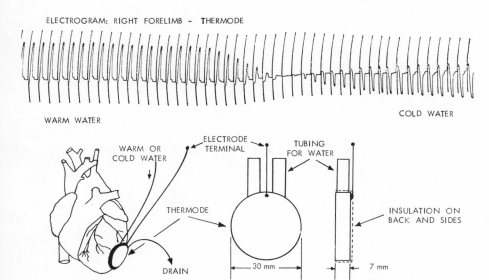

Fig. 2-32. Semidirect cardiac electrogram obtained with warm and cold water flowing through the thermode (From H. E. Hoff, L. A. Geddes, A. G. Moore, and J. Vasku, *J. Electrocardiol.* 1970, **3**:333–335. By permission.)

potential of mammalian myocardium by placing the thermode between the pericardium and apex of the left ventricle of a dog. The electrode consists of a silver-plated copper chamber, insulated everywhere except the surface applied to the myocardium. Figure 2-32 illustrates the semidirect cardiac electrogram taken with the thermode (paired with an electrode on the right forelimb), recorded first with warm water (42°C) flowing through the thermode. Cold water (0°C) was then caused to flow, and the progressive delay in ventricular repolarization inder the thermode is shown by the gradual reduction in amplitude and inversion in the *T* wave.

The thermode can be arranged to be part of a suction electrode to create a very practical self-adhering electrode whose temperature can be easily controlled. Figure 2-12*c* illustrates one such arrangement; here, when a slight negative pressure is applied to the annular space surrounding the thermode, the electrode adheres well to a wetted surface.

3

Subintegumental Electrodes

It is often desirable to locate an electrode close to an aggregate of irritable cells; frequently the cells are below the integument or even below the surface of an exposed organ. In such a situation it is convenient to employ a surgical implantation technique; if short-term recording is desired, it is usually permissible to advance an electrode into or near the structure of interest. Although an electrode used for this purpose may have a small area, it is still large with respect to the dimensions of the cells from which electrical activity is to be recorded.

Needle Electrodes

The hypodermic needle electrode entered electrophysiology via the studies of Adrian and Bronk (1929), who wanted to record the electrical activity of a few skeletal muscle fibers in man and the cat. To do so they made monopolar and bipolar hypodermic needle electrodes, somewhat like those in Fig. 3-1a. The monopolar electrode consisted of a 36-gauge enameled copper wire (193-μ diam.) and the bipolar electrode employed two 44-gauge enameled copper wires (81-μ diam.) centrally mounted in hypodermic needles. Although the investigators described the bipolar electrode, they did not use it; parenthetically it must be noted that since the differential amplifier had not been developed at that time, a shielded bipolar electrode was of little practical use.

With the central conductor of the monopolar electrode (Fig. 3-1a) connected to the grid and the outer conductor connected to the grounded side of their amplifier, Adrian and Bronk recorded the action potentials of the human triceps, both at rest and with increasing contraction. Their capillary electrometer records showed that the frequency of the action potentials, which were obviously recorded from a few muscle fibers adjacent to the electrode, was related to the force of contraction.

The use of bare needles to record the ECG was described by Straub (1922),

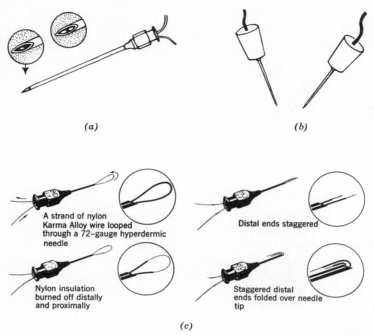

(a)   *(b)*

A strand of nylon
Karma Alloy wire looped
through a 72–gauge hyperdermic
needle

Distal ends staggered

Nylon insulation
burned off distally
and proximally

Staggered distal
ends folded over needle
tip

(c)

Fig. 3-1. Subintegumental electrodes: (*a*) hypodermic electrode; (*b*) needle electrodes; (*c*) steps in making new bipolar electrode assembly before sterilization (from J. V. Basmajian and G. Stecko, *J. Appl. Physiol.* 1962, **17**:849. By permission).

who observed that if the area of each electrode in contact with the tissue fluids was small, the result is distortion in the ECG, as recorded with the string galvanometer; this phenomenon is discussed on page 120. Perhaps for this reason, needle electrodes were infrequently used for electrocardiography until much later.

In clinical electromyography it is customary to use a steel needle inserted directly into the muscle, paired with another larger electrode on the surface of the skin. Sometimes two needle electrodes are placed into a muscle. Figure 3-1*b* illustrates a frequently employed needle electrode, as described by Jasper (1945). When the shaft of the needle electrode is coated with insulating varnish, the area of the electrode in contact with the active tissue is quite small; in the case of Jasper's electrode, the tip area was about 0.2 mm².

During the last several years the need has arisen for small-area electrodes that may be left in place for prolonged periods, and many interesting electrodes have been developed to meet this requirement. The main goals have been ease of insertion, freedom from pain during insertion and while the electrode is in place, mechanical and electrical stability with movement, and minimal restraint. Although few electrode types have been perfected, some very promising electrodes have been constructed; most have been used for electromyography. For example, Basmajian and Stecko (1962) developed a bipolar, fine-wire (25-μ diam.) electrode that is easily inserted and remains well anchored in the muscle. The steps in construction of this electrode appear in Fig. 3-1*c*. The sketch at the lower right hand corner of the illustration shows the configuration of the electrode when it is ready for placement in the muscle. When the desired depth has been reached, the needle is withdrawn, leaving the electrode wires in the muscle; the bent ends serve as hooks preventing the wires from easy withdrawal.

A similar fine-wire subcutaneous electrode was described by Scott (1965), who employed an insulated wire (Karma, Driver-Harris Co., Harrison, N.J.) passed through the lumen of a hypodermic needle and bent back to pass along the outside of the needle (Fig. 3-1*d*-1). The needle and wire are inserted to the desired depth in the muscle (Fig. 3-1*d*-2) and then, with the outside wire held firmly, a pair of forceps is applied to the inner wire. Winding the wire on the forceps (Fig. 3-1*d*-3) causes it to be cut by the sharp edge of the hypodermic needle; whereupon the needle is withdrawn, leaving the outer wire in the

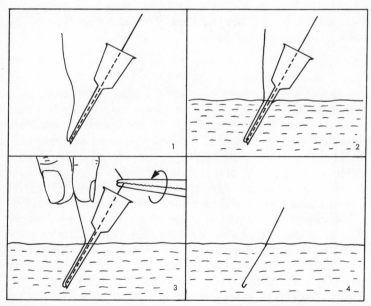

Fig. 3-1*d*.   Scott's electrode. (Redrawn from R. N. Scott, *IEEE Trans. Bio-Med. Eng.* 1965, **BME-12**:46–47.)

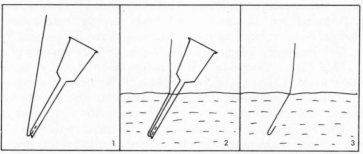

Fig. 3-1e. Parker's electrode. (Redrawn from G. Parker, *Amer. J. Phys. Med.* 1968, **47**:247–249.)

tissue (Fig. 3-1*d*-4). The active surface of the electrode is the cross-sectional area of the wire.

A simpler method of inserting a fine-wire electrode was developed by Parker (1966). With his technique, a short length at the end of a fine wire is bent back upon itself. The bent- back portion is then inserted into the tip of the hypodermic needle (Fig. 3-1*e*-1) and the needle and wire are advanced into the muscle (Fig. 3-1*e*-2). At the desired depth, the needle is withdrawn, leaving the electrode hooked into the muscle (Fig. 3-1*e*-3). With this technique, monopolar or bipolar electrodes can be installed. The active surface of the electrode is the cross-sectional area of the wire.

Cutting Electrodes

Occasionally it is necessary to record bioelectric events (most frequently the ECG) from animals with thick, dry, and sometimes hairy and oily hides. In this situation, conventional plate electrodes do not establish adequate ohmic contact, even when the integument is cleaned or shaved and electrode paste is rubbed in. To solve this problem, and to provide an easily applied electrode that remains tenaciously in place and establishes a low-resistance contact with such animals, Hermann (1962) developed cutting electrodes from battery clips with phonograph needles mounted between the jaws (Fig. 3-2, sketches 1–4). The clips were applied to a fold of skin and a rubber sleeve covered all but the biting end. Hermann reported that the electrodes have been used to take ECGs on anesthetized baboons, dogs, rats, and a variety of other experimental animals. No electrical data were given for these very practical electrodes, which will undoubtedly see considerable service.

The goal described by Hermann can be attained without soldering phonograph needles to a battery clip. There is a clip called the "cable-piercing" clip (Type 50 C, Mueller Electric Co., Cleveland, Ohio which is equipped with a sharp point, but Hermann found that it was too far back in the jaws to pierce a fold of skin. However, the point can be brought forward easily by sawing off about 1/8 in. from the jaws of the clip.

The author has also used alligator clips fitted with an injection needle mounted between the jaws to permit flooding the area under the electrode with an electrolyte of choice. This type of electrode (Fig. 3-2, sketch 5) has been used with an insulated rubber covering to record the action potential from an

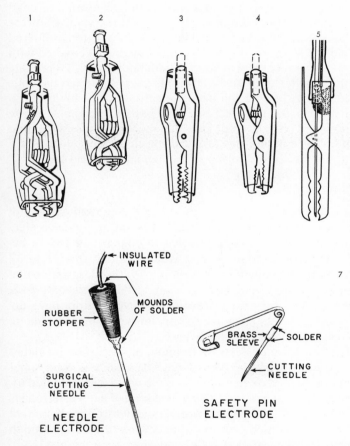

Fig. 3-2. Various types of cutting electrodes: 1–4, cutting electrodes made by soldering phonograph needles inside the jaws of spring clips (from G. Hermann, *Circ. Res.* 1962, **11**:736–738); 5, injection electrode (unpublished); 6 and 7, safety-pin and needle electrodes made with surgical cutting needles (from Geddes et al., *Southwest. Vet.* 1964, **18**:56–57. By permission.)

atrial appendage in the dog in order to demonstrate shortening of the refractory period by the local application of acetylcholine.

In many instances it is difficult to insert needle electrodes below the thick hides of animals. To solve this problem, the author (Geddes et al., 1964) developed two types of cutting electrodes (Fig. 3-2, sketches 6 and 7) made from surgical cutting needles, which have beveled sharpened shanks that permit easy insertion through the hide. The size of the needle can be selected by the investigator.

When cutting electrodes are used, movement artifacts can be minimized by inserting the needles so that the area of the bare metal electrode in contact with the tissues is constant. The safety-pin electrode should be inserted through a pinch of skin and fastened. When the pinch is released, the skin will press against the head and spring of the safety pin. The connecting wire is soldered to the brass sleeve, and the sleeve and solder connection are all covered with insulation to prevent them from coming in contact with body fluids. To provide strain relief for the solder joint, the connecting wire is passed through the coils of the spring and tied. Similarly with the needle electrode, it is advisable to insulate the soldered portions of the electrode and part of the shank above the cutting edge and to insert the electrode into the animal far enough so that no bare needle protrudes. The author has used cutting electrodes successfully to record ECGs from horses, cows, dogs, armadillos, snakes, and lizards.

Depth Electrodes

In many circumstances—particularly in studies of the electrical activity of neuron clusters deep within the brain, when it is necessary to advance electrodes into the regions of interest, it is convenient to employ what have become known as depth electrodes. Many types, which contain from 1 to 37 recording sites, have been developed since the mid-1940s; descriptions of typical electrodes were presented by Chartrain et al. (1959), Delgado (1955, 1964), and Ray (1963, 1965). In most instances the investigators have developed their own electrodes; the two most popular versions are due to Delgado and Ray (Figs. 3-3 and 3-4, respectively).

Delgado's electrode (Fig. 3-3a) consists of a bundle of six Teflon-insulated stainless steel wires (120 μ diam. and 100 mm long). The insulation is removed from the first millimeter of the end of each wire and the wires are cemented together with the tips staggered 3 mm apart. The active area of each electrode is about 0.4 mm². The other ends of the wires are connected to a multicontact connector equipped with two additional stainless steel wires, which are used to anchor the connector to the skull after the electrode has been installed.

The method of application (Fig. 3-3b-c) consists of advancing the electrode into the brain via a burr hole in the skull. At the desired depth, the electrode wires are fixed to the skull with cement, bent, and run parallel to the skull

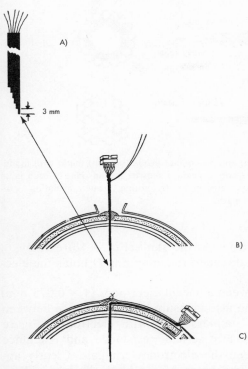

A)

3 mm

B)

C)

Fig. 3-3. Delgado's depth electrode (Redrawn from J. M. R. Delgado in *Physical Techniques in Biological Research,* Vol. V, Part A, W. Nastuk, Ed. New York: Academic Press, 1964. By permission.) © Academic Press.

(under the scalp). The electrical connector (whose undersurface has been sealed) is then anchored to the skull by the two stainless steel wires attached to it. These anchor wires can be used as a reference electrode or to ground the subject.

Ray's electrode (Fig. 3-4a), which is commercially available (Medical Applications Dept., IBM, Rochester, Minn.), consists of a bundle of up to 37 insulated wires bonded together or to a length of 24-gauge stainless steel needle tubing using a high-temperature varnish. Each wire is platinum-iridium (90/10) and is 0.0035 in. diam.; the active electrodes are made by scraping the varnish from the wires at the desired places. The scraped areas are then platinized (see p. 32) to reduce the tissue-electrode impedance by about one hundred-fold. The contact area employed by Ray was 0.075×1.00 mm.

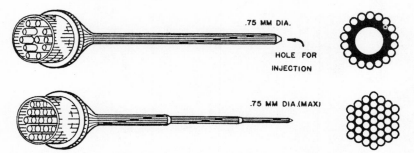

Fig. 3-4a. Diagram of 18-contact probe (top) and 37-contact probe (bottom) made with laminations of wires; hollow-tubing core and "self" core types are shown. To the right of each is an enlarged cross section. Contacts are made at specific points by cutting through insulation as indicated. Probes can be made any convenient length.

The method of securing the electrode to the skull (Fig. 3-4b) employs a socket B threaded into the skull. A threaded retaining ring D holds the electrode connector C securely in place.

Ray measured the impedance between adjacent electrodes (1×0.075 mm) separated by 1 cm in 0.9% saline; he also measured the impedance between 1-mm bare 39-gauge (80-μ diam.) stainless steel wire electrodes. Figure 3-5 presents the equivalent (presumably series), resistance and reactance characteristics for the circuit between the electrode terminals. Clearly apparent are the high resistance and reactance of the bare platinum-iridium

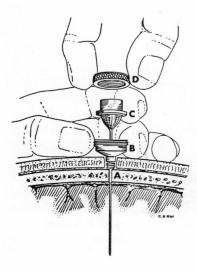

Fig. 3-4b. Parts used for chronic implantation of Ray's depth electrode: a threaded-end, stainless steel cranial pin A has been driven into a hole predrilled in the calvarium; the retaining cup B is then screwed onto the threaded pin; the electrode C is inserted into the cup and locked into position by the retaining ring D. (From C. Ray, et al. Proc. Staff Meet. Mayo Clin. 1965, **40:** 771–780. By permission.)

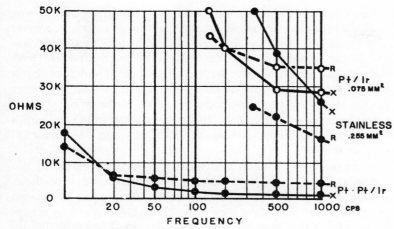

Fig. 3-5. Equivalent resistance R and reactance X of platinum-iridium, stainless steel and platinized platinum-iridium depth electrodes in 0.9% saline. Spacing between electrodes = 1 cm. (From C. D. Ray, et al. *Proc. Staff Meet. Mayo Clin.* 1966, **40**:771–804. By permission.)

and stainless steel electrodes and the increase in both resistance and reactance as frequency is reduced. Equally conspicuous is the remarkable reduction in resistance and reactance produced by platinizing the platinum-iridium electrodes. This highly desirable alteration permits using Ray's small-area depth electrode with EEG amplifiers, which serve routinely and have an input impedance of 2 to 5 MΩ.

The depth electrodes described by Delgado and Ray have been implanted into the brains of animals and man and left there for prolonged periods to record the electrical activity of subcortical structures under a variety of normal and abnormal states. Ray reported that the central stainless steel needle support, in one of his electrode types, could be used for the injection of materials into the brain or for the passage of a guarded microelectrode. He also stated that his electrode could be used to measure localized impedance changes and that by the application of the proper polarizing voltage, it is suitable for the continuous polarographic recording of oxygen tension.

Although a considerable amount of valuable information has been derived from depth recording in the brain, there has always been some concern about the possible damage to nerve cells and tracts as the electrode is advanced into the brain substance. Soon after depth recording was introduced, Dodge et al. (1955) investigated this problem and reported that "pertinent information has been scarce since few of the patients studied by this technic [depth recording] have later died". He stated that "the surgeons who have actually inserted the depth electrodes in the now well over 85 patients from whom more than 3000 contacts have been recorded, concurred in the opinion that little damage was done to cerebral tissue, and that the attendant surgical risk due to intracerebral electrography done properly was indeed minimal." Further documentation came as a result of the opportunity to examine the brain of a patient who died of pneumonia nineteen months after he had been studied with four depth electrodes (each consisting of 6 strands of #38 Formvar-insulated copper wire) left *in situ* for 6 days. Microscopic examination of the brain showed almost no damage. (In more recent studies, the same investigators used depth electrodes consisting of 6 strands of 46-gauge (37.5-μ diam.) Formvar-insulated stainless steel wire.)

In a later report by Fischer-Williams and Cooper (1963), substantially the same information was obtained. These investigators made 302 recordings using depth electrodes made from 4 to 7 strands of insulated nickel-plated copper wire 100 μ diam.; the recording area for each wire was made by scraping off the insulation for a distance of 1 mm from the tip. They had the opportunity to conduct histological examinations of the brains of 14 subjects, and in only one of these was there evidence of histologic changes attributable to the electrode. It must be pointed out that this electrode had been used for stimulation as well as recording.

Delgado (1964) made extensive studies on the response of monkey brains to installation of his stainless steel depth electrode (ca. 0.5 mm diam.) Recognizing that stainless steel is the material that produces the least tissue reaction, he reported that the first response was due to mechanical factors (hemorrhage and trauma) and was confined to the periphery of the electrode. A few days later, edema and an inflammatory reaction were noted; after a few weeks, encapsulation of the electrode was common. At a distance of 100 to 200 μ beyond the capsule, the brain was found to be normal. (This sequence of events is also described in the section "Tissue Response to Electrode Metals" (see p. 129). With the 0.5-mm diam. electrode, the width of the track formed was about 1 mm. Delgado also reported that the risk of infection was slight.

The Distributed-Area Electrode

Gerstein and Clark (1964) desired to record the electrical activity of neuron pools with a single, rather than a multicontact depth electrode. They therefore constructed a 10-μ tungsten electrode using Hubel's electropointing technique (see p. 138) and completely covered the electrode with vinyl insulation; then the insulation was punctured at different places in the region extending 100 μ from the tip. The puncture method consisted of advancing the tip of the electrode into a hanging drop of clean mineral oil for protection. Using a 100-power magnification to view the procedure, the tip of another microelectrode was brought up at right angles to it and positioned 5 to 10 μ distant from the region to be punctured. The two electrodes were then connected to a Tesla coil and a short series of sparks was created. The authors reported that they could control the size of the tiny holes in the insulation by varying the intensity and number of sparks. The effectiveness of the puncturing technique was verified by the electrolytic gas-bubble method described by Hubel for measurement of the area of an exposed electrode.

After making several distributed-area electrodes having many holes randomly located within 100 μ of the tip of the electrode, Gerstein and Clark successfully recorded action potentials in the cochlear nuclei. By obtaining recognizable action potentials and moving the electrode, they were able to observe that it had a "capture area" equivalent to a sphere of 200 μ diam.

Since it is difficult to control the size and location of the holes in the insulation on a small electrode using the Gerstein–Clark technique, Mela (1966) described the fabrication of similar electrodes using a ruby laser beam to make holes in the insulation. Puncture of the insulation was carried out in fluid under microscopic examination. The size of each hole was measured by Hubel's electrolytic gas-evolution technique. Holes of 2 μ diam. were made in vinyl-coated tungsten electrodes of 10 to 15 μ diam.; the method permitted making holes of controllable diameters ranging in size from 2 to 9 μ.

It is too soon for evaluation of the distributed-area electrode. No comparisons have been made using several single electrodes with the same geometrical arrangement as a distributed-area electrode. While *in vivo* comparisons may be difficult to conduct, *in vitro* studies could be made easily and undoubtedly will be carried out.

Impedance of Subintegumental Electrodes

The impedance-frequency characteristics of a subintegumental electrode depend on the type and area of the electrode material, the kind of electrolyte-tissue environment, and the current density used for measurement. Although many of the types described have seen extensive service, there is only a limited amount of quantitative descriptive information. Nevertheless, the available data can be used to characterize the properties of these small-area electrodes.

Perhaps the most frequently used subintegumental electrode is the needle electrode (Fig. 3-1*b*). The range of areas employed, of course, depends on the construction. Even when a needle is coated with an insulating compound, it is often difficult to measure the exposed area by microscopic examination. A method described by Hubel (1957) may be employed in many cases to determine how much of the metal is not covered by the insulating compound. The method consists of placing a drop of saline solution on a microscope slide which is viewed along with a size calibration mark. An indifferent electrode and the tip of the needle electrode to be measured are placed into the drop.

With the needle electrode as the cathode, sufficient current is passed to produce electrolysis. From the exposed area of the needle electrode, hydrogen
gas bubbles are evolved and, when the electrode is rotated, the dimensions
of the exposed tip can be estimated and the area calculated.

Using the area-measuring procedure just described, Geddes and Baker
(1966) measured the impedance-frequency characteristics appearing between
the terminals of pairs of stainless steel needle electrodes inserted in a lead-II
configuration in a series of dogs. A low value of current was employed for all
measurements (10 μA rms); the corresponding current density range varied
from 1 to 0.001 mA/cm^2. The data obtained (Fig. 3-6) clearly reveal the inverse
relation between electrode impedance and frequency for a particular electrode;
also apparent is the inverse relation between electrode impedance and surface
area. These findings are in agreement with the nature of the circuit formed
between a metal electrode and an electrolyte. Curves for similar electrodes
have been reported by Gray and Svaetichin (1951), Tasaki (1952), Gesteland
(1959), Plutchik and Hirsch (1963), and Schwan (1963).

A better idea of the impedance of needle electrodes can be obtained from
loading studies (see p. 39). Such data were acquired for needle electrodes
ranging in area from 73 to 1 mm^2 inserted subcutaneously in the dog to obtain
lead-II ECGs. Similar studies were carried out with smaller-area needle

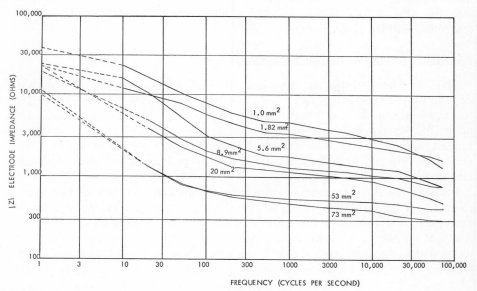

Fig. 3-6. Impedance-frequency curves for needle electrodes. (From L. A. Geddes and L. E.
Baker, *Med. Biol. Engng.* 1966, **4:**439–450. By permission.)

electrodes (125,00–500 μ^2) placed in the gastrocnemius muscles of frogs to record the EMG evoked by the application of single stimuli to the sciatic nerve. In the electrocardiographic studies, the amplitudes of the *P*, *R*, and *T* waves obtained with a 4.4 MΩ input impedance were taken as 100%; the amplitudes obtained with the different resistive input impedances (loads) were expressed as percentages, and the three percentages for each input impedance were averaged. Figure 3-7 presents the loading data obtained for the needle electrodes used to record the dog ECG. Since the resistance required to reduce the open-circuit voltage to one-half is approximately equal to the resistance of the generator (i.e., the animal and the two electrode-tissue interface impedances), the inverse relation between electrode area and impedance can be recognized in Fig. 3-7 by the lower value of resistance that is required for the larger-area electrodes to decrease the bioelectric signal to one-half.

In the case of the small-area needles used to record frog EMGs, the upward amplitude recorded with the highest available input impedance (ca. 1000 MΩ) was taken as 100%; the amplitudes obtained with the various resistive input impedances were expressed as percentages of this initial value. The data obtained (Fig. 3-8) illustrate that the larger the area of the electrode, the lower the value of resistance required to reduce the bioelectric signal to one-half.

It is well known that the larger the area of an electrode, the lower its impedance. However, because an electrode-electrolyte impedance contains both resistive and reactive components, and because different bioelectric

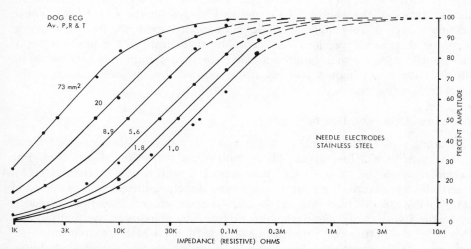

Fig. 3-7. Average loss of amplitude for the *P*, *R*, and *T* waves of the canine ECG recorded with stainless steel needle electrodes of various areas as amplifier input impedance (resistive) is reduced.

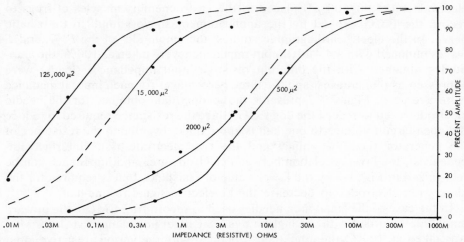

Fig. 3-8. Loss of amplitude of the frog EMG recorded with stainless steel electrodes of different areas as amplifier input impedance (resistive) is reduced.

events have different sinusoidal (Fourier) spectra, it is virtually impossible to develop a simple relation between electrode area and the resistive impedance that will reduce the amplitude of a bioelectric event to one-half. The truth of this statement is illustrated by the differences in the loading curves in Fig. 2-8, 13 for the *P*, *R*, and *T* waves of the human ECG, in Fig. 3-7 for the *P*, *R*, and *T* waves of the canine ECG, and in Fig. 3-8 for the frog EMG. These loading curves merely indicate the level of input impedance for a bioelectric recorder that would produce unacceptable distortion (50% amplitude loss); the input impedance that would not cause significant amplitude loss would be 100 to 1000 times higher. Further discussion of this subject appears in the following section.

Waveform Distortion Due to Loading

Loading a pair of electrodes while a bioelectric event is being recorded can cause more than a loss of amplitude; waveform distortion also occurs. For example, when the canine ECG was recorded with a pair of stainless steel electrodes approximately 1 mm² in area and a rectangular pulse was placed in series with the electrode-subject circuit, the waveform distortion produced by loading was easily demonstrable. Figure 3-9*a* illustrates the method and Fig. 3-9*b* shows the control record taken with a 4.4-MΩ resistive input impedance. As the input impedance was lowered (by placing successively lower values of resistance across the amplifier input terminals), the loss in amplitude and the alteration in waveform of the *P-QRS-T* waves and square pulse are clearly apparent (Fig. 3-9*c–f*). With 300 KΩ across the input terminals, there were recognizable *P*, *S*, and *T* wave changes along with a noticeable tilt on the square pulse. With 100 KΩ and especially with 30 KΩ across the input terminals, not only is there an overall loss in amplitude, but the *P* and *T* waves and the square pulse were dramatically changed, all becoming diphasic. There was a continued loss of overall amplitude as loading was increased. The ECG in Fig. 3-9*f*, recorded with 10 KΩ across the input terminals and with twice the amplification, is vastly different from that in Fig. 3-9*b*. The outstanding differences are the loss of *P*, *R*, and *T* wave amplitudes and the addition of

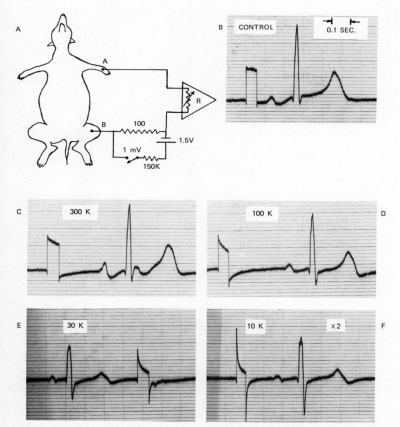

Fig. 3-9. Distortion produced by progressive reduction in amplifier input impedance (resistive). (From L. A. Geddes and L. E. Baker, *Med. Biol. Engng.* 1966, **4**(5):439–450 By permission.)

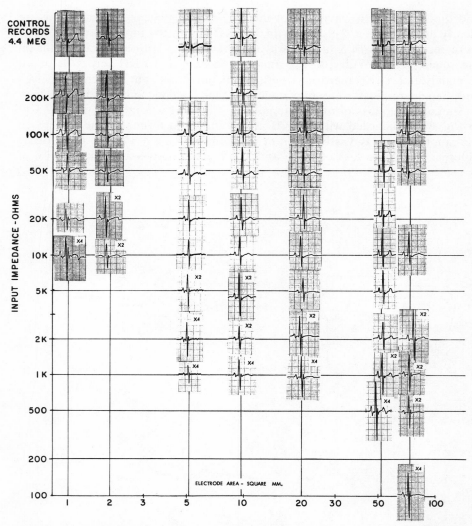

Fig. 3-10. The relationship between input impedance and electrode area for the canine ECG using stainless steel needles. (From L. A. Geddes and L. E. Baker, *Med. Biol. Engng.* 1966, **4**:439–450. By permission.)

diphasic components to each. In addition, loading produces a noticeable shift in the *S-T* segment (Fig. 3-9*d–f*).

The manner in which distortion produced by the use of an amplifier having too low an input impedance is related to electrode area is shown in Fig. 3-10. In this illustration the ECGs of anesthetized dogs were recorded with pairs of stainless steel electrodes of various areas connected in lead-II configuration. Each control record, which appears at the top of each column, was recorded with an input impedance of 4.4 MΩ. Inspection of the ECGs from top to bottom of each column reveals the changes that occurred as the input impedance of the amplifier was lowered. For example, for the 1-mm² electrodes, noticeable distortion has occurred when the input impedance was reduced to 100 KΩ. For the 10-mm² electrodes, detectable distortion occurred when the input impedance was 50 KΩ. With the 73-mm² electrode, distortion was apparent when the input impedance was 10 KΩ.

The distortion produced by loading various small-area needle electrodes used to record the frog EMG is shown in Fig. 3-11. Along the top of this illustration are the various control evoked-muscle-action potentials recorded with a rectangular pulse placed in series with the indifferent electrode; these recordings were made with an input impedance of about 1000 MΩ. For any given electrode area, the distortion can be recognized by inspecting the recordings from the top to bottom of that column. For example, with the 500-μ² electrode, distortion is just perceptible with an 88-MΩ input impedance. For the 125,000-μ² electrode, distortion becomes apparent at about 0.5 MΩ. The type of distortion imposed on the action potential is a decrease in the amplitude of the upward phase and an increase in the amplitude of the downward phase. The distortion in the rectangular pulse consists of a slope on the top of the wave and an undershoot on the descending portion.

The distortions evident in Figs. 3-10 and 3-11, which are the result of electrical differentiation of the signals, have been noted by several investigators. For example, in the early days of electrocardiography when string galvanometers with their relatively low resistance (ca. 5 KΩ) were used, Lewis (1915) showed that a normal ECG recorded with platinum electrodes was distorted (Fig. 2-10). In such cases attenuated *P* and *T* waves and enhanced *S* waves were obtained. Similarly, Pardee (1917), using a string galvanometer and German-silver electrodes applied to a bandage soaked in saline, showed that the rectangular-wave calibration signal was distorted when the area of each electrode was decreased from 300 to 8 cm². With the smaller electrodes, the calibration signal, instead of rising rapidly and exhibiting a flat top, showed a sharp overshoot and an RC-type decay to a sustained plateau. On turning off the calibration signal, there was an undershoot and an RC decay to the baseline. Pardee (1917) observed a similar type of distortion when electrodes were applied to patients with thick dry skin or when the blood vessels under

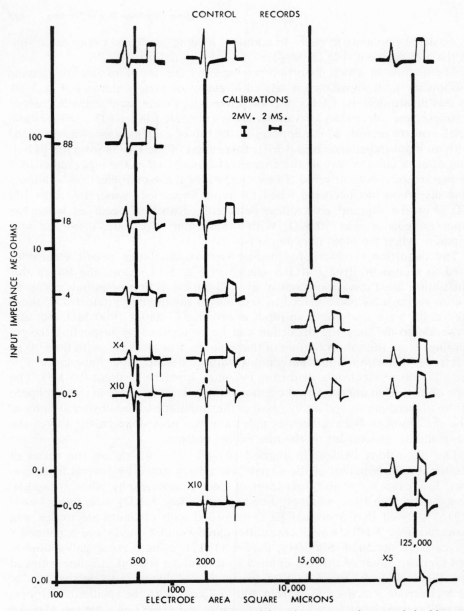

Fig. 3-11. Distortion in frog muscle action potentials and a square pulse recorded with needle electrodes of various areas (*X* axis) with decreasing amplifier input impedance (*Y* axis). To examine the effect of reduced input impedance, refer to the records in any column reading from top (control records) to bottom. (From L. A. Geddes, and L. E. Baker, *Med. Biol. Engng.* 1967, **5**:561–589. By permission.)

the electrodes were constricted. The distortion disappeared when an electrolyte was rubbed into the skin; it often disappeared as time passed and the electrolyte penetrated the dry horny layers of the skin.

Although Pardee did not investigate the phenomenon thoroughly, his observations were in agreement with those of previous and later workers, notably Straub (1922), who attempted to record the ECG with needle electrodes and the string galvanometer and Einthoven (1928), who showed that the electrode-subject interface impedance introduced a time constant into the circuit which electrically differentiated the *P* and *T* waves. The practical importance of these facts in recording the ECG was again demonstrated by Sutter (1944), using needle electrodes and a vacuum-tube amplifier, and by Roman and Lamb (1962). The latter investigators attached miniature recessed electrodes to the chest of a human subject and used them first with an amplifier having a high input impedance to obtain control records; they then lowered the input impedance of the amplifier by connecting different resistance values across it. The distortions in the ECG were what would be called clinically significant, consisting of a displacement of the *S-T* segment and a slight depression in the latter part of the *T* wave.

The clearly recognizable distortions reported by these observers and demonstrated by Figs. 3-9 through 3-11 are those expected in a system that is deficient in low-frequency response. Such a situation can be predicted on the basis of the general capacitive nature of the interface impedance between a metal electrode and an electrolyte. The sine-wave impedance-frequency curves add further confirmation to the reactive nature of electrode impedance.

The relation between amplifier input impedance and electrode area that will serve to avoid distorting a bioelectric event is not the same for all metals. Cooper (1963) compared the distortions encountered when silver-silver chloride, platinum, silver, copper, gold, and stainless steel electrodes, all 0.1 mm² in area, were connected to an amplifier with a 750 KΩ input impedance. The test consisted of passing a square wave of current through a saline bath in which the electrode pairs were immersed. The types of waveforms detected by the various electrode pairs appear in Fig. 3-12. The same kind of distortion represented in Figs. 3-9 through 3-11 is clearly evident namely, electrical differentiation or, stated differently, a loss of low-frequency response. Interestingly enough, under these circumstances, the silver-silver chloride electrodes reproduced the waveform best and the stainless steel electrodes provided the poorest reproduction of the test signal. The author hastens to point out that this does not mean that stainless steel electrodes cannot be used with success; it does mean that a high input impedance amplifier should be used with small-area stainless steel electrodes.

These results, which are consistently reproducible, call attention to the need to use an input stage with an input impedance many times larger than that of the bioelectric generator-electrode system.

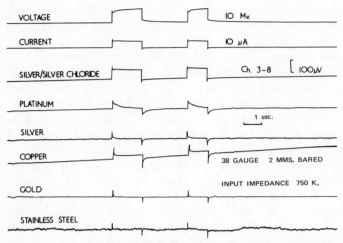

Fig. 3-12. Distortion produced by pairs of electrodes of different metals in the detection of a square wave of current passing through saline. Electrode area in each case = 0.1 mm²; amplifier input impedance = 750,000 Ω. (From R. Cooper, *Amer. J. EEG Technol.* 1963, **3:**91–101. By permission.)

The distortion illustrated in Figs. 3-9 through 3-12 is primarily due to two factors. First, with loading (i.e., the use of an amplifier input circuit that is not high with respect to the electrode-bioelectric generator system), the electrode impedance becomes a dominant part of the input circuit and the voltage across the input terminals of the bioelectric recorder is reduced. In addition, the amount of phase shift is different for the various frequency components of the bioelectric event. With loading, too, the electrode current density is increased and the resistive and reactive components of the electrode polarization impedance become nonlinear, therefore the magnitude of the electrode impedance becomes a function of the amplitude of the bioelectric signal. Thus small- and large-amplitude signals will encounter different impedances. The exact contribution of each of the two sources of distortion for the various electrodes employed in recording bioelectric events is still unknown.

Not only does an increase in current density alter the electrode impedance, it changes the half-cell potential. Even with such a relatively nonpolarizable electrode as the calomel cell, current flow alters the half-cell potential. Rothschild (1938) showed that the maximum current density for this type of electrode was 15 $\mu A/cm^2$ before the half-cell potential was noticeably altered. The exact limits of current densities permissible for the various types of electrodes have not been investigated adequately.

In summary, the best way to avoid encountering the distortions just described is to employ a bioelectric recorder that has an input impedance many times (e.g., 100 to 1000) higher than that of the electrode-tissue circuit for the frequency spectrum of the bioelectric event of interest. A simple way of estimating the impedance of the tissue-electrode system is to conduct loading tests such as those described earlier in this chapter.

Sometimes the loss of low-frequency response due to loading can be used for filtering purposes. For example, placing a resistive load across a pair of electrodes used to record the ECG from a moving subject results in a dramatic reduction in baseline wandering (as well as a loss in P and T wave fidelity). If the channel is used to count heart rate by counting the R waves, this simple type of filtering is attractive, but it should never be used if ECGs of high fidelity are desired.

Electrode Area and Amplifier Input Impedance

It was pointed out previously in this chapter that the circuit between a pair of electrode terminals can be considered as a generator (the bioelectric event) in series with an impedance (that of both electrodes and the volume-conductor matrix between the electrodes and the source of the biopotential). It is well known that measurement of the potential presented to the electrodes requires a device with an input impedance many times (e.g., 100 to 1000) higher than the impedance measured between the electrode terminals for the sinusoidal frequency spectrum of the bioelectric event being measured. It is also well known that as electrodes are made smaller (i.e., their area is decreased), their impedance increases. As shown in this chapter, an easy way of determining the order of impedance represented by the electrodes and the bioelectric generator is to conduct loading tests by connecting successively lower values of resistance across the electrode terminals and plotting the reduced amplitudes versus the resistance values. The resistance that reduces the amplitude (measured under open-circuit conditions) to one-half gives a rough idea of the impedance of the electrode-bioelectric generator system. The use of an input impedance about 1000 times this value will allow measurement of a bioelectric event with insignificant distortion.

When large-area electrodes are employed, the impedance between the electrode terminals is fairly low and it is not difficult to obtain an adequately high input impedance in the biopotential recorder. However, when the electrode area is small, it becomes increasingly difficult to obtain an adequately high input impedance. To illustrate the order of magnitudes encountered in practice, the data obtained from the loading tests using the recessed electrodes (to record the human ECG), the needle electrodes (to record the canine ECG), and the needle electrodes (to record the frog EMG) are plotted in Fig. 3-13. This illustration shows the resistance values that decreased the amplitude of each particular bioelectric event to one-half versus the area of the electrodes

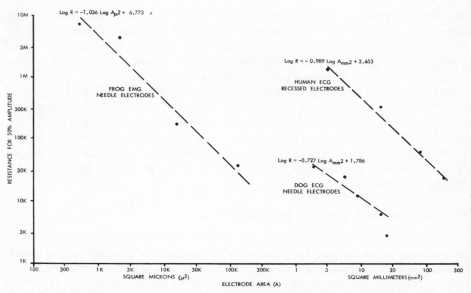

Fig. 3-13. The relation between input impedance and electrode area: plots of the resistive values of input impedance required to reduce the open-circuit amplitudes of the frog EMG and the dog and human ECGs by 50% when recorded with different types of electrodes.

used to measure the bioelectric event. It is quite apparent that, with the large-area recessed and needle electrodes, the resistance values for 50% amplitude ranged from 3000 Ω to about 1 MΩ. Attainment of an input impedance 100 to 1000 times these values does not present difficult electronic problems. However, when electrodes with areas ranging from 500 to 100,000 μ^2 were used, the resistance values for 50% amplitude were 8 MΩ and 45 KΩ, respectively. It is quite easy to obtain an input impedance adequate for the larger-area electrode (i.e., 45 K$\Omega \times 1000 = 45$ MΩ); it is rather difficult, however to accommodate the requirements for the 500-μ^2 electrode, which ideally requires an input impedance on the order of 8000 MΩ.

The discussion, relating amplifier input impedance and electrode area, is presented to alert the reader to the necessity of using an adequately high input impedance when small-area metallic electrodes are employed. Figure 3-13, which was compiled from trials with different bioelectric events recorded with electrodes of different area indicates the order of electrode-subject impedances found in practice. In the section on intracellular electrodes, additional data are presented to show the impedances encountered with micron-sized electrodes, which are small enough to be advanced into single cells without damage.

Tissue Response to Implanted Electrodes

In many recording situations it is necessary to place metallic electrodes in direct contact with body tissues and fluids for prolonged periods. For example, electrodes are often implanted in muscle and brain tissue to record the bio-electric signals of these structures for periods of months. In the early days of depth-electrode recording in the brains of humans, Dodge et al. (1955) had the opportunity to study the tissue response to two electrodes, each consisting of 6 strands of Formvar-insulated copper wire (97.5-μ diam.) which had been *in situ* for six days. Nineteen months later the brain was examined histologically. Tissue changes were seen at the points of entry of the electrodes; minimal tissue changes were found along the tracks of the electrodes.

Faced with the problem of recording the electrical activity of structures deep within the brains of humans and animals, Fischer et al. (1957) studied the responses of the brains of cats to 1-cm lengths of 24-gauge wires, left *it situ* for periods up to four weeks. The wires employed were of chlorided silver, bare copper, and stainless steel. Both bare and insulated* wires were employed. After one week, histologic studies showed tissue responses to all the materials. The responses were dependent on the types of metals employed—the insulating compounds were virtually without tissue response. Silver and copper wires proved to be the most toxic to brain tissue. After three weeks, a narrow ring of necrotic tissue surrounded the silver wire. Surrounding this ring was a circular edematous region 2 mm diam. The reaction to the copper wire at the same time was similar, except that an increase in vascularity had occurred. The copper wire was encircled by necrotic tissue and the adjacent tissue was edematous. The diameter of the lesion varied between 1.5 and 7 mm. With the stainless steel wire, the size of the lesion was determined by the extent of the mechanical trauma produced by introduction into the brain. Only minimal edema was found. These investigators concluded, therefore, that stainless steel is the electrode material of choice for such studies.

In another series of experiments, Collias and Manuelidis (1957) inserted bundles of six stainless steel electrodes (125-μ) into the brains of cats and described the histologic changes that occurred over periods extending up to six months. They found that an orderly sequence of changes took place in the tissue surrounding the electrode track. At the end of 24 hours, there was a zone of hemorrhage, necrosis, and edema extending to about 1 mm from the electrode. After three days, there was less hemorrhage and necrotic debris, and by the seventh day, a layer of capillaries (0.1 mm) occupied the necrotic zone. By the fifteenth day, the capillaries had almost completely replaced the necrotic region, and connective tissue had started to form. After the passage

* Tygon, Formvar, Thermobond M472, and polyethylene.

of a month, the necrotic debris had disappeared and a well-defined capsule surrounded the electrode track. After four months, a thick dense capsule completely surrounded the electrode track.

Robinson and Johnson (1961) performed studies similar to those just described. Into cat brains they implanted 125-μ diam. wires of gold, platinum, silver, stainless steel, tantalum, and tungsten and examined the tissue responses at different times over a period extending to six months. After about one week, the differences between the metals began to be detectable. Gold and stainless steel produced the least tissue reaction; tantalum, platinum and tungsten produced more. Silver precipitated a vigorous tissue response. Encapsulation of all electrodes was evident at 15 days; the thickest capsules were formed around the metals that provoked the greatest tissue response.

The successful use of 1-mm carbon rods as recording electrodes implanted in the brains of monkeys was reported by Roth et al. (1966). Six months after implantation, no histological changes were seen in the cortical regions surrounding the electrodes. An additional advantage of these electrodes was high electrical stability; the variation in potential in a 5-minute period was less than 1 μV as measured between two electrodes separated by 11 mm in 0.9% saline. The electrode pair exhibited a low impedance, amounting to 1100 Ω from 10 to 120 Hz and falling to lower values at higher frequencies. Linearity of the voltage-current characteristic extended to a dc current density of about 10 mA/cm². All these characteristics would indicate that the use of carbon for recording electrodes merits further investigation.

The author hastens to emphasize that in most of these studies the tissue changes were observed in the brains of animals with electrodes that carried no current. Results of research on the response of other tissues to other electrode materials and the tissue responses to current-carrying (stinulating) electrodes are beginning to appear. It is to be anticipated that current-carrying electrodes will produce their own special category of response, since so many electrolytic reactions can occur.

Marking Metal Electrode Sites

In many studies in which an electrode is advanced deep within tissue, it is desired to know the location of the electrode tip. Inferences can be made from initial measurement with reference to anatomical landmarks. When studies are made in the brain, the various stereotaxic instruments, and the atlases composed for them, provide an even better identification of the location of the tip of an electrode. Often the track formed by the electrode is invisible in histological sections; sometimes it can be made more apparent by appropriate preparation of the tissue. For example, Guzman et al. (1958) reported being able to identify electrode tracks in rat brains when the electrodes were left *in situ* and the brains were perfused with 20 ml of 40% formalin solution injected into the carotid arteries. When a ferrous metal electrode is placed within the nervous system, it is customary to mark its location by electrolytic deposition of iron in the tissue and later perfuse it with potassium cyanide to obtain a blue spot; this technique is called Prussian-blue marking. Obviously it should be possible to obtain marks using nonferrous metal electrodes—if there can be found a perfusate containing a suitable compound that reacts with the metal ions to produce a visible spot. Consideration must also be given to selection of a marking solution that will not alter the fixing process which permits histological manipulation of the tissue. In addition, the metallic ions deposited must not facilitate or inhibit cell function. It must also be noted that there is a shrinkage factor associated with each tissue-fixing procedure; with many, the factor is 30 to 40%. Sometimes electrode sites are marked by purposeful tissue damage made by creating a small thermal lesion with direct or radio-frequency (alternating) current.

Marking With Prussian Blue   The Prussian-blue marking method has been used with considerable success when iron-containing electrodes are placed within the central nervous system. Marking is achieved by electrolytic deposition of iron ions (i.e., electrode positive) into the tissue; the deposit is then made visible by perfusion of the tissue with formalin (to harden it) and with potassium ferrocyanide (to produce the blue spot). When the tissue is sectioned and viewed microscopically, the Prussian-blue spot is clearly identifiable. The size of the spot is controlled mainly by the size of the electrode, the intensity of the current, and the time the current is allowed to flow.

Adrian and Moruzzi (1939) reported that Hess was the first to use the Prussian-blue method to mark the sites of steel electrodes; in their own studies, employing rabbits and monkeys, Adrian and Morruzzi connected a steel microelectrode (40-μ diam.) to the positive pole of a 4 to 6 V battery, the negative pole connected to a reference electrode) and allowed the current to flow for 10 to 20 sec. They then perfused the brain with a formalin solution, sectioned it, and stained the sections with cyanide solution, which produced an easily recognizable blue spot at the site of the electrode tip. Marshall (1940) applied Prussian-blue marking to the frozen-section technique of preparing cerebral tissue for slicing. For marking, he connected the steel microelectrode to the positive pole of a 3 to 4.5 V battery for 5 to 10 sec, which resulted in a current flow of 10 to 20 μA. The tissue (cat brain) was perfused with a solution of 10% formalin, 1% potassium ferrocyanide, and Ringer's solution, whereupon it was treated with alcohol, frozen, and sectioned to show well-defined blue spots where the electrode tips had been located. Similarly, Schiebel and Schiebel (1956) marked the locations of flexible steel wire electrodes (10-μ diam.) in brain tissue by passing 100-μA (microelectrode positive) of current for 5 sec. With the electrodes in place at the end of the experiment, the brain was perfused with 50 to 75 ml of a 1–1 mixture of 4% potassium ferrocyanide and 4% acetic acid. The brain was then cut into blocks and fixed in a 1–1 mixture of 4% formalin and 95% alcohol for 5 to 6 hours; then the electrodes were removed and the blocks were sectioned. The blue spots obtained occupied spherical regions about 200 μ in diameter. In a similar study, Green (1958) reported success in marking the site of the tip of an 0.5-μ stainless steel electrode that had been made by the electropointing method (see p. 138). He passed 2 μA for 15 to 30 sec (microelectrode positive) and then fixed the tissue with a solution of 10% formalin containing 1% potassium ferrocyanide; extremely small Prussian-blue spots were produced. He reported that a much higher current (100 μA) and longer time (10 sec) yielded a colored spherical region 200 μ diam.

From the foregoing, it is apparent that the Prussian-blue marking technique is practical and controllable when used in conjuction with iron-containing electrodes. There is a fair agreement on the current and time needed to obtain a recognizable spot. From the literature cited, it appears that, with the microelectrode positive, 500 to 1000 μA-sec of charge (current × time) will produce a spherical colored region about 200 μ diam. The current levels employed range from 2 to 100 μA, the smaller values being used for the smaller-area electrodes. It must be remembered that although these currents are low, because of the very small electrode areas, they represent extremely high current densities. Since electrode resistance decreases with increasing current density (see p. 19), the electrode resistance values encountered during current flow are much lower than would be estimated on the basis of electrode area. For this reason it is unwise to use a constant-voltage source to obtain the current because it will not be possible to initially select a current and maintain it at a constant level while the mark is being made.

The use of a constant-current supply will permit selection of the current prior to its application to the electrodes; such a source can be approximated by using a high resistance connected to a high-voltage source. For example, a typical extracellular electrode-subject circuit resistance during current flow would be on the order of about 0.1 to 0.5 MΩ; connection of a resistance of 2 MΩ in series with a voltage source of 100 V would limit the current to 50 μA with the electrodes short-circuited. With the short circuit removed and the current flowing through the electrode-subject circuit, the current would be slightly less. With this technique the desired current can be preselected by adjustment of the voltage used to drive current through the high series resistor. (In the example chosen, 2 MΩ was suggested, but a higher value can be employed with a higher voltage.)

Marking With Direct-Current Lesions   If it is permissible to use a small lesion to identify the location of the tip of a microelectrode, then marking by the use of direct current can be employed with electrodes of any metal. The method appears to have originated around the turn of the twentieth century when Sellier and Verger (1898, 1903) and Horsley and Clarke (1908) reported on studies carried out virtually simultaneously. Sellier and Verger made pea-sized lesions in the thalami of dogs using bipolar needles having exposed tips of a few millimeters, currents of 9 to 15 mA were passed through these needles for periods ranging from 7 to 10 minutes. Horsley and Clarke's contributions were the quantification of the size of the lesion and the introduction of a device—the stereotaxic instrument—that permitted precise location of a single lesion-making electrode by the use of X, Y, and Z coordinates based on skull landmarks.

The characteristics of the lesion produced by direct current are best described in the words of Horsley and Clarke, who wrote

Electrolytic lesions of the brain, especially anodal ones, are quickly and easily produced with very slight injury to any other parts; their size can be accurately regulated, their form depends on the nature of the electrode, they are precisely defined and the necrosed tissue passes in all directions almost abruptly into the uninjured tissue which does not appear to be even temporarily affected by the lesion. . . . During an excitation experiment [i.e., when the electrode was used for stimulation], a restricted electrolytic lesion has been made to mark some spot from which a definite response had been attained, we have found on advancing the needle another millimeter, that the uninjured tissue immediately adjoining is normally excitable.

Horsley and Clarke were also concerned with the nature and size of the lesion when the monopolar brain electrode was made positive (anodal lesion) and negative (cathodal lesion). Anodal lesions showed evidence of bleaching due to the formation of chlorine gas and coagulation of the protein. Cathodal lesions were characterized by more tissue destruction, and therefore they chose to use anodal current (i.e., the lesion-making electrode was connected to the positive pole of the dc supply).

Subsequent studies have merely validated the work of Horsley and Clarke. Both bipolar (concentric and side-by-side) and monopolar electrodes have been employed. Unfortunately, not all of those who used the technique provided quantitative data regarding polarity, current, time, and lesion size. Nonetheless, the literature contains sufficient data for characterization of the lesion in terms of the parameters employed to produce it.

As shown by Horsley and Clarke (1908), the size of a lesion depends on the exposed area of the electrode and the product of the current and the time for which it flows, providing the current employed was less than 5 mA. Using an 0.19-mm diameter platinum needle with an 0.5 to 1 mm surface exposed to brain tissue, they reported "Combining many observations together it is evident that for a unit of time, e.g., 1 minute, there will result about 1 mm breadth of destruction for each mA of current employed." It must be remembered that they used anodal current; with cathodal current the lesion will be slightly larger. In the well-known studies of hyperphagia in rats, Hetherington and Ranson (1940) made lesions about 1 mm diam. by passage of 2 mA for 20 secs. Kreig (1946), who devised his own stereotaxic instrument for the rat brain, was able to produce lesions of about 1.5 mm diam. by passing a current of 2 mA for 15 secs. Whittier and Mettler (1949) made lesions of about 1 mm diam. in monkey brains by passing a current of 2 mA for 15 sec between the core and sheath of a concentric electrode 0.63 mm diam. the core was made negative to the sheath. In a second series of studies the same investigators, with Carpenter (1950), produced 1-mm diam. lesions in monkey brains using a monopolar (positive) electrode and passing 5 mA for 15 secs. In a similar study using cats, Hendley and Hodes (1953) made lesions 2 mm diam. by passing a current of 3 mA for 20 sec through a monopolar electrode 1.2 mm diam. In another series of studies on the effect of hypothalamic lesions on food intake, Anand et al. (1955) made lesions of about 1 to 1 1/2 mm diam. in cat and monkey brains by passing 3 mA for about 30 sec through a unipolar electrode located in the hypothalamus.

All the evidence cited can serve as a semiquantitative guide to the size of a lesion that will be produced by the passage of direct current through the tip of a needle electrode. Although most of the lesions studied were produced in animal brains, the results in other soft tissues are not expected to be very different. Most of the studies employed needle electrodes about 0.5 to 1 mm diam. insulated down to a small tip. Without regard to polarity (which is of some importance in the studies that were performed carefully), the range of current multiplied by time varies between 20 to about 90 mA-sec; these parameters produced lesions having diameters of about 1 to 1 1/2 mm. There is ample evidence that with high currents (i.e., greater than about 5 mA) and long times (in the range of minutes), the lesion diameter is not proportional to the product of milliamperes and seconds; it is somewhat smaller. From these data it should be possible to estimate the range of current and time required for making less destructive lesions suitable for electrode marking.

It is to be noted that the resistance of the electrode-tissue interface decreases during the initial passage of the current. If a constant-voltage dc supply is used, a high current will flow and a small amount of gas and perhaps steam will be produced; under this condition the electrode resistance will rise. To prevent such electrode-tissue resistance changes from altering the current, a constant-current source should be used.

Creation of a lesion by the passage of direct current through a small-area electrode-electrolyte interface can erode the electrode tip. Loucks et al. (1959) studied the erosion produced in a variety of electrode metals following the passage of currents of the magnitude and for the times used in lesion-making. With anodal current applied to stainless steel, chromel, platinum-iridium, silver, and tungsten in saline, and in a few instances in brain tissue, they observed cup-like depressions in all but the silver and stainless steel electrodes; the silver electrode developed a mushroom-shaped cap (presumably of silver chloride), and the tip of the platinum-iridium electrode was unaltered.

The observations of Loucks et al. are not particularly surprising, since lesion-making with direct current is the same as the process that is used in the electropointing method (see p. 138), carried out at reduced current. For this reason, an inert metal should be used to avoid the liberation of electrolytic products; from the studies reported it appears that either platinum (as used by Horsley and Clarke) or platinum-iridium (as advocated by Loucks et al.) would be a good choice. Although other metals may suffice, preliminary tests in saline or in tissue employing the current levels used in lesion-making will produce useful information.

Simplicity is the attractive feature of electrode marking by the creation of small lesions produced by direct current; adequate control of size can be achieved by appropriate choice of current, time, and electrode area. The major drawbacks to the method are the local deposition of the products of electrolysis and erosion of the electrode. If the electrolytic products do not alter cell function, then the method can be used with confidence for marking.

Marking With Radio-frequency Lesions   The probability that lesions made with direct current had two undesirable effects—namely, erosion of the electrodes and deposition of the products of electrolysis in adjacent tissue—led investigators to employ radio-frequency (200 KHz to several megahertz) current to make purely thermal lesions. If it is permissible to mark an electrode site by a small thermal lesion, then the use of radio-frequency current becomes attractive. Carpenter and Whittier (1952) showed that radio-frequency lesions were quite similar to dc lesions and that their size was similarly related to the current intensity, the time of flow, and the electrode area. They pointed out that with radio-frequency current, the quality of the insulation is especially important because tiny cracks or thin spots will also cause lesions to be produced. Zervas (1965) took advantage of this property to produce asymmetrical lesions.

Quantitative prediction of the size of lesions made by radio-frequency current was slow in coming, principally because when the investigations were being carried out, there were few instruments for the generation and measurement of low-intensity radio frequency current.

One of the first studies attempting to quantify the size of a lesion in terms of the electrical parameters employed was that of Brown and Henry (1934). They applied up to 20 V of 3-MHz current for up to 2 1/2 minutes to a monopolar electrode (0.01 in. diam.) bare only at the tip, and obtained brain lesions 0.2 mm diam. Hunsperger and Wyss (1953) applied a constant power (310—330 mW) for 10 sec to a monopolar electrode (1 mm² in area) placed in the brain of a cat; the sizes of the lesions depended on where the electrode was located. In the thalamus, the volume of the lesions ranged from 1.5 to 3 mm³; in the internal capsule the same parameters produced lesions in the range of 0.5 to 1.5 mm³.

Aronow (1960), using up to 500-mA of 2-MHz radio-frequency current applied to electrodes ranging from 0.2 to 15 mm diam., produced lesions measuring 0.2 to 16 mm diam. With a constant time he found that the lesion diameter was proportional to the current.

Von Bonin et al. (1965) investigated the effect of electrode size, current intensity, and duration on the size of lesions made in cat brains using 200-KHz and 1.75 MHz radio-frequency current; they also monitored the temperature at the lesion site. The lesions they obtained were ellipsoidal in shape, with an average length-to-diameter ratio of 1.5 and a range of values extending from 1.1 to 3.7. The size of a lesion was found to be linearly proportional to the temperature rise and the electrode area. When the temperature was held constant, the size of a lesion was proportional to the duration of current flow and electrode area.

Using a microthermocouple to monitor the temperature at the lesion site, Dieckmann et al. (1965) showed that lesions of fairly predictable size could be produced with temperature rises to 70°C by manipulation of the current, time of flow, and electrode size. They also pointed out that the nature of the tissue at the electrode tip (e.g., gray or white matter) also influenced the size of the lesion. Szekely et al. (1966) applied 20 to 30 mA of 2-MHz current for 2 to 10 sec to 0.2-mm diam. electrodes in various animal brains and obtained ovoid lesions measuring $6.26 \pm 0.66 \times 3.19 \pm 0.79$ mm; the temperature rose to 56 to 58°C during passage of the current.

Alberts et al. (1966) took a different approach to the problem of predicting the size of a lesion made by radio-frequency current; they showed that the size of lesions produced with a given area of electrode was almost linearly related to a temperature maintained for 2 minutes. The shape of the lesions so produced was ellipsoidal and remarkably uniform, having an average length-to-diameter ratio of 1.8. Choosing a temperature of 60°C maintained for 2 minutes as optimum, they showed that the length of the lesion was proportional to the length of the exposed tip of the electrode; the width of the lesion was unaffected by the length of the exposed tip.

From the foregoing presentation of the parameters employed to make discrete brain lesions using radio-frequency current, the reader should be able to estimate the parameters required to form lesions of a predetermined size for marking electrode sites.

4

Intracellular Electrodes

When it is necessary to investigate the characteristics of the source of a bioelectric phenomenon that exists across the membrane of a single cell, transmembrane electrodes are required. One electrode is placed inside the cell, the other on the surface or nearby in the fluid environment. An intracellular electrode need only be small with respect to the size of the cell in which it is placed; thus penetration by the electrode will not cause damage. Just how small the electrode should be is not always easy to establish. Engineers use a rule of thumb that states that anything which is one-tenth to one-hundredth as large as something else, is considered to be small. In the case of intracellular electrodes, making an electrode that is one-hundredth the diameter of the cell of interest sometimes presents practical difficulties and often imposes very severe electrical requirements in the measuring instrument. Nonetheless, the tip of an intracellular electrode must be small enough to penetrate a cell membrane without allowing leakage of the intracellular ions. If the electrode tip is too large, the cell is injured, cytoplasm leaks out, and the potential measured is the injury, not the membrane potential.

Intracellular electrodes are of two types, metallic and nonmetallic. The former type usually consists of a slender needle of a suitable metal sharpened to a fine point or formed by electrolytic etching (electropointing). Another type of metal intracellular electrode is made of a glass micropipet filled or coated with a metallic substance. A nonmetallic intracellular electrode consists of a glass micropipet filled with an electrolyte. Because of the fragility of intracellular electrodes, they are usually made by the investigator. For this reason, the details of construction are presented along with descriptions of the various types.

It is to be noted that the electrical characteristics of metal microelectrodes and micropipets are not at all similar, and for this reason they are often used for different purposes. Both are high in impedance. The metal microelectrode resembles a leaky electrolytic capacitor, having a very high low-frequency impedance and a low high-frequency impedance. The micropipet, on the other hand, has a high resistive impedance. Thus the metal microelectrode is more suitable for measuring action potentials (usually extracellularly) than membrane potentials. The micropipet is ideally applicable to the measurement of resting membrane potentials; with suitable precautions (i.e., reducing the effect of shunt capacitance), it can easily serve to record action potentials. The full meaning of these statements will become clear from the following presentation.

METAL MICROELECTRODES

Mechanical Sharpening   Because of the small tip sizes desired, it is difficult to mechanically sharpen metal rods or wires. Nonetheless, Grundfest and Campbell (1942) described a method of grinding steel needles to produce 5 to 10 μ points. The needle is mounted in a rotating chuck and the tip is sharpened with a high-speed counter-rotating grinding wheel, which permits attainment of a high cutting speed, despite the small tip diameter.

ELECTROPOINTING   In most instances it is possible to point the end of a metal wire by electrolytically removing material; this technique was introduced by Grundfest et al. (1950), who employed it to fabricate stainless steel electrodes having 1-μ tips. They placed the end of a stainless steel wire (0.25 mm diam., hard-drawn 18-8 type) in an acid bath consisting of "34 cc concentrated sulfuric acid (sp. gr. 1.84), 42 cc of *ortho*-phosphoric acid (sp. gr. 1.69), and 24 cc of distilled water to total 100 cc of solution. A portion of this solution is placed in a small metal beaker which is connected to the negative side of a 6-V dc source. The wire to be "electropointed" is made the anode. A switch, variable resistance, and a milliammeter are included in the circuit. A microscope of low magnification is fitted in front of the beaker so that the stage of the electrolytic pointing may be observed at any time by raising the needle out of the acid solution."

A current of 30 mA was first employed to form the taper of the needle. During this part of the procedure the end of the wire was gradually withdrawn. When the taper was formed, the current was reduced and the needle was withdrawn more slowly to allow the etching process to provide the final taper and tip diameter desired. Following the pointing process, the authors recommended dipping the tip in 10% hydrochloric acid, then washing and thoroughly drying it and giving it a coat of insulating compound.

Use of the electrolytic technique to point tungsten wire was also described by Hubel (1957), who fabricated needles having tip diameters ranging from 0.5 to 0.05 μ. Starting with tungsten wire (0.125-mm diam.) he found that the slope of the taper could be controlled by raising and lowering the wire during all but the final stage in fabrication. After electrolytic pointing was completed, a final polishing step was added in which the terminal few milli-meters are immersed in a saturated aqueous potassium nitrate (KNO_3) solution and a 2 to 6 V alternating current is passed between the wire and a nearby carbon rod; the current may be conveniently obtained from a 6.3-V filament trans-former fed by a Variac (General Radio Co.). The optimum voltage is not crit-ical, but currents that are too low or too high tend to cause pitting. Hubel stated

If the polishing is allowed to continue until all bubbling ceases, a rather abrupt pencil-like point is obtained which has a tip of ultramicroscopic dimensions (from 0.5 to 0.05 μ in diameter). Such a result is explained by the fact that the meniscus height depends on the diameter of the wire, which decreases as the polishing proceeds.

An ingenious automatic method of using the electropointing technique to produce a batch of 20 stainless steel or tungsten electrodes with tip diameters ranging from 1 to 10 μ was described by Mills (1962). The only difference in the fabrication techniques for the two metals was the type of solution used for the etching.

Mill's method consisted of mounting 20 lengths (1 1/2-2 in.) of the material to be pointed on a metal shaft that rotated at about 5 rpm. As the shaft turned, the electrodes dipped into and out of the electropointing solution, which was composed of equal parts of sulfuric acid, phosphoric acid, and water for stain-less steel needles and 0.5M sodium hydroxide for tungsten wires (0.010 in. diam.). The rotating shaft was connected to the positive pole of a 3-V battery with the negative terminal connected to a 2-cm² platinum electrode mounted to the side of the electrolytic bath, which contained 500 ml of solution. The current that flowed when the electrodes were in the solution was about 200 mA. Mills started with stainless steel pins having tip diameters of 30 to 50 μ; after about 6 hours of electropointing, during which the direction of rotation was reversed every half hour, 85% of the electrodes had tip diameters ranging from 1 to 6 μ. They were made ready for recording by washing, drying, and coating with several layers of insulating varnish (Formvar). After each coat was applied, the electrode was allowed to air dry (tip upward); the insulation was then hardened by baking in an oven at 150°C for 30 minutes.

The use of alternating current for electropointing was described by Wol-barsht et al. (1960). These investigators placed the end of a length of 8 to 10 mil platinum-iridium (70%, 30%) wire "in a solution of 50% sodium cyanide and 30% sodium hydroxide, added to prevent the formation of hydrogen cyanide." Alternating current (60 Hz) was passed between the wire to be pointed and a nearby carbon electrode in the solution. The taper was formed by connection of a 6 to 10 V (rms) source, and while current flowed, the solution was stirred and the wire was slowly withdrawn. The tip, which measured about 1 μ, was formed by reducing the voltage to 0 to 8 V; agitation of the solution was not necessary at this point in the procedure. The electrode was coated with glass insulation (see p. 141) and the tip was platinized to lower its resistance. (Making the microelectrode negative with respect to a platinum wire in a 1% platinous chloride solution accomplished the platinization). The current was provided by a 15-V power supply in series with a 1-MΩ resistor; the platinization process required passage of the current for 15 to 30 sec.

The method of using alternating current for electropointing was further developed by Guld (1964), who etched 0.25-mm platinum-iridium (70%, 30%) wires to produce electrodes having tip diameters of 1 μ. Guld presented considerable quantitative data on his technique for obtaining a desired taper and tip diameter. In general, an electrode was etched by the application of 12 V (rms) to the platinum-iridium wire and a nearby carbon electrode (10 cm² in area); the solution employed was 100 to 500 ml of sodium cyanide ($8M$); enough sodium hydroxide was added to prevent the formation of hydrogen cyanide. Current was passed for about 3 minutes; the rate of withdrawal determined the taper. The tip was formed by etching at 12 V (rms) in a slightly unsaturated sodium cyanide solution and stopping the current when it reached 30% of its initial value. The electrode was then coated with glass using Wohlbarsht's method (see p. 141) and the tip was exposed by the platinizing process, employing 0.1% platinous chloride solution. Platinization was accomplished by using 10-msec current pulses; the electrode was made the negative pole of an 8 to 15 V source. After each current pulse, the electrode impedance was measured and the process was stopped when the "desired impedance" was reached (1–20 MΩ).

The first thought that occurs when alternating-current etching is proposed is that the method will not work. With alternating current, theoretically, as much current flows in one direction as in the opposite; and according to Faraday's law of electrolysis, the net effect on each electrode would be zero. However, if the deposits laid down on the electrodes with current flow in one direction have conductivities different from those deposited with current flow in the other direction, the average current in one direction will not be the same as that in the opposite direction; hence, the system is nonlinear and rectification occurs.

Electrolytic rectifiers have been known for some time; the most prominent example employs an aluminum electrode that becomes coated with aluminum oxide (a good insulator) during one half-cycle; the coating is removed during the other half-cycle. That rectification occurs in the use of an alternating voltage for etching platinum-iridium electrodes was documented by Guld, who showed that although the alternating voltage waveform is sinusoidal, the current waveform is not. In fact, his use of a capacitor across the electrolytic etching bath guaranteed smoothing out of the current pulses to produce an average direct voltage which, in turn, produces the current for the electrolytic processes.

Insulating Metal Electrodes   No matter how metal electrodes with small-diameter tips are fabricated, it is necessary to place insulation on all but the tip. Grundfest et al. (1950) advocated withdrawing the needle from the insulating compound and allowing drying to take place with the tip upward. Hubel (1957) coated his electrodes with E53 Insul-X (Insul-X Co., Ossining, N.Y.) or with clear vinyl lacquer (S-986S, Stoner-Mudge, Inc., Pittsburgh, Pa.). An electrode was slowly withdrawn from the insulating compound, and when the tip emerged from the insulating material, a bead ran up the tapered portion of the electrode. At this time the tip was quickly reimmersed and withdrawn slowly until no bead formed. Then the insulating compound was allowed to dry at room temperature; no baking was advocated.

A technique for placing glass insulation on a small metal electrode was described by Wohlbarsht et al. (1960). Starting with an electropointed platinum-iridium (70%, 30%) electrode, they were able to insulate it by advancing it through a melted bead of glass (Corning solder glass 7570) having the same thermal coefficient of expansion as the metal [8.5×10^{-6} cm/(cm)(°C)]. The bead was formed by heat in an electrically energized V-shaped platinum-iridium element. During passage of the electrode through the melted glass, the tip was also covered by a very thin layer of glass; this was ruptured by passage of current through it by the platinizing process. The appearance of gas bubbles at the tip indicated that the thin glass covering had been removed.

A glass covering provides a very durable low-leakage insulation. However, successful use of it requires that the chosen glass have exactly the same thermal coefficient of expansion as the metal electrode. If the coefficients are different, the glass will crack during cooling and the insulation will be imperfect.

Measurement of Exposed Area of Metal Microelectrodes   With the insulating procedures just described, it is often difficult to see the extent of the exposed tip, because the layer of insulating material is usually so thin near the tip that microscopic examination usually fails to uncover the boundary of the insulation. Under such circumstances, the tip area can often be determined by using the method described by Hubel (1950). A drop of saline on a glass slide is viewed with a microscope. A small wire is then placed in the drop and connected to the positive pole of a battery; the other pole is connected to the metal microelectrode. The electrode is advanced into the drop, and the tip is viewed in the microscope. The active area can be estimated by observation of the area from which hydrogen bubbles are evolved. This method is very convenient to apply and has been used successfully by the author to measure the area of a variety of large and small needle electrodes.

Heat-Pulled Metal Microelectrodes

Taylor (1925) reported that a 35-gauge platinum wire placed in a quartz capillary could be heat pulled in a microburner to produce a glass-insulated platinum electrode having an exposed tip of 1 μ. Presumably the type of quartz used had nearly the same thermal coefficient of expansion as the platinum wire ($8.99 \times 10^{-6}/°C$); because Taylor did not mention any breakage during cooling. Interestingly enough, in the same paper he reported making a micro-magnet by heat pulling an iron wire (thermal coefficient $12.1 \times 10^{-6}/°C$) in a quartz capillary. Because of success with metals having different coefficients of expansion, it would appear that a slight mismatch in coefficients can be tolerated.

A low-cost, high-quality metal microelectrode was developed by Svaetichin (1951), who first melted silver solder in a 1-cm diam. soda glass tube, then pulled the tube to obtain a long length of 5-mm diam. glass-covered conductor of silver solder. Because silver solder and glass melt at nearly the same temperature and both have similar thermal expansion coefficients, a uniformly insulated and mechanically stable structure was obtained. Segments of this material were then heat pulled in a microburner to produce glass-insulated electrodes with tip diameters in the range of 1 to 90 μ. To reduce the impedance of the small metal tip, it was first plated with rhodium and then covered with platinum black (see p. 32). Considering the tip diameters obtained, the resistance of the electrodes was remarkably low (see Fig. 4-3).

A slightly different procedure for producing metal-filled glass capillary electrodes was described by Dowben and Rose (1953). These investigators melted a mixture of gallium (50%) and indium (50%), or pure indium, and sucked the melted metals (melting points 110 and 150°C, respectively) halfway up into a 5-in. length of 0.5-mm i.d. Pyrex tube having a thermal coefficient of expansion of $3.3 \times 10^{-6}/°C$. The alloy, which wetted the glass tubing, was then allowed to cool. At a point 2 mm beyond the level of the alloy, the glass tube was heat-pulled to form the empty glass taper and the tip of the electrode. (The authors stated that a glass micropipet puller could be used to advantage.) After the tip had formed, the electrode was heated on a hot plate until the alloy melted, whereupon a sewing needle was inserted into the tubing and used as a piston to force the melted metal into the empty end and out the tip so that a bead (ca. 30–50 μ diam.) formed beyond the tip. When the electrode cooled, the bead was blown off and the exposed metal tip (2–4 μ diam.) was gold plated, then platinized. For gold plating, they stated

A gold cyanide solution (0.2% Au) is used for plating, with a 1.5 V dry cell in series with a 5–10 megohm resistor as a source of current. The microelectrode is connected to the negative pole. For an electrode with a 2–4 μ tip, the plating time is 30–45 sec.

On the chemically inert and impervious gold base, platinum black is electrodeposited from a bath of 0.4% platinum chloride. The plating time and current are the same as those for gold plating.

The technique of heat pulling a metal in a quartz or glass tube is simple to use and produces insulated electrodes with exposed metallic tips ranging from 1 to 90 μ diam. The application of platinum black to such electrodes is highly desirable to reduce the electrode-electrolyte impedance. The electrical characteristics of these electrodes are described later.

The two requirements that should be recognized for successful use of this technique are the use of a metal and insulation with about the same melting points, and the same thermal coefficients of expansion. From the reports presented, it would appear that not all investigators took care to use a metal with the same thermal expansion coefficient as the insulating covering. Omitting to do this causes stresses to develop as the electrode cools. There is thus the danger of producing microcracks and sudden fracture if the electrode is tapped gently. However, since such failures did not occur, some degree of mismatch in thermal coefficients of expansion may be tolerable in practice. The data in Table 4-1 may be of value to those who are contemplating use of this simple process.

Electrodeposited Metal-Filled Micropipet   To overcome the high resistance of the electrolyte-filled micropipet (see p. 154), Weale (1951) devised a method of filling a micropipet with silver, which was deposited electrolytically. Starting with an empty glass micropipet, he advanced a pointed silver wire into it. The silver wire (60 μ) was pointed by drawing it until it broke whereupon a piece of platinum wire was soldered to the opposite end. With the pointed silver wire in the taper of the micropipet, the large end was sealed to the platinum wire by melting the glass. Then the electrode was placed in boiling distilled water and some of the air in the micropipet was expelled by the increased temperature. The submerged electrode was then allowed to cool, which caused enough water to enter the capillary tip and establish contact with the silver wire. Then the tip of the micropipet was filled with silver by electroplating with a saturated solution of silver nitrate. The platinum wire (soldered to the silver wire in the micropipet) was connected to the negative pole of a 12-V battery in series with a variable resistance and a microammeter; the positive side of the battery was connected to a silver electrode in the silver nitrate solution. The plating process was carried out using 30 μA for about one-half hour, the time necessary to fill the tip with silver. Weale employed an automatic device to determine when this point has been reached—a sudden drop in the resistance, which was continually measured during the plating process, indicated that the tip was filled with silver.

Table 4-1 Melting Points and Thermal Expansion Coefficients

Material	Melting Point (°C)	Expansion Coefficient $\times 10^{-6}$/°C
Pyrex glass	1245	3.3
Vycor glass	1550	0.8
Soda glass	1000	12.0
Quartz crystal		5.21
Quartz fused	1500	0.256
Silver solder		
(10% Ag, 52% Cu, 38% Zn)	870	
(45% Ag, 15% Cu, 15% Zn, 25% Cd)	620	
Corning 7570 solder glass		8.5
Silver	960	17.04
Platinum	1773	8.99
Platinum-iridium		8.84
Gallium	30	
Indium	156	41.7
Tin	232	22.57
Antimony	630	12.0
Gold	1063	13.2
Iron (soft)	1535	9.07
Iron (stainless steel)	1400–1450	10–20
Flame Temperatures		
Natural gas–Air	1950	—
Acetylene–air	2325	—
Acetylene–oxygen	2900–3000	—

Weale stated that he was able to make electrodes with tip diameters of 10 μ and resistances on the order of 1500 Ω. Smaller-diameter electrodes exhibited considerably higher resistance. No additional processing was carried out.

Vapor-Deposited Metal-Film Microelectrodes

Two interesting types of small-size vapor-deposited metal electrodes are available commercially (Transidyne General Corp., Ann Arbor, Mich. 48103). One type is solid and the other is hollow, like a micropipet; both are made from glass stock 0.75 or 1 mm diam. In both types (called Microtrodes), the glass is covered with a thin layer of platinum applied by the vapor-deposition technique similar to that used by Einthoven to make quartz filaments conducting in his string galvanometer (1903).

Several of the geometric arrangements available are shown in Fig. 4-1. The exposed platinum tip beyond the insulation is about 1 to 2 μ; tip diameters are in the range of 1 to 20 μ. Because these solid and hollow Microtrodes have only been available for a short time, it is not possible to comment on their

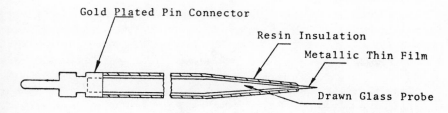

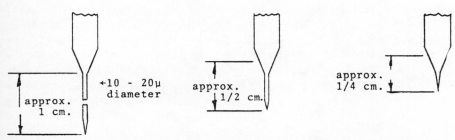

Fig. 4-1. Vapor-deposited metal film microelectrodes. (Courtesy of Transidyne General Corp., Ann Arbor, Mich. 48103.)

performance characteristics. However, their low cost, specified dimensions, and reusability would seem to indicate that they will see considerable service.

Integrated-Circuit Technique Microelectrodes

Another method of constructing small-area electrodes was developed by Wise and Starr (1969), who applied integrated-circuit technology to make single and multiple electrodes from gold strips that are bonded to a film of silicon dioxide on a substrate of silicon (Fig. 4-2). The gold was then covered with an 0.4 μ layer of silicon dioxide, which was etched off to form the exposed tip. Wise and Starr have fabricated electrodes with exposed tips having areas of 15 μ^2. An illustration presented by the authors shows a linear array of three electrodes which occupies less than 50 μ. The capacitive coupling to the silicon support (ground) is given as 0.5 pF/mm of electrode length.

The electrodes described by Wise and Starr have the advantage of precise control of dimensions. However, in their present state of development, they can only be used extracellularly or placed into relatively large cells. Nonetheless, there is no doubt that a slight improvement in technology will permit fabrication of much smaller electrodes .

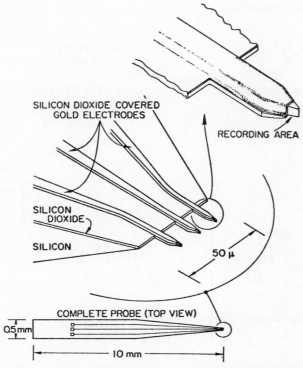

SILICON DIOXIDE COVERED
GOLD ELECTRODES

RECORDING AREA

SILICON
DIOXIDE

SILICON

50 μ

COMPLETE PROBE (TOP VIEW)

0.5 mm

10 mm

Fig. 4-2. Integrated-circuit technology microelectrodes. (From K. D. Wise and A. Starr, *Proc. 8th Int. Conf. Med. Biol. Eng., 1969*, Sect. 14-5. By permission.)

Electrical Properties of Metal Microelectrodes

Impedance The high impedance of a metal microelectrode is due to the characteristics of the small-area metal-electrolyte interface. By knowing the electrical parameters of such a structure, it is possible to predict the impedance for an electrode of a known area. It was pointed out in Chapter 1 that, over a considerable frequency range, the series-equivalent reactance X and resistance R for many metal-electrolyte interfaces are approximately equal and that each varies approximately as $1/f^\alpha$ where f is the frequency and α is approximately 0.5. Assuming that $R = X$, then the impedance Z will be

$$Z = \sqrt{R^2 + X^2} = \sqrt{2}\, X$$
$$= 1.41X = \frac{1.41}{2\pi fC} \tag{4-1}$$

Therefore, if the series capacitance (C)-frequency (f) relationship is known for a given area of electrode-electrolyte interface, it will be possible to calculate the approximate impedance.

Because electrode capacitance varies with area, it is possible to use the data in Fig. 1-6 to estimate the impedance-frequency characteristic for a small-area metal microelectrode within the constraint that reactance and resistance are equal and both vary as $1/f^a$. In order to use the data in Fig. 1-6 and equation (4-1) to calculate the approximate impedance-frequency relationship for a metal microelectrode of area a, it is necessary to modify the constant K (which was derived using the units of microfarads and square centimeters), to allow its use with electrodes (which are square microns in area—1 $\mu^2 = 10^{-8}$ cm² and farads (1 μF $= 10^{-6}$ F). Performing this adjustment gives the following expression for the series capacitance C (infarads) for an electrode of area a (μ^2):

$$C = 10^{-14} \frac{Ka}{f^\alpha} \qquad (4\text{-}2)$$

The impedance-frequency relationship can be obtained by substitution of this expression into equation (4-1); performing this substitution gives the following for the impedance Z (Ω):

$$Z = \frac{1.41}{2\pi f(10^{-14}Ka/f^\alpha)}$$
$$= \frac{0.225 \times 10^{14}}{Kaf^{(1-\alpha)}} \qquad (4\text{-}3)$$

It is of interest to use this expression to calculate the impedance-frequency relationship for a particular electrode area and to compare the data with those in the published literature. A useful electrode size would be 10 μ^2, which is equivalent to the surface of a conical electrode 1-μ diam. and 6.35 μ high (i.e., the tip angle is 9°). Inspection of Fig. 1-6 reveals that the highest impedance would be expected with bare metals because of the low values for the capacitance-frequency relationship; conversely, the lowest impedance would be expected from platinum black because it exhibits the highest value for the capacitance-frequency relationship. Using an average value of $C = 350/f^{0.5}$ for bare metal and $C = 149,700/f^{0.366}$ for heavy platinum black, the expected impedance-frequency characteristics for a 10-μ^2 electrode of each of these surfaces have been calculated. The plots obtained (Fig. 4-3) also show the values obtained by Gray and Svaetichin (1951–1952) for their 5-μ diam. glass-insulated platinum-black electrode (19.4 μ^2), by Gesteland et al. (1956) for their 3-μ platinum-black electrode (28 μ^2) and 50 μ^2 silver-silver chloride electrode, and by Hubel for his 8-μ^2 tungsten electrode.

Inspection of Fig. 4-3 reveals that the impedance-frequency values reported

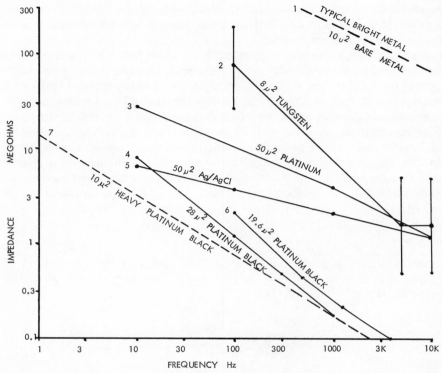

Fig. 4-3. Impedance-frequency characteristics of metal microelectrodes: (1) bare metal (calculated from $C = 350/f^{0.5}$ and $X = R$); (2) tungsten (Hubel, 1959); (3) platinum (Gesteland et al., 1956); (4) platinum black (Gesteland et al., 1956); (5) silver-silver chloride (Gesteland et al., 1956); (6) platinum black (Gray and Svaetichin, 1951–1952); (7) heavy platinum black (calculated from $C = 149000 f^{0.366}$ and $X = R$).

in the literature cited are in fairly good agreement with those predicted by equation (4-3) using the data in Fig. 1-6. Perhaps the most important point made by Fig. 4-3 is that electrolytically prepared electrodes (e.g., platinum black and silver-silver chloride) have quite low impedances. Likewise, Hubel's $8-\mu^2$ electrolytically pointed electrode has an impedance considerably lower than was predicted on the basis of a bare metal surface slightly larger in area, probably because electrolytically formed surfaces have quite large areas.

From the foregoing it can be seen that the deposition of a coat of platinum black offers an excellent means of reducing the impedance at all frequencies to values well below those exhibited by bare platinum. Although the platinizing process (see p. 32) has been specified for large-area electrodes, the values obtained may not be directly scaled for the platinization of microelectrodes. Robinson (1968) pointed out that the current density value obtained by scaling on an area basis, are such that platinization will not occur; he therefore advocated the use of a current density 10 times that found to be optimum for large-area (cm^2) electrodes.

The impedance-frequency characteristic for a small-area (50-μ^2) silversilver chloride surface in contact with saline illustrates that this surface has a much lower impedance than bare silver, but somewhat higher than platinized platinum. Attractive as the relatively low impedance of the silver-silver chloride surface may be (see p. 25), it is not well suited mechanically for metal microelectrodes; the chloride coat is friable and easily damaged. In addition, there are no data available regarding the current density and time for optimum chloriding of silver-silver chloride microelectrodes; existing data can only serve as a starting point. The curves in Figs. 1-12 and 1-13 for an optimum chloride coat for large-area silver chloride electrodes indicate that it may well be necessary to increase the chloriding current density to achieve an adequate chloride deposit.

Because of its mechanical strength and inertness in living tissue, stainless steel is frequently used for microelectrodes. It should be noted that the price paid for these two desirable qualities is high impedance, since stainless steel exhibits a low value of capacitance at all frequencies.

From the foregoing it is apparent that metal microelectrodes exhibit a very high low-frequency impedance. Thus is very difficult to use them to measure resting membrane potentials. On the other hand, their lower high-frequency impedance permits their easy use to measure action potentials; with amplifiers with only moderately high input impedances, they are particularly well suited for the extracellular measurement of action potentials. This subject is discussed in the following section.

Equivalent Circuit Having established that the impedance of a metal microelectrode is high, it is appropriate to develop a simple equivalent circuit to identify all the voltages and impedances that exist between the terminals of a metal microelectrode A and its reference electrode B; Fig. 4-4a shows a metal electrode located in the environment of an irritable cell. Tracing out the path between terminals A and B, the first item encountered is the resistance of the connecting wire and the metal used for the microelectrode; the former is obviously negligible. Many metal microelectrodes are made from metal rods about 0.25 mm diam.; the resistance of a 1-cm length depends on the resistivity of the metal employed (most frequently, platinum, tungsten, and stainless steel). The resistances per centimeter of an 0.25-mm rod of these metals are 20, 11, and 130 Ω; respectively; thus the resistance of the metallic portion of the microelectrode is negligible.

Because the shaft of a metal microelectrode is covered with an insulating material (e.g., glass or insulating varnish) which is in turn in contact with a

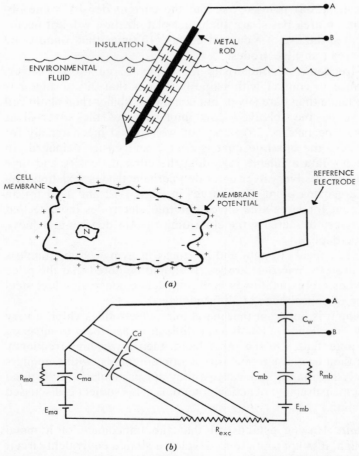

Fig. 4-4. Extracellular metal microelectrode: (*a*) extracellular metal microelectrodes; (*b*) equivalent circuit.

conducting fluid environment, a coaxial capacitor is created. With an insulating coating amounting to 10% of the diameter of the shaft of the microelectrode, the capacitance C_d of the shaft can be calculated from the expression for the capacitance of a coaxial cable with inner and outer conductors of diameters d and D, respectively, and the space between filled with material of dielectric constant k. The capacitance C_d per millimeter is given by

$$C_d = \frac{0.024\,k}{\log_{10}D/d} \text{ pF/mm} \qquad (4\text{-}4)$$

With 10% insulation, $D/d = 110/100 = 1.1$, and with glass or insulating varnish as the dielectric, the dielectric constant will be between 3 and 5, with perhaps 4 as an average value. Under these conditions the capacitance per millimeter of shaft is

$$C_d = \frac{0.024 \times 4}{\log_{10} 1.11} = 2.1 \text{ pF}$$

Note that although the capacitance depends on the length of the microelectrode surrounded by fluid, it does not depend on the diameter of the microelectrode, but rather on the ratio of the diameter (plus insulation) to the shaft diameter. The distributed capacitance C_d exists between the shaft of the metal microelectrode and the extracellular fluid R_{exc} and forms a circuit as shown in Fig. 4-4b.

The next item to be encountered in tracing out the circuit between terminals A and B is the small-area metal-electrolyte interface at the tip of the microelectrode; this can be represented by a series or parallel capacitance C_{ma} and resistance R_{ma} and a potential E_{ma}. The magnitude of the resistance and capacitance depend on the exposed area and the type of metal and electrolyte which make up the extracellular environment; the smaller the area, the higher the impedance of the series or parallel combination of R_{ma} and C_{ma}. The electrode potential E_{ma}, is not determined by exposed area, it is dependent on the type of metal and electrolyte and the temperature (see Chapter 1).

Because the microelectrode is outside the cell, the next item to be encountered is the resistance of the extracellular fluid. Ignoring the distributed capacitance C_d for the moment, the reference electrode potential E_{mb} and impedance, consisting of C_{mb} and R_{mb}, are encountered in completing the circuit from A to B. Because the area of the reference electrode B is much larger than that of the tip of the microelectrode, the parallel impedance of C_{mb} and R_{mb} is small and can be neglected. However, the potential of electrode B, E_{mb}, is not area dependent, nor can it be neglected.

Thus it can be seen that with the metal microelectrode outside a cell, the circuit between the electrode terminals consists of a series circuit of two electrode potentials and impedances and the resistance of the extracellular fluid. The distributed coaxial capacitance of the microelectrode is in parallel with a part of the circuit. Another capacitance C_w is due to the wires connected to the reference and microelectrode; it exists in parallel with the electrode terminals and its value depends on the length and spacing of the wires used to connect the electrodes to the measuring instrument. Because some of the components in the equivalent circuit are more important than others, simplification of the circuit can be carried out. (A description of such a simplification is reserved for the case of the microelectrode in the cell.)

When the cell becomes active and recovers with the microelectrode outside the cell, the cyclic depolarization and repolarization cause current to flow in the extracellular fluid. Hence, a potential will appear across R_{exc} in the equivalent circuit of Fig. 4-4b. The amount of this voltage that appears across terminals A and B depends on the distance of the electrodes from the cell, their orientation with respect to the direction of spread of excitation, the electrode impedances (R_{ma}, C_{ma}, R_{mb}, C_{mb}, C_d, and C_w), and the magnitude of the impedance of the measuring instrument connected across terminals A and B. As stated previously, the circuit is simplified to identify the most important of these impedances.

When a metal microelectrode is advanced into a cell as in Fig. 4-5a, the potential appearing between terminals A and B, connected to the microelectrode and the reference electrode, is the sum of three potentials, the metal-electrolyte junction potential of the microelectrode tip E'_{ma},* the cell membrane potential E_{mp}, and the reference electrode-electrolyte junction potential E_{mb}. When the membrane potential is being measured, it is assumed that the sum of the first and third is constant. Note that the potential of the microelectrode has now changed from that which existed when it was in contact with extracellular fluid E_{ma} to that when in contact with intracellular fluid (E'_{ma}). Although small, this difference should be noted; it is usually ignored. Because metal microelectrodes are seldom used for the measurement of resting membrane potentials, a discussion of the difference between E_{ma} and E'_{ma} is not presented. (A similar situation exists with the micropipet discussed on p. 167.)

In addition to the potentials just identified, there are several impedances of importance, and the complete equivalent circuit for a metal microelectrode in a cell can be assembled by tracing out the circuit between electrode terminals A and B. Referring to Fig. 4-5b, the first item encountered is the impedance C'_{ma}, R'_{ma} of the tip of the metal microelectrode, which is in contact with intracellular fluid; in conjunction with the tip there is the electrode potential E'_{ma}. Neglecting the distributed capacitance for the moment, the next to be encountered is the resistance of the intracellular fluid R_{inc} and then the membrane with its potential E_{mp}, along with the components that account for it (which appear in Fig. 4-20). The circuit is completed by the resistance R_{exc} of the extracellular fluid and the potential E_{mb} and impedance of C_{mb} and R_{mb} of the reference electrode B. Between terminal A and the extracellular fluid appears the distributed (coaxial) capacitance C_d of the shaft of the metal microelectrode. In addition, there is capacitance C_w between the wires used to connect the electrodes to the instrument that measures the cell potentials.

Fortunately some of the components in the equivalent circuit in Fig. 4-5b are small with respect to others and can be neglected. Therefore it is possible

---

* The prime notation is used to identify a different value for the same component shown in Fig. 4-4.

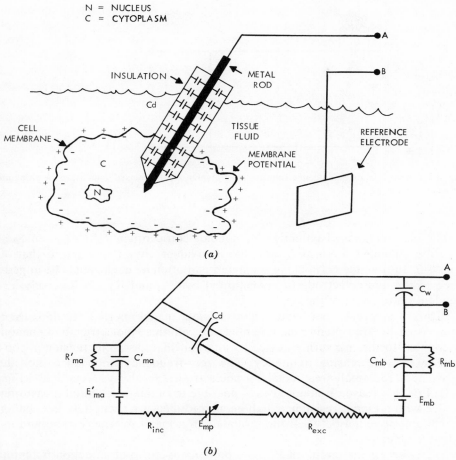

N = NUCLEUS
C = CYTOPLASM

INSULATION

METAL ROD

Cd

CELL MEMBRANE

TISSUE FLUID

MEMBRANE POTENTIAL

REFERENCE ELECTRODE

A

B

C

N

(a)

A

C_w

B

C_d

R'_{ma} C'_{ma}

C_{mb} R_{mb}

E'_{ma}

E_{mb}

R_{inc} E_{mp} R_{exc}

(b)

Fig. 4-5. Intracellular metal microelectrode: (*a*) metal microelectrode inserted into a cell; (*b*) equivalent circuit.

to synthesize a simple electrical circuit that mimics the electrical behavior of the actual circuit. For example, in comparison to the impedance of the microelectrode tip C'_{ma}, R'_{ma}, the resistance of the intracellular and extracellular fluids R_{inc} and R_{exc} can be neglected. Because the area of the reference electrode is many times greater than that of the microelectrode tip, the impedance of the former (C_{mb}, R_{mb}) can be neglected. The distributed capacitance C_d between the microelectrode shaft and the extracellular environment can be collected together with that of the wiring C_w to form a single capacitance $C_d + C_w$.

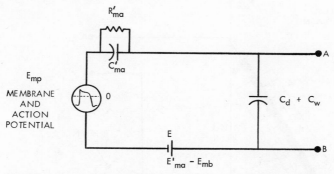

Fig. 4-6. Approximate equivalent circuit for a metal microelectrode used to measure the potential of a cell membrane.

The electrochemical potentials of nonbioelectric origin E'_{ma}, E_{mb} can likewise be collected to form E, and the equivalent circuit reduces to that of Fig. 4-6. In practice $C_d + C_w$ is small and can often be neglected. The impedance of the microelectrode tip (constituted by C'_{ma} and R'_{ma}) is inversely dependent on the area of the tip.

Figure 4-6 reveals that, when an intracellular metal microelectrode is used to measure the transmembrane potential of a cell, this bioelectric phenomenon is coupled to the measuring device by a circuit that has an increasing impedance with a decreasing frequency. The zero-frequency impedance (i.e., dc resistance) for small-area metal microelectrodes is always very high (Fig. 4-3). For this reason it is not always possible to obtain a potential-measuring device with an input resistance high enough to measure accurately the resting membrane potential between the terminals of a metal microelectrode and its reference electrode.

The effect of the circuit of Fig. 4-6 on the waveform of an action potential is determined by the input impedance of the amplifier to which it is connected. With an adequately high input impedance (e.g., 100 to 1000 times that of the electrode circuit for the frequency spectrum of the action potential), no waveform distortion will be encountered. However, if the input impedance is not high enough, the low-frequency components of the action potential will be attenuated as in Figs. 3-9 through 3-11. (The relation between electrode area and input impedance is discussed on pp. 120 and 127.)

THE ELECTROLYTE-FILLED MICROPIPET

When transmembrane resting and action potentials are to be measured, it is customary to employ the micropipet (Fig. 4-7), which resembles a small medicine dropper with a tip drawn down to 1 μ diam. A terminology has been developed to describe the various parts of a micropipet. The upper part is called the stem or shaft. The region where the stem encounters the tapered portion is sometimes called the shoulder; often the tapered portion is called the shoulder or shank. The term tip is usually reserved for the region of the pipet that enters the cell. In this book, the terms stem, taper, and tip along with the taper angle θ and tip angle ϕ, as in Fig. 4-7, are employed.

A suitable electrolyte is used to completely fill the tip, taper, and stem, and a well-chlorided silver electrode dipping into the filling solution serves as the electrical terminal. The important consideration regarding the micropipet (in addition to use of the proper measuring instrument) is its tip diameter, which should be small with respect to the size of the cell into which it is placed.

An interesting study relating the magnitude of the measured membrane potential with the tip diameter of the intracellular electrode was conducted by Woodbury et al. (1951). Using the frog ventricle, which is composed of

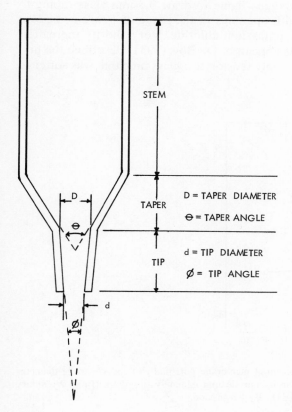

Fig. 4-7. The micropipet.

cells approximately 20 μ diam., they measured the membrane potential with micropipets having different tip diameters. Their data (Fig. 4-8) illustrate that the larger the tip diameter, the less the potential that was measured. The voltage measured with a tip diameter on the order of 1 μ (ca. 5% of the cell diameter) was taken as 100%, since this was the smallest diameter electrode used. Extrapolation of their data (Fig. 4-8) indicates that an electrode with a slightly smaller tip diameter might have measured a slightly larger membrane potential. Therefore, their data lend support to the rule-of-thumb that 1 to 10% identifies the appropriate range of an electrode diameter with respect to the size of a cell.

Micropipets suitable for measuring transmembrane potentials have been available since 1925, when Ettisch and Peterfi described the use of a 2-mm-wide flame to pull 1-mm glass tubing to points about 10 μ diam. These electrodes were filled with 0.1N potassium chloride in agar and were employed with calomel half-cells to measure the (membrane) potentials of plant and animal cells in conjunction with a polarized electrostatic electrometer. In the same year Gelfan used an oxygen flame to draw 0.5-mm glass tubing to make micropipets having tips with openings in the range of 1 to 2 μ; he also filled his micropipets with 0.1N potassium chloride. Very shortly thereafter, machines for making micropipets appeared. Du Bois (1931) described the first of these; it employed a spring to apply tension to a glass tube that was softened

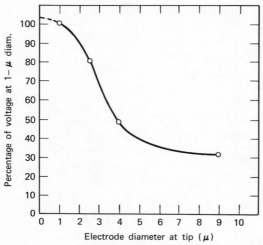

Fig. 4-8. The relation between the measured membrane potential (%) and electrode diameter (μ). The data were obtained from frog cardiac muscle approximately 20 μ diam. (From Woodbury et al., *Amer. J. Physiol.* 1951, **164**:307–318. By permission.)

by an electrically energized platinum heater. As the glass softened, the tension was increased considerably, and the tube was automatically removed from the vicinity of the heater during the final pull. This technique of using an initial gentle pull followed by a sudden forceful one, while the heat is reduced, has been employed in almost all subsequent micropipet makers; descriptions of some of these were presented by Livingston and Duggar (1934), Benedette-Pichler and Rachelle (1940), Alexander and Nastuk (1953), Winsbury (1956), Lux (1960), Byzov and Chemystov (1961), and Chowdhury (1969). In the instrument made by Alexander and Nastuk, an initial force of 100 g, followed by 1700 g, produced 10° tips with outside diameters of 0.4 to 0.5 μ. In Wins-bury's machine, an initial force of 0.8 lb was followed by 1 to 5 lb to produce 3° tips with diameters in the range of 0.5 to 10 μ. Lux found that cooling a small area of the glass while it was being drawn permitted attainment of a sharper taper. The instrument devised by Byzov and Chemystov, which employed a sharp blow to provide the second force, produced 10 to 15° tips with diameters in the range of 0.1 to 0.08 μ. Chowdhury's instrument yielded micropipets having tip diameters as small as 0.05 μ (i.e., 500 Å) by the use of a two-stage pull combined with cooling from two tiny jets of compressed air controlled by solenoid valves. This process yielded tip angles in the range of 11°.

Not all who employed micropipets used automatic machines for their fabrication; Hodgkin and Huxley (1939) and Curtis and Cole (1940) used simple techniques to construct micropipets 10 to 100 μ diam. which were used with squid giant axons 500 to 1000 μ in diam. Similarly, Ling and Gerard (1949) were able to draw micropipets down to less than 1 μ diam. for recording transmembrane potentials in frog skeletal muscle cells.

From the foregoing it is apparent that the basic steps in the fabrication of a small-tip micropipet consist of the circumferential application of heat to a small area of suitable tubing which is placed under an initial tension. When the glass softens, the tension is increased as rapidly as possible and the heat is turned off; sometimes active cooling is added. Careful timing and adjust-ment of the amount of heat, as well as the initial and final tensions and cooling, permit the production of micropipets with controlled dimensions. An excellent account of how these factors interact to control tip diameter and taper was presented by Frank and Becker (1964). Although many micropipet makers are commercially available, and most provide a means for controlling tensions and temperature, each investigator must learn the technique of adjustment to obtain micropipets with the desired tip characteristics.

The Floating Micropipet

The tip of a conventional micropipet usually breaks during the first con-traction when transmembrane potentials from contracting muscle cells are being recorded. To solve this problem, Woodbury and Brady (1956) developed a micropipet that moves along with contracting and relaxing muscle cells. Their micropipet was constructed by placing an 0.001-in. tungsten wire in a conventional $3M$ KCl-filled micropipet and breaking the stem to leave a segment about 1 cm long attached to the tip. If the movement is slight, a straight wire a few inches in length is adequate; if the movement is large, it is advisable to use a 3 to 4 in. length of tungsten wire, later bending it with a right angle near the middle. The broken part is slipped off the tungsten wire, which is then affixed to the preamplifier mounted in a convenient manipulator (Fig.

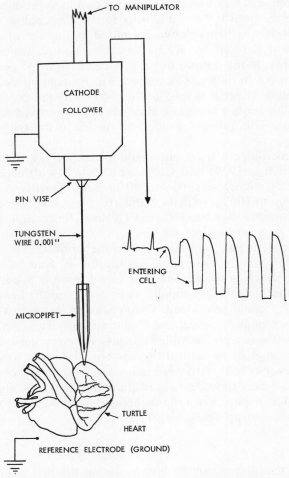

Fig. 4-9. Penetration of a ventricular cell by a floating micropipet.

4-9). The micropipet is gently lowered so that the tip comes in contact with the muscle. When the muscle contracts, the springiness of the tungsten wire allows the micropipet to move along with the muscle. The record shown in Fig. 4-9 was made with such an electrode, placed on the surface of a turtle ventricle; in this position the conventional ECG was recorded. As the ventricle continued to contract and relax, the tip of the micropipet entered a cell, revealing the membrane and monophasic action potential.

Using a short tungsten wire, Woodbury and Brady reported success in recording resting membrane and action potentials from the cardiac pacemaker of the frog, the turtle, the rat, the guinea pig, the rabbit, the monkey, the canine ventricle, frog skeletal muscle, and the uterus of the pregnant guinea pig and rabbit. The author has had considerable success using this electrode in the teaching laboratory to record the membrane and action potentials of frog and turtle atria and ventricles.

Filling Micropipets

Filling a micropipet with a chosen electrolyte does not come about by filling the shaft or by placing the micropipet tip downward in the electrolyte; special techniques must be used to coax the fluid into the microtip. In general, the difficulty in filling increases as the tip diameter decreases. Frank and Becker (1964) reported that micropipets having tip diameters larger than 5μ can be filled by applying pressure by a syringe and a blunt-ended needle placed snugly in the tapered portion of the micropipet. Micropipets with tip diameters as small as 0.5μ or less have been successfully filled using a variety of the techniques to be described. However there also appears to be a finite time during which micropipets can be stored, once filled. (Storage of dry empty micropipets is inadvisable because dust particles tend to gain access and block the tip.) If micropipets are filled, but not immersed in the electrolyte (usually $3M$ KCl), crystals soon appear at the microtip, rendering the micropipet useless. Kennard (1958) reported that even with filled micropipets stored by immersion in the filling electrolyte, deposits and growths capable of occluding the tip often appear after about one week. To solve this problem, he advocated storage in the dark at reduced temperature and the addition of a bacteriostatic agent to the surrounding solution. Prolonged storage is also inadvisable because repeated handling of the pipets (to remove one or several for an individual experiment) damages the microtip. To avoid these difficulties most investigators fabricate a batch of about 10 to 50 micropipets, fill them immediately, check several for resistance and electrical stability, and use them all as soon as possible.

When a micropipet is immersed tip upward in an electrolyte, the fluid enters the stem but does not fill the tip; when immersed tip downward, only a very small amount of fluid enters the tip. To cause the electrolyte to fill the tip, taper, and shaft, investigators have replaced entrapped air bubbles with vapor from the electrolytic solution or removed them by the use of positive or negative pressure.

The technique of filling by boiling seems to have been introduced by Ling and Gerard (1949), who immersed their micropipets in a dilute solution of potassium chloride contained in a vessel covered by a heavy lid. A blowtorch was then used to boil the electrolyte vigorously for a half-hour; the vigorous bubbling caused the lid of the vessel to rise and fall intermittently, producing a fluctuating pressure. During this time, the air in the micropipets and above the salt solution was replaced by potassium chloride vapor and the volume of the solution was decreased and therefore concentrated. At the end of the boiling period, cold potassium chloride solution was added; during the cooling period the micropipets became filled.

Nastuk (1953) found that micropipets filled by the boiling method of Ling and Gerard exhibited appreciable junction potentials; he therefore omitted the boiling procedure. He first fabricated his micropipets from cleaned glass tubing and placed them in well-filtered $3M$ potassium chloride solution. The air in the shafts of the micropipets was drawn off with a fine pipet and, after about 2 hours of immersion, the tips of the micropipets were filled by capillary action. Any remaining air bubbles were removed by poking a thin, pointed glass fiber down the shaft of the micropipet; a few thrusts were usually adequate to remove a bubble.

The technique used at present by many investigators was developed by Mrs. Tasaki (Tasaki et al., 1954), who mounted the micropipets on a holder so that they could be immersed tip downward in a vessel of either ethyl or methyl alcohol. The vessel was capped and the alcohol within it was boiled gently for 10 minutes under reduced pressure (boiling temperature 25–40°C). The micropipets were then allowed to cool in the alcohol at atmospheric pressure. During this time (5–20 minutes) the bubbles disappeared. After the cooling period, the micropipets were immersed in distilled water for about 2 to 2.5 minutes and transferred to $3M$ potassium chloride (below 45°C), whereupon the pressure was reduced for about 30 minutes. The micropipets were then stored overnight immersed in $3M$ potassium chloride at atmospheric pressure.

Caldwell and Downing (1955) described a simple filling method that can be employed if it is permissible to allow passage of about 2.5 days between fabrication and use of the micropipets. They reported that immediately after fabrication, the micropipets were placed tip downward in $3M$ potassium chloride and, within one-half hour, 0.5 mm of the tips had become filled. The shaft of the micropipet was then filled with distilled water and the tips were left immersed in the $3M$ potassium chloride solution. Caldwell and Downing reported that filling occurred over a 60-hour period, because of the difference in the vapor pressure of water and the solution. When each micropipet was filled, the distilled water in the shaft was removed and replaced by $3M$ potassium chloride by means of a fine pipet. Micropipets filled in this way were allowed to stand immersed in the salt solution until, by diffusion, they were filled uniformly with $3M$ potassium chloride.

Kennard (1958) described a simple procedure that must be applied immediately after fabrication of a micropipet. This method consisted of placing the micropipet (tip downward) in a closed flask containing $3M$ potassium chloride. The filling procedure was completed by applying negative pressure derived from a water-faucet pump, to the flask for a few hours.

Two reports have described the preparation of prefilled micropipets; in both the procedure consisted of heat pulling with the electrolyte in the standard tubing used for fabrication of micropipets. In the first report by Kao (1954), a length of capillary tubing (0.7 mm o.d.) was selected and right-angle bends were made a few centimeters from each end. The tubing was then filled with the desired electrolyte and the tubing was clamped in a horizontal (gamma) micropipet puller with the heating coil around the center of the electrolyte-filled tube and the ends of the tube dipped into wide reservoirs of the chosen electrolyte. When heat and the pulling were applied to the fluid-filled capillary tube, electrolyte was expelled into the reservoirs and, as the capillary tube was being drawn, electrolyte was drawn back into the two micropipets from the reservoirs. The filled micropipets were then removed and held tip downward to permit any gas bubbles to rise. Kao stated that bubbles remained in about 40% of the electrodes; however, these bubbles were removed by storing the micropipets for a few days in a rack placed in a container filled with the electrolyte. Kao's prefilled technique was developed further by Crain (1956), who filled a length of Kimble glass tubing (0.7–1.0 mm o.d.) with $3M$ potassium chloride, leaving an air space 2 to 3 cm long at each end. One end was then flame sealed and the tube was placed in the micropipet puller with the heating coil located several millimeters away from the termination of the liquid column and facing the open end of the tube. When the heat and drawing force were applied to form the micropipet, the tip became automatically filled.

From the foregoing descriptions the reader will be able to recognize the important steps to be taken in filling micropipets. In summary it is perhaps advantageous to call attention to some of the important details which, if attended to, will increase the yield in fabricating usable micropipets. It should be recognized that, since micropipets have tip diameters of a micron or less, absolute cleanliness must be maintained at every step in the fabricating and filling procedure. The capillary tubing used for making the micropipets must be scrupulously cleaned so that no particulate matter remains in the lumen. The tubing should be washed and rinsed thoroughly with distilled water. In addition, the filling procedure should be commenced very soon after fabrication to minimize the chance of airborne dust particles finding their way into the tiny lumen of the tip. The electrolyte used for filling must be completely free from particulate matter. Chemically pure salts dissolved in triply distilled water are necessary. In addition, the electrolyte should be filtered employing negative pressure. When filtered, the electrolyte should be stored in a covered container that has also been carefully cleaned and well rinsed with distilled water.

The filling procedure developed for the author by Cantrell consists of mounting a batch of about 50 micropipets tip downward in a Teflon rack (Fig. 4-10), which is mounted to a cover that forms an airtight seal on the top of a thick-walled glass vessel 6 in. high, 4 in. diam.; in the cover there is a tube for the application of negative pressure. The vessel is two-thirds filled with ethyl alcohol and the cover is carefully placed on it, thereby totally immersing the electrodes in alcohol. Negative pressure is applied using a water-faucet suction device and maintained for 24 hours; the negative pressure can be maintained by evacuating the vessel and clamping off the tube leading to the suction pump. If any of the electrodes contain air bubbles, after 24 hours they can easily be filled by applying and releasing the negative pressure a few times. The micropipets are then removed and the alcohol in the vessel is replaced by distilled water; when the cover is replaced, the micropipets are immersed in distilled water. Vacuum is again applied and the micropipets are allowed to stand for 24 to 48 hours to replace the alcohol by distilled

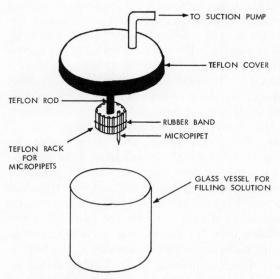

Fig. 4-10. Filling apparatus for micropipets.

water. The distilled water in the vessel is then replaced by the filling electrolyte (usually $3M$-KCl) and the vacuum is reapplied and maintained for 24 hours. The electrodes are now filled and ready for use. A final test of adequacy of filling is the failure of a hard vacuum to produce bubbles in the micropipets. Filling can also be judged from the measurement of resistance; an unusually high resistance for a known micropipet tip diameter indicates the presence of either a bubble or particulate matter.

Resistance

Because the tip of a micropipet is filled with an electrolyte, it constitutes an electrolytic conductor of small cross-sectional area; it is this property that gives a micropipet its high resistance, not the electrode-electrolyte interface in the stem (which is low in impedance compared with tip resistance); the resistance of the electrolyte in the tapered portion of the micropipet contributes very little resistance. Actually the tip of the micropipet is a truncated conducting cone in which the internal tip diameter d, the angle ϕ (radians) of the sides (at the tip), and the resistivity ρ (ohm-cm) of the fluid used to fill it, constitute the most important factors determining the resistance of a micropipet. By simple integration it is easy to show that the tip resistance R_t is given by $R_t = 4\rho/\pi\phi d$. If the angle ϕ is larger than 0.17 rad. (i.e., 10°) the following general expression should be used: $R_t = 4\rho/\pi d \cot \phi$. Examination of these expressions shows that the use of an electrolyte with low resistivity ρ and a steep angle ϕ at the tip favor holding the resistance down if a small diameter d must be used.

It was soon learned that the use of filling solutions that were isotonic with intracellular fluids produced micropipets with high resistance. Nastuk and Hodgkin (1950) showed that if their micropipets were filled with isotonic potassium chloride (118 mM), they exhibited resistances in the range of 50 to 150 MΩ; the values encountered when $3M$ potassium chloride was used varied between 10 and 30 MΩ, a five fold decrease. This reduction is not consistent with the resistivity values for 0.118-mM potassium chloride ($\rho = 75$ Ω-cm at 20°C) and $3M$ potassium chloride ($p = 3.7$ Ω-cm at 20°C); here the ratio is about 20—slightly larger than the approximate values reported. Nonetheless, the remarkable reduction in impedance is worthwhile.*

Various investigators have measured the resistances of their micropipets, and there is general agreement among the values presented for the various tip diameters and electrolytes used to fill the micropipets. For example, Curtis and Cole (1940), using isotonic potassium chloride as the filling electrolyte, listed the resistance range for 10 to 60 μ electrodes as 1 to 25 MΩ. Renshaw

---

* Filling micropipets with potassium chloride solution has another advantage; namely, minimization of liquid-junction potentials (see p. 9).

et al. (1940) reported a resistance range of 0.5 to 1 MΩ for their 40-μ electrodes filled with mammalian Ringer's solution. Graham and Gerard (1946) measured resistance values between 7 and 10 MΩ for tip diameters slightly smaller than 5 μ, filled with potassium chloride isotonic with frog muscle. Ling and Gerard (1949), using electrodes about 1 μ in tip diameter, found the resistance to be in the vicinity of 20 MΩ when filled with KCl isotonic with frog muscle. Nastuk and Hodgkin's (1950) micropipets of 0.4-μ tip diameter, when filled with $3M$ potassium chloride exhibited a resistance of 50 to 150 MΩ. The automatic machine developed by Alexander and Nastuk (1953) produced 0.5-μ micropipets with resistances in the range of 10 to 13 MΩ when filled with $3M$ potassium chloride. The electrodes used by Frank and Fuortes (1956), with tip diameters of 0.5 μ and filled with $3M$ potassium chloride, were found to have resistances in the range of 15 to 50 MΩ. Grundfest (1955) reported that the resistance range of KCl-filled micropipets, having tip diameters from a few microns to 0.1 μ, is from a few megohms to 100 MΩ; Eccles (1955) gave values of 10 to 20 MΩ for $3 M$ KCl-filled micropipets 1 to 0.5 μ diam. Hence it is seen that a micropipet with a tip diameter of about 1 μ, filled with $3M$ potassium chloride, would be expected to have a tip resistance of about 10 MΩ. This value is not inconsistent with that predicted by the foregoing expression. For example, substituting these values [0.5-μ internal tip diam., $p = 3.7$ Ω-cm for $3M$ KCl at 20°C, and tip angle $\phi = 1°$ (0.017 rad.)] into the expression gives a tip resistance of 5.5 MΩ.

Thus the use of a concentrated potassium chloride solution to fill a micropipet reduces its resistance considerably and thereby imposes less stringent requirements on amplifier input impedance for measuring membrane and action potentials. However, it must be remembered that there exists a concentration gradient and a small liquid-junction potential at the tip of the micropipet when it is in contact with anything other than $3M$ potassium chloride, which almost never happens; this matter is discussed subsequently.

When a micropipet is filled with $3M$ potassium chloride and the tip is advanced into a cell, there exists a concentration gradient (e.g., $3M$ KCl versus intracellular ions). Thus the electrolyte will diffuse from the micropipet into the cell. Initially the diffusion rate will be high and according to Nastuk and Hodgkin (1950), with an 0.4-μ tip in a frog muscle cell, the diffusion rate is on the order of 6×10^{-14} mole/sec. Coombs et al. (1955) reported that the diffusion of $3M$ potassium chloride from a micropipet is on the order of 4×10^{-14} mole/sec. Whether such diffusion fluxes significantly after cellular function during a measurement remains to be seen.

Measurement of Resting Membrane Potential

When the potential difference that exists across a cell membrane is measured with a micropipet, it is customary to call the potential between the electrode terminals A and B zero before the micropipet is advanced into the cell; this situation is diagrammed in Fig. 4-11a. When the micropipet is advanced through the cell membrane (Fig. 4-11b), the potential that suddenly appears between electrodes A and B is usually taken as the membrane potential. In fact, the sudden appearance of this potential indicates that the tip of the micropipet has entered the cell. Because an apparent zero potential has been established as described previously, it is possible to identify the overshoot or reverse polarization that occurs when the cell becomes active and recovers; a typical representation appears in Fig. 5-27.

The procedure just described is eminently practical, but it incorporates a small error that should be identified, particularly in view of the serious efforts now being made to account for the magnitude of membrane and action potentials. The error can be found by examination of Fig. 4-11a for the purposes of identifying all the potentials between the electrode terminals A and B. Starting with electrode terminal A, there is a potential difference between the metallic electrode in contact with the electrolyte filling the micropipet; this potential will be designated as E_{ma}. If the fluid in the micropipet is different from that

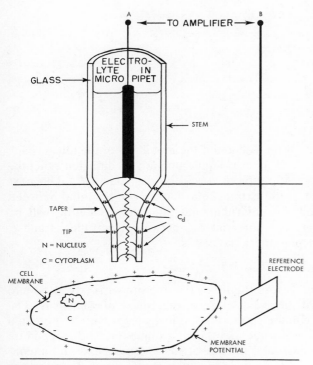

Fig. 4-11a. Micropipet tip located outside a cell.

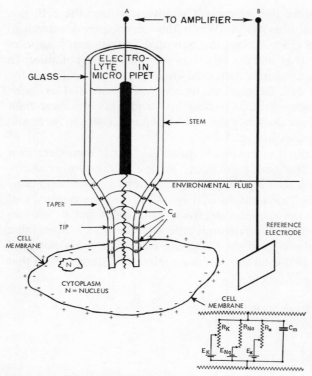

Fig. 4-11b. Micropipet tip located inside a cell.

constituting the environment of the cell and if the mobilities of the ions in each solution are different, there will arise a liquid-junction or diffusion potential at the tip of the micropipet; this potential will be designated E_j. The next potential to be encountered in the extracellular path is the potential between the reference electrode and the environmental fluid; this potential will be designated as E_{mb}. Therefore the potential E_{AB} between electrodes A and B before the tip of the micropipet enters the cell is

$$E_{AB} = E_{ma} + E_j - E_{mb}$$

The potential E_{AB}, measured before the tip of the micropipet is advanced into the cell, is usually assigned the value of zero, regardless of what it is. The potential E_{ma} between the metal electrode in the micropipet and the potential E_{mb}, between the reference electrode and the environmental fluid, depend on the type of metal used for each and the activities (which depend on the concentration and temperature) of the ions in the fluids they contact. Fortunately, these potentials remain the same whether the tip of the micropipet is outside or inside the cell. The junction potential E_j, between the fluid filling the micropipet and the environmental fluid, depends on the concentrations and mobilities of the ions in each solution. When the tip of the micropipet penetrates the cell membrane (Fig. 4-11*b*), not only does the membrane potential E_{mp} appear in the circuit, but the liquid-junction potential changes because the fluid filling the micropipet is now in contact with that which constitutes the cytoplasm; this junctional potential is designated as E_j'. Thus the potential E_{AB}' between electrodes *A* and *B* after the tip of micropipet has passed through the cell membrane is

$$E_{AB}' = E_{ma} + E_j' + E_{mp} - E_{mb}$$

By this procedure, what has been called the membrane potential is the difference between the potential between the electrode terminals with the tip of the micropipet inside the cell E_{AB} and that measured before the tip penetrated the cell membrane E_{AB}. Stated mathematically: the measured membrane potential *MMP* is:

$$\begin{aligned} MMP &= E_{AB}' - E_{AB} \\ &= E_{ma} + E_j' + E_{mp} - E_{mb} - (E_{ma} + E_j - E_{mb}) \\ &= E_j' - E_j + E_{mp} \end{aligned}$$

As long as the difference between the junction potentials $E_j' - E_j$ at the tip of the micropipet is small, the potential difference that suddenly appears between the electrode terminals (when the micropipet tip enters the cell) can be taken as the membrane potential. (The use of KCl as a filling solution for micropipets aids in minimizing the liquid-junction potential because K^+ and Cl^- ions have very nearly equal mobilities.)

Only a few investigators have concerned themselves with the error in membrane potential that can be contributed by junction potentials; this subject is discussed presently. Because of the difficulty and uncertainty in calculating liquid junction potentials, Curtis and Cole (1942) wrote "it seems advisable to report only the measured values of potential." With the techniques routinely employed it would appear that the measured resting membrane potentials are slightly low due to the liquid-junction potential, which may amount to about 5 mV.

Electrode Potentials

Although the potentials of the electrode in the micropipet E_{ma} and the reference electrode E_{mb} remain the same in the procedure described previously, it is illuminating to consider their magnitudes. These electrode-electrolyte junction potentials depend on the type of metal used and the type and concentration of the electrolytes they contact. In most instances, the silver-silver chloride surface is employed. The potential of a silver-silver chloride $E_{Ag/AgCl}$ electrode is dependent on the activity of the chloride ions in its fluid environment and is given by

$$E_{Ag/AgCl} = 0.2225 - \frac{RT}{F} \ln A_{Cl^-}$$

where R is the gas constant, T is the absolute temperature, F is the Faraday, and A_{Cl^-} is the activity of the chloride ions in the environmental solution. At 25°C the expression reduces to

$$E_{Ag/AgCl} = 0.2225 - 0.05915 \log_{10} \gamma m$$

where γ is the activity coefficient and m is the molality of solution in contact with the silver-silver chloride electrode.

From the foregoing relation it is possible to calculate the electrode potentials of the silver-silver chloride electrode in contact with $3M$ potassium chloride and environmental fluid, which often corresponds to 0.6 and 0.9% sodium chloride, the molal concentrations for these three solutions are 3.30, 0.102 and 0.155, respectively. At 25°C, the activity coefficients are 0.578, 0.780, and 0.755. Hence the potential of the silver-silver chloride electrode in contact with the $3M$ potassium chloride in the micropipet is 206 mV; the potentials for the reference electrode in contact with 0.6 and 0.9% sodium chloride are 288 and 278 mV, respectively. Neglecting the junction potential at the micropipet tip for the moment, the fact that the silver-silver chloride electrodes are in contact with different solutions guarantees that the potential measured between the electrode terminals is not zero; the difference amounts to -82 mV when the $3M$ potassium chloride electrode is compared with the reference electrode in 0.6% sodium chloride and -72 mV when the latter is in 0.9% sodium chloride.

Liquid-Junction Potential of a Micropipet

The liquid-junction potentials E_j and E_j' are of special importance when it is desired to know the true magnitude of a particular membrane potential and the amount of overshoot of its action potential. As pointed out previously, a junction potential can arise when the electrolyte in the micropipet is of a type or concentration different from the fluid in contact with the tip of the micropipet. Before the micropipet is advanced through the cell membrane, the tip is in contact with the fluid environment of the cell. Therefore, the junction potential E_j' is constituted by the potential difference existing between $3M$ potassium chloride and environmental fluid (usually NaCl of concentration C); therefore

$$E_j = E_{3MKCl/NaCl(C)}$$

When the tip of the micropipet penetrates the cell membrane, the junction potential E'_j is constituted by

$$E'_j = E_{3MKCl/cytoplasm}$$

It must be pointed out that the important consideration relating to the junction potential at the tip of the micropipet is not its value E'_j when in contact with the environmental fluid, but its difference $E'_j - E_j$ when it enters the cell and comes into contact with its ionic interior. It is extremely difficult to calculate the true values of this difference because determining the liquid-junction potential, when the tip is in the cell, requires an accurate knowledge of the species of ions and their activities and mobilities in the cytoplasm. Unfortunately, this information is not available. However, to shed light on this situation Adrian (1956) conducted a series of useful measurements using a 3MKCl-filled micropipet and measured its potential in solutions of sodium chloride and potassium chloride of various concentrations; his results are presented in Fig. 4-12. The importance of this study illustrates that if the external environment of a cell were a solution of 0.125M sodium chloride

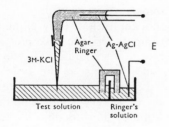

(a)

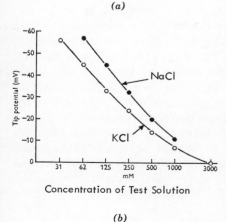

Concentration of Test Solution

(b)

Fig. 4-12. The potential of a micropipet: (*a*) arrangement for measurement of the potential of a micropipet $E = $ Ag/AgCl + Agar-Ringer + 3M KCl + test solution + agar-Ringer + Ringer's + Ag/AgCl; (*b*) Tip potential (corrected for other potentials) of a 3M KCl-filled micropipet versus concentration of test solutions of sodium chloride and potassium chloride. (From R. H. Adrian, *J. Physiol.* 1956, **133**:631–658. By permission.)

(0.7% NaCl), the junction potential would be 45.3 mV. If the interior of a cell is equated to $0.125M$ potassium chloride, the junction potential would be 32.9 mV. Therefore, when the tip passed from the external to the internal solution, a difference in potential of 12.4 mV would accompany the appearance of the membrane potential.

Tip Potential

By checking the junction potentials predicted by use of the Planck equation,* Adrian found not only that the measured tip potential was excessively large, but that it correlated with electrode resistance; this indicates that the smaller the tip diameter, the larger the discrepancy in potential from that predicted from the ionic gradient, which remains the same regardless of tip diameter. Agin and Holtzman (1966) and Agin (1969) continued the study and pointed out that since the glass wall at the tip of a micropipet is only about 0.1 μ thick, it has the properties of a thin membrane. In fact, they showed that what is now known as the tip potential is related to the ionic concentration of the environment and that the potential virtually disappears when the tip is removed by breakage. (It must be recognized that what remains is the liquid-junction potential.) Starting with $3M$ KCl-filled micropipets, they measured the potential in a series of test solutions of varying concentration; they eliminated the tip potential by breaking the tip of the micropipet. Their results, plotted in Fig. 4-13, show that when dilute potassium chloride or sodium chloride was used, the tip potentials were large, increasing with decrease in the ionic concentration of the environment. They further showed that the tip potential could be reduced considerably by the addition of a very small amount of calcium chloride and even reversed by the addition of thorium chloride to the test solutions of potassium and sodium chloride.

The tip potential has come under considerable scrutiny because it imposes a small error in the measurement of the membrane and the overshoot in action potentials. The error is due to an ionic gradient that forms at the inner and outer surfaces of the tip of the micropipet; Agin's concept of it appears in the inset of Fig. 4-13. It has not been determined whether the ionic gradient is the result of adsorption or selective permeability of the glass used for the micropipet. It is known, however, that the type of glass affects the magnitude of the gradient; Lavallée and Szabo (1969) reported that the tip potential is pH dependent if the tip diameter is small (i.e., the resistance is greater than 100 mΩ). In this situation the wall at the tip is very thin and, presumably, the action resembles that of a glass pH electrode. Kurella (1969) also reported that the type of glass used strongly influences the tip potential and that "microelectrodes made of quartz have practically negligible potentials."

---

* See Mac Innes (1961) for calculation of junction potentials.

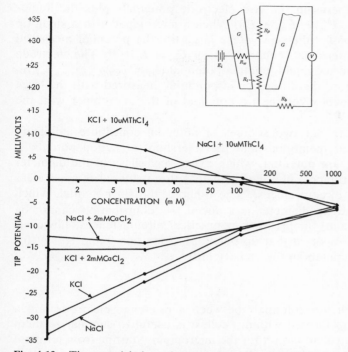

Fig. 4-13. Tip potentials in various solutions. (Drawn with data and insert from D. Agin and D. N. Holtzman, *Nature.* 1966, **211**:1194–95.) Equivalent circuit illustrating possible origin of tip potential: E_i = interfacial potential; G = glass wall of the microelectrode; R_p = resistance of bulk electrolyte in the pipet; R_t = resistance of the tip; R_w = resistance of the wall; R_b = resistance of the bath electrolyte; V = voltage-measuring device. (Redrawn from D. P. Agin, in *Glass Microelectrodes*, Lavallée, Schanne, and Hebert, Eds. New York: John Wiley & Sons, 1969.

From the foregoing it is apparent that the tip potential depends on tip diameter, the type and concentration of the fluid in the micropipet and in the environment, and the type of glass that forms the micropipet. As shown by Agin (1966, 1969), tip potential can be reduced by the addition of a small quantity of thorium chloride. However, this technique is not yet widely used. Because tip potential varies with electrode preparation, investigators are starting to report the selection of micropipets with low tip potentials. The tip potential can be determined by measuring the total potential between a micropipet with a large tip diameter (or with a broken tip) and the reference electrode, with the tip of the micropipet in the fluid environment of the cell. Because the tip potential attains an appreciable magnitude only when the tip diameter is small, the potential appearing between the electrode terminals equals the difference between the two electrode potentials plus the liquid-junction potential (i.e., $E_{ma} - E_{mb} + E_j$). When a micropipet with a small tip diameter is substituted for the one with a large tip, the potential measured between the electrode terminals becomes $E_{ma} - E_{mb} + E_j + E_t$. The electrode potentials E_{ma}, E_{mb} and the liquid-junction potential ($E_j = E_{3MKCl/environment}$) remain the same; therefore the difference in potential measured with the large and small tip micropipets gives the tip potential of the micropipet with the small tip diameter.

Thus it can be seen that two sources of potential contribute ambiguity to the measurement of membrane and the overshoot of action potentials. One is the liquid-junction potential, which exists regardless of the diameter of the tip of the micropipet; it is believed that this potential is small (less than 10 mV in most circumstances). The other is the tip potential, which becomes larger as the tip diameter is reduced. Its contribution cannot be ascertained with certainty at present; in practical circumstances, however, Schanne et al. (1968) reported that tip potential is between 0 and 10 mV for electrodes with tip diameters in the vicinity of 0.5 μ.

Equivalent Circuit

Having identified all the potentials between a reference electrode and a micropipet with the tip located within a cell, it is useful to summarize them and synthesize an equivalent circuit for the micropipet. Starting from terminal *A* and tracing out the circuit (Fig. 4-14), the first potential encountered is E_{ma}, the potential between the electrode metal and the electrolyte filling the micropipet. If the fluid in the cell is different from that in the microelectrode, there will be a liquid-junction potential at the tip E_j'. If the tip diameter is small and the ionic concentration within the cell is low, there may be a tip potential E_t.* Next to appear in the circuit is the membrane potential, E_{mp} and then the potential between the reference electrode and the extracellular fluid E_{mb}. Thus the potential E_{AB} between the electrode terminals *A* and *B* consists of the sum of five potentials

$$E_{AB} = E_{ma} + E_j' + E_t + E_{mp} - E_{mb}$$

Neglecting for the moment the potentials and examining the impedances in traversing the circuit, there are encountered the resistance of the connecting wire, which is negligible, then the impedance of the electrode-electrolyte junction in the stem of the micropipet R_{ma}, C_{ma}, followed by the resistance R_t

---

* It is to be noted that the tip potential may not be the same when the tip is outside of the cell. Thus, the measured membrane potential *MMP* when the tip of the micropipet enters the cell will be equal to $\Delta E_j + \Delta E_t + E_{mp}$, where ΔE_j is the change in liquid-junction potential (see p. 167) and ΔE_t is the change that occurs in tip potential when the tip enters the cell.

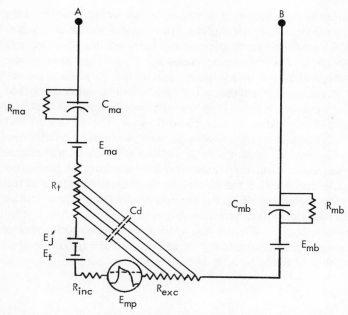

Fig. 4-14. Approximate equivalent circuit for a micropipet used to measure the potential of a cell membrane.

of the electrolyte filling the tip of the micropipet, the resistance of the electrolyte within R_{inc} and without R_{exc} the cell, the impedance of the reference electrode-electrolyte interface R_{mb}, C_{mb}, and finally the negligible resistance of the wire connecting the reference electrode to terminal B. These impedances form a series circuit in conjunction with the five potentials previously identified; they are collected along with the potentials and presented in Fig. 4-14.

In addition to the components just identified, there is a distributed capacitance C_d existing between the fluid in the micropipet and the extracellular fluid. The magnitude of this important capacitance, along with the tip resistance, determines the response time of the micropipet. As early as 1950, Nastak and Hodgkin recognized that the important part of the capacitance is that which is outside the cell, because there is a potential difference across it.

Since in many measurement situations the relative magnitudes of the various impedances are known, it is possible to simplify the equivalent circuit considerably. For example, the resistances of the connecting wires amount to a fraction of an ohm and therefore can be neglected. In the whole circuit, the resistance R_t of the tip of the micropipet is by far the highest, amounting to about 10 to 200 MΩ for typical micropipets. The area of the electrode-electrolyte interface in the stem is usually large, or can be made large, which results in the impedance of R_{ma}, C_{ma} being negligible with respect to the 10 to 200 MΩ tip resistance. Likewise, the resistance of the intracellular R_{inc} and extracellular R_{exc} fluids and the reference electrode-electrolyte impedance R_{mb}, C_{mb} can also be ignored; the distributed capacitance C_d cannot be neglected, however. With reasonable accuracy the circuit can be reduced to that of Fig. 4-15a. This first simplification shows that the membrane potential E_{mp} is connected to the amplifier input terminals A and B via R_t, the 10 to 200 MΩ resistance of the tip of the micropipet; the resistance is shunted by the distributed capacitance C_d, which amounts to about 0.5 pF/mm of tip length. A voltage then appears across the terminals which is the sum of the membrane potential E_{mp}, the potential of the electrode-electrolyte interface in the stem E_{ma}, the liquid-junction potential E_j', the tip potential E_t, and the potential of the reference electrode-electrolyte junction E_{mb}.

In many circumstances, E_{ma}, E_j', E_t, and E_{mb} are stable and can be corrected for in determining the membrane potential. Therefore, they can be summed to form a single potential E. In addition, it is customary to replace a circuit

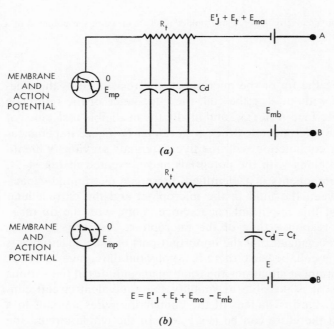

Fig. 4-15.   Equivalent circuits for a micropipet used to measure membrane and action potentials: (*a*) simplified equivalent circuit; (*b*) equivalent circuit with lumped parameters.

consisting of distributed parameters by one with lumped parameters in accordance with the known electrical behavior of the circuit; thus the distributed capacitance C_d and tip resistance R_t can be represented as in Fig. 4-15b, in which R_t' and C_d ($\simeq C_t$) are the single resistance and capacitance values that simulate the response of R_t with its distributed capacitance C_d.

In Fig. 4-15b, the membrane potential is coupled to the amplifier terminals A and B via a high series tip resistance R_t' and a moderate tip shunt capacitance C_t. The effect of this particular combination of circuit elements places a limit on the ability of the micropipet to respond to a sudden change in potential; this characteristic is described as the response or rise time. Failure to achieve a response time shorter than the rising or falling phase (whichever is shorter) of the action potential will prevent faithful reproduction of the action potential.

Because it is not possible to calculate accurately the resistance and distributed capacitance of the tip of the micropipet, two simple techniques are used to measure them; the circuit arrangements are diagrammed in Fig. 4-16.

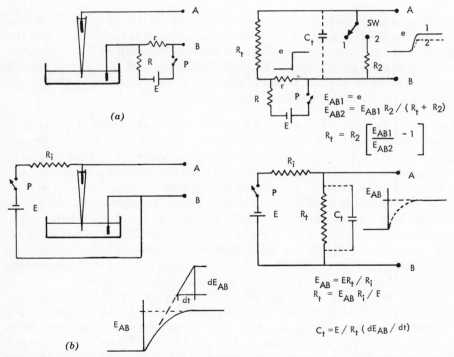

Fig. 4-16. Measurement of tip resistance and capacitance: (a) constant-voltage method; (b) constant-current method.

Both methods can be applied with the tip of the micropipet in the environmental fluid or in the cell.

Measurement of Tip Resistance and Capacitance

When a micropipet is used with its reference electrode to measure transmembrane resting and action potentials, it is necessary to know the resistance and capacitance of the tip. The resistance, which is determined by the diameter and the angle of the tip and the type of electrolyte filling it, identifies the magnitude of the amplifier input impedance required; that is, for measurement of the true potential difference between the micropipet tip and the reference electrode, the amplifier input impedance must be many times larger than the tip resistance (see p. 39). For a given size of micropipet, tip resistance is often used as a measure of the acceptability of the electrode. For example, if the tip is clogged, the resistance becomes high; if it breaks, the resistance is low. Tip resistance and capacitance, which also depend on tip dimensions, determine the rise time or the ability of the micropipet to reproduce the rapidly changing portion of an action potential.

Tip Resistance   Two methods are used to measure tip resistance; with one, a constant voltage is applied in series with the reference electrode, and the voltage across the electrode terminals is measured without and with a known resistance placed across them. With the other method, a constant-current source is placed in parallel with the micropipet and reference-electrode terminals, and the voltage increase is measured.

CONSTANT-VOLTAGE METHOD   As stated previously, with the constant-voltage method, a known voltage is suddenly inserted in series with the reference electrode; the technique for carrying this out (Fig. 4-16a) is used in many commercially available headstages. Often the inserted voltage serves to calibrate the sensitivity of the measuring system. In Fig. 4-16a, a step of voltage e is caused to appear across a low value of resistance r by closing the switch P. The resistors r and R together constitute a voltage divider which reduces the large voltage E such that $e = Er/(R + r)$. The actual value for e is chosen so that the cell will not be stimulated when applied with the electrode tip in the cell. Another reason for using a low voltage is to avoid decomposition of the electrolytes or heating the fluid in the tip of the micropipet.

With the switch SW across the electrode terminals A and B in position 1, the voltage E_{AB_1}, which appears when switch P s closed, will be equal to e if measured by an amplifier with a very high input resistive impedance. With the switch in position 2, a resistor R_2 is placed across the terminals A and B and the voltage E_{AB_2}, measured when switch P is closed, will be less. Voltage E_{AB_2} is equal to iR_2; the current i is equal to the voltage applied e divided by the total circuit resistance $R_t + R_2$. If R_t is greater than all the other resistances in the electrode circuit, which is nearly always the case; then

$$E_{AB_2} = iR_2 = \frac{eR_2}{R_t + R_2}$$

$$E_{AB_1} = e$$

$$E_{AB_2} = \frac{E_{AB_1}R_2}{R_t + R_2}$$

Manipulation of these equations gives a value for the tip resistance R_t in terms of the voltage without (E_{AB_1}) and with (E_{AB_2}) the resistor R_2 across the terminals AB; thus

$$R_t = R_2 \left(\frac{E_{AB_1}}{E_{AB_2}} - 1 \right)$$

Note that the magnitude of the voltage e is unimportant and that, in this derivation, the tip capacitance and electrode impedances are neglected. The values E_{AB_1} and E_{AB_2} are those that persist as long as the switch P is depressed. Note also that if $R_t = R_2$, $E_{AB_2} = 0.5\ E_{AB_1}$.

CONSTANT-CURRENT METHOD   With the constant-current method, both tip resistance R_t and effective capacitance C_t can be measured; the former will be discussed first. Referring to Fig. 4-16*b*, the procedure consists of passing a very small step of constant current through the micropipet; this is accomplished by connecting a voltage source E, which is in series with a resistor R_i, across the electrode terminals. The value for R_i is chosen to be many times larger than the highest value expected for the tip resistance R_t; in practice, resistance values of 1000 to 10,000 MΩ are not uncommon for R_i. Thus the current i in the circuit when the switch P is depressed will be $E/(R_i + R_t)$. Because R_i is much greater than R_t, the current will be very close to E/R_i and virtually independent of R_t. When P is depressed, the voltage appearing across the electrode terminals will increase by a value equal to the current ($i = E/R_i$) multiplied by the tip resistance R_t (i.e., $E_{AB} = ER_t/R_i$). Therefore, the voltage across the electrode terminals is proportional to the resistance of the tip of the micropipet. By knowing the value of the voltage E and the resistance R_i and by calibrating the sensitivity of the measuring apparatus connected to the electrode terminals, the tip resistance can be calculated as follows:

$$R_t = \frac{E_{AB}}{E} R_i$$

Tip Capacitance   After tip resistance has been measured, it is possible to determine the effective capacitance of the tip by the procedure described for the constant-current method. The method consists of determining the initial rate of rise of the voltage dE_{AB}/dt when switch P is depressed (Fig. 4-16b). Solving the differential equation to obtain the manner in which E_{AB} rises with time to its final value $ER_t/(R_i + R_t)$ gives

$$E_{AB} = \frac{ER_t}{R_i + R_t} \left[1 - e^{-t(R_i + R_t)/R_iR_tC_t} \right]$$

If R_i is much larger than R_t, the slope of the curve $\dfrac{dE_{AB}}{dt}$ becomes

$$\frac{dE_{AB}}{dt} = \left[\frac{E}{C_tR_i} \right] e^{-t/C_tR_t}$$

The initial slope (at $t = 0$) therefore is

$$\frac{dE_{AB}}{dt} = \frac{E}{R_iC_t}$$

Rearrangement of the equation isolates C_t to give

$$C_t = \frac{E}{R_i \text{ (initial slope)}}$$

Thus by knowing E and R_i, and measuring the initial slope (V/sec), the tip capacitance can be calculated.

In summary, the constant-current method, is approximated by applying a voltage in series with a resistance (many times higher than the resistance of the micropipet tip) across the terminals of the reference electrode and micropipet. This produces a voltage that rises exponentially to a final value E_{AB} from which the resistance of the tip can be calculated as follows:

$$R_t = \frac{E_{AB}R_i}{E}$$

From the initial rate of rise of the voltage dE_{AB}/dt, the capacitance of the tip of the micropipet can be calculated as follows:

$$C_t = \frac{E}{R_i \text{ (initial slope)}}$$

Negative-Capacity (Positive-Feedback) Compensation

The effective distributed capacity at the tip of a micropipet and at the input of a headstage can be markedly reduced by the use of a negative-capacity (positive-feedback) amplifier. Figure 4-17 diagrams a micropipet and reference electrode connected to a typical negative-capacity amplifier created by using a field-effect transistor (FET) as a source follower (a cathode follower can also be used). The micropipet signal is connected to the noninverting terminal (+) of

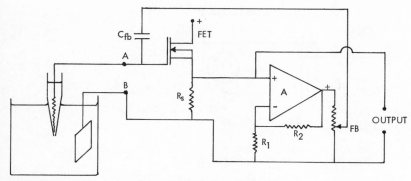

Fig. 4-17. Negative-capacity (positive-feedback) amplifier.

an operational amplifier A, whose gain is specified by the ratio of the feedback resistors R_2/R_1; in practice the gain should be only slightly greater than unity. The output, which reflects the bioelectric event detected by the micropipet, is obtained from the ends of the source resistor R_s. Capacity neutralization is provided by feeding a selected fraction of the output of the operational amplifier to the input, in phase, via a coupling condenser C_{fb}, which is a few picofarads in size. The feedback fraction is selected by use of a potentiometer FB across the output of the operational amplifier; an alternate method consists of connecting the feedback coupling capacitor C_{fb} directly to the output of the operational amplifier and controlling the gain of the latter by varying the ratio of the feedback resistors R_2/R_1. Considerable care must be exercised in adjusting the amount of feedback so that the loop gain—the ratio of the input to the source follower to the output signal fed back—is less than 1.0; sustained oscillations will occur with no input if the feedback fraction exceeds 1.0. Even with slightly less feedback, distortion will be added to a bioelectric event.

In practice, it is never possible to calculate the required feedback because the values of tip resistance and total input capacitance are unknown and, moreover, they vary from micropipet to micropipet. For this reason, a test signal is applied and the feedback is adjusted for best reproduction of the test signal. At present there are three ways to apply a test signal; with all, the optimum feedback is that which provides the best reproduction of a square wave. The three methods—the square-wave constant-voltage and constant-current methods and the capacitively coupled triangular-wave method—are illustrated in Fig. 4-18.

With the square-wave constant-voltage method (Fig. 4-18a), a low resistance r is placed in series with the lead wire to the reference electrode B; across the resistor, a square wave of amplitude e is caused to appear by connecting a square-wave generator (of voltage E) in series with a resistance R that is many

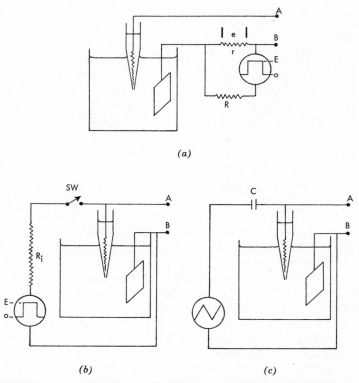

Fig. 4-18. Three methods of applying a test signal for adjustment of negative-capacity (positive-feedback) headstages used with micropipets: (*a*) Square-wave constant-voltage method; (*b*) square-wave constant-current method; (*c*) triangular-wave capacitively coupled method.

times larger than *r*. The voltage *e* is therefore equal to $Er/(r + R)$ as shown in Fig. 4-18*a*. The frequency of the square wave is chosen so that its period (1/frequency) is long with respect to the rise time of the input circuit, which is determined by the tip resistance and total capacitance at the input circuit. On application of the test signal, an exponentially rising and falling wave (Fig. 4-19*b*) appears at the output of the source follower; this is due to the tip resistance and the total capacitance of the input circuit. With an increasing amount of feedback, the rising and falling phases of the reproduced wave become steeper (Fig. 4-19*c*); with more feedback, a slight overshoot appears (Fig. 4-19*d*), and with excessive feedback, the overshoot becomes much larger and a series of damped oscillations (Fig. 4-19*e*), i.e., ringing, appears. Obviously the optimum feedback adjustment is shown in Fig. 4-19*d*.

With the square-wave constant-current method (Fig. 4-18*b*) the test signal

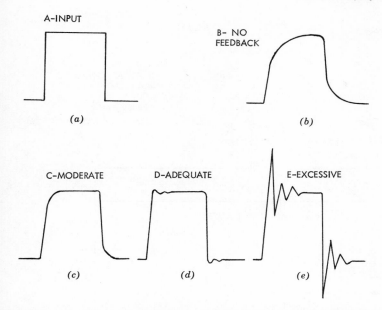

Fig. 4-19. Adjustment of the negative-capacity (positive-feedback) amplifier: (*a*) input; (*b*) no feedback; (*c*) moderate; (*d*) adequate; (*e*) excessive.

(provided by a square-wave generator) is connected to a resistance R_i, which is much higher in value (e.g., 100–1000 times) than the micropipet tip resistance; the series combination (R_i and the square wave-generator of voltage E) are connected via a switch across the micropipet and reference electrode terminals. The procedure for adjustment of the feedback is the same as with the square-wave constant-voltage method; namely, increase the feedback to obtain the best reproduction of the square-wave test signal.

The triangular-wave, capacitively coupled method is illustrated in Fig. 4-18c. This method, suggested to Freygang (1958) by Lettvin, employs a triangular-wave generator coupled via a capacitor C which is a few picofarads in value. The amount of feedback is increased until the best square wave is reproduced at the output of the headstage. A square wave results when the proper feedback is obtained because the resistance of the tip of the micropipet, in series with C, constitutes a differentiating circuit. True differentiation of a triangular wave (i.e., one in which the slope is constant) produces a square wave.

The ability of each of the three methods to provide perfect compensation for all the capacitances associated with the micropipet and input circuit of the headstage is subject to question because in each case the circuit facing the bioelectric event at the tip of the micropipet differs from that for which the

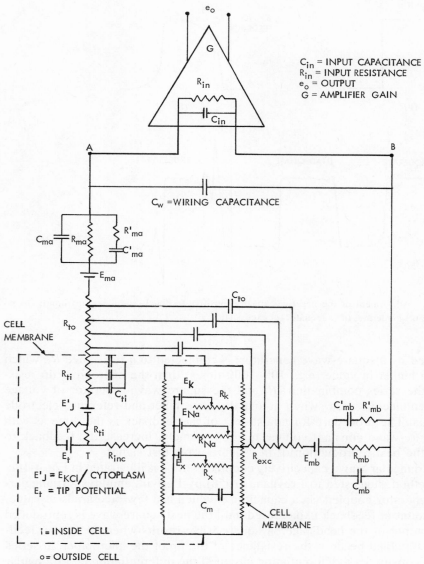

C_{in} = INPUT CAPACITANCE
R_{in} = INPUT RESISTANCE
e_o = OUTPUT
G = AMPLIFIER GAIN

C_w = WIRING CAPACITANCE

CELL MEMBRANE

$E'_J = E_{KCl}$ / CYTOPLASM
E_t = TIP POTENTIAL

i = INSIDE CELL

CELL MEMBRANE

o = OUTSIDE CELL

Fig. 4-20. Complete equivalent circuit for micropipet in a cell.

compensation methods are adjusted. Nonetheless, the three methods offer the best means of adjusting the feedback to compensate for input capacitance. However, as pointed out by Amatneik (1958), the input circuit is more complex and consists of many distributed, rather than lumped, components; complete neutralization is therefore impossible. To clarify this point, the equivalent circuit for a micropipet in a cell is presented in Fig. 4-20 and discussed in the following section. Even casual inspection reveals that the event to be measured, the transmembrane potential and its excursion during activity, is separated from the electrode terminals A and B by several potentials and impedances. Furthermore, the use of an amplifier to measure transmembrane potential places another impedance $R_{in}C_{in}$, across the electrode terminals; the greatest care must therefore be exercised so that the impedance of the measuring apparatus (amplifier) will not in any way alter the potential E_{AB} appearing between the electrode terminals.

Complete Equivalent Circuit of the Micropipet

The equivalent circuit in Fig. 4-20 is complex, and in many instances certain of its components can be neglected, however, it is of value to identify all the components to emphasize their existence. Starting with terminal B, connected to the reference electrode, (and neglecting the resistance of the connecting wire), the first impedance encountered is that of the reference electrode and the fluid that constitutes the environment of the cell; this impedance was previously represented as a parallel circuit consisting of R_{mb} and C_{mb}. It has been suggested that a more accurate representation of the impedance-frequency characteristic is obtained when a series circuit R'_{mb}, C'_{mb} is placed in parallel with R_{mb}, C_{mb}. Another property of the electrode-electrolyte interface is its half-cell potential (dependent on the type of metal, the type and concentration of the electrolyte it contacts, and the temperature); this potential is assigned the value E_{mb} for the reference electrode.

Continuing in a clockwise direction from electrode B to A, the next circuit component encountered is R_{exc}, the resistance of the fluid in the environment of the cell. Next to appear is the cell membrane with its capacitance C_m (amounting to about 1 $\mu F/cm^2$ of cell membrane area). C_m is in parallel with the potentials E_K, E_{Na}, E_X (which are a reflection of the ionic gradients) and their conductances, g_K, g_{Na}, g_X which describe the manner in which the membrane becomes active and recovers; (see Chapter 5). Within the cell, the resistance of the cytoplasm R_{inc} is the next element. Finally, the tip T of the micropipet is reached; at this point there are two potentials, the tip potential E_t and the liquid-junction potential E'_j. The former depends on the type of glass used, the thinness of the wall at the tip, and the ionic components within the micropipet and the cytoplasm. The liquid-junction potential E'_j depends on the type and concentration of the fluid filling the micropipet and the type and concentration of the ions in the cytoplasm.

The resistance of the tip of the micropipet is divided into two portions, that inside the cell R_{ti} and that in the outside environment R_{to}; because the tip diameter is small, both are high. Between the fluid filling the tip and that in contact with the outside of the tip there is capacitance because the micropipet material constitutes a dielectric. The capacitance of the portion inside the cell C_{ti} is relatively unimportant because there is virtually no potential difference across it; the capacitance of the portion outside the cell C_{to}, which is in contact with the environmental fluid, is extremely important because it is charged to the membrane potential. Both capacitances are distributed along the high-resistance tip of the micropipet.

In the taper of the micropipet the resistance of the electrolyte is usually low, but in the stem there is an electrode-electrolyte impedance and a potential. As in the case of the reference electrode, the electrode-electrolyte impedance is represented by a parallel R_{ma}, C_{ma} combination paralleled by a series circuit R'_{ma}, C'_{ma}. The potential of the electrode-electrolyte junction in the stem is designated as E_{ma}, and its magnitude is dependent on the same factors as those for the reference electrode E_{mb}.

Because wires must be used to connect the micropipet and the reference electrode to the amplifier that measures the membrane and action potentials, there is a capacitance between them; this has been represented as C_w in Fig. 4-20.

As stated previously, Fig. 4-20 reveals that the event to be measured (the membrane potential and its excursion) is far removed from the terminals used to measure it. Only by careful choice of a measuring instrument, which must have a high input resistance R_{in} and a low input capacitance C_{in}, and which must be located as close as possible to the electrode system (to make C_w as small as possible), it is possible to have confidence in the magnitude of a resting membrane and action potential.

Electrode Noise

Although electrodes can be efficient detectors for bioelectric events, they are not without their imperfections—random voltages, often called "noise," are the most serious. Noise becomes an important consideration when electrode area is small (i.e., when the resistive component of electrode impedance is high). Noise also becomes prominent when amplification is increased to examine details of action potentials and fluctuations in membrane potentials. The noise level sets the ultimate limit for the identification of a signal.

Noise arises as the random movement of charge carries. In electrodes, this action takes place primarily at the electrode-electrolyte interface, where a spontaneous reaction may be in progress as the electrode comes into equilibrium with its environmental electrolyte; the passage of current also, may cause noise. In a well-designed electrode, these sources of noise are avoided. However, the random movement of charge carriers at an electrode-electrolyte interface, or within an electrolyte, endows an electrode with a resistive component of its impedance and the resistive component is the noise generator.

All resistors are generators of noise voltages at all frequencies. The root-mean-square (rms) voltage was defined by Johnson (1928) as

$$E = \sqrt{4kTR\,(f_h - f_l)}$$

Where E is in volts, k is the Boltzmann constant (1.38×10^{-23} J/°K), T is the absolute temperature (°K), R is the magnitude of the resistance (Ω) generating the noise voltage, and f_h and f_l are the upper and lower cutoff frequencies of the measuring system; $f_h - f_l$ is therefore its bandwidth.

It can be seen that the actual noise voltage of a resistance varies as the square root of the magnitude of the resistance R, the absolute temperature T, and the bandwidth of the measuring apparatus. When using electrodes to measure bioelectric events, it is only possible to exercise control of resistance and bandwidth: increasing either increases the noise voltage. How important noise voltage is in the measurement of bioelectric events with high-resistance electrodes can be demonstrated by calculating the rms noise voltage for a 1-MΩ resistor at room temperature (20°C) connected to a measuring system with a bandwidth extending from 0 to 10 KHz. Substitution of these values into the equation for E gives an rms noise voltage of 12.8 μV.

In a practical circumstance the noise voltage (Johnson noise) appears as a thickening of the baseline of the recording instrument and is composed of randomly occurring voltages of all frequencies in the bandwidth of the measuring system. Because of its characteristic appearance, it is often called grass, and because it contains all frequencies, it is sometimes designated white noise. When monitored aurally, such noise resembles a hissing sound.

The practical importance of electrode noise has been reported for both metal microelectrodes and electrolyte-filled micropipets. For the former, Gesteland et al. (1959) showed the applicability of the Johnson noise concept by first measuring the resistive and reactive components of the impedance of a conical platinum electrode (10 μ diam. 5 μ high) in physiological saline. Next they determined the rms noise voltage of the electrode system by connecting it to a measuring instrument in which the bandwidth ($f_h - f_l$) could be specified. Then a variable resistance was substituted for the electrode system and adjusted to obtain the same noise voltage. Comparison of the resistive component of the electrode impedance with the substituted resistor gave excellent agreement.

The resistive noise associated with the high resistance of the fluid-filled micropipet tip was emphasized by Renshaw et al. (1940) when they measured this nonbioelectric signal and expressed it as a widening of the baseline of the recorder (oscilloscope). A 10-MΩ micropipet gave a 150-μV peak-to-peak noise voltage when connected to a system with a bandwidth extending from 0 to 10 KHz. The investigation correctly pointed out that the amplifier input (grid) resistor is also a noise generator and that all efforts should be made to attain a low electrode resistance; thus the resistance at the input will be as low as possible, which means, in turn, that the minimim noise voltage will be present.

Ohmic noise voltages do not usually constitute a serious problem when transmembrane potentials are measured because the signal is so large (ca. 100 mV). However, when a high-gain recording system having a wide bandwidth is used, electrode noise can become objectionable; this case is often encountered when action potentials are recorded with small-area electrodes located at a distance from the active cells. In this circumstance it is unwise to use a bandwidth wider than is necessary for faithful reproduction of the event. Just what bandwidth should be employed depends on the bioelectric event under investigation. An adequate bandwidth may be identified by reducing the bandwidth while observing the bioelectric event. If even a slight reduction in bandwidth alters the waveform of the bioelectric event, the investigator should be suspicious of the fidelity capabilities of his measuring system. A more scientific method to evaluate the measuring system consists of subjecting a recorded event to frequency analysis to determine its harmonic spectrum. Then the frequency response of the measuring system is measured. If its bandwidth turns out to be coincident with that revealed by the frequency analysis, then the system is probably limiting the response of the event. If the system bandwidth far exceeds that of the bioelectric event, it can be believed that the event has been faithfully reproduced.

Marking Micropipet Sites

Several investigators (Rayport, 1957; Bultitude, 1958, and Tomita et al., 1959) have been able to adapt the Prussian-blue marking technique (see p. 130) to identify the location of the tip of a micropipet; others have used different methods. For example, MacNichol and Svaetichin (1958) succeeded in depositing crystal violet, and Oikawa et al. (1959) ejected silver ions to identify the locations of their micropipets. In all these studies a dc potential was applied to the micropipet to eject the appropriate ions electrophoretically, and the tissue was perfused with an appropriate solution to create the chemical reaction that would produce a visible spot; the size of the spot depended on the quantity of electricity employed, expressed as the product of current and time.

Prussian-Blue Marking   A ferric-ion salt brought together with a ferrocyanide (usually alkali) produces the insoluble iron cyanide which is the Prussian-blue pigment. Given this underlying principle, it is clear that two choices exist: one of the substances can be placed in the micropipet, the other can be used to perfuse the tissue; or the solutions can be reversed, using the opposite polarity to eject the appropriate ions. Both techniques have been employed.

Rayport (1957) was able to record membrane and action potentials with an 0.5-μ diam. micropipet filled with a solution of $3N$ $FeNH_4$ SO_4. Identification of the tip site was accomplished by making the microelectrode positive with respect to an indifferent electrode and allowing a current of 0.05 μA to flow for 45 to 55 minutes, which resulted in ion ejection. The site was made visible by soaking the brain in a solution of potassium ferrocyanide [$K_4Fe(CN)_6$] and formalin for not more than 72 hours. Paraffin sections were prepared and stained with Kernechtrol. Although Rayport stated that the electrode sites were clearly visible in 25-μ tissue sections, he did not report on the sizes of the blue spots.

Bultitude (1958) and Tomita (1959) used slightly different methods to obtain Prussian-blue spots at the tips of their micropipets. Bultitude filled his micropipets with $2.5M$-KCl to which $0.5M$ $Na_4Fe(CN)_6$ had been added. After recordings had been made, the Ringer's solution that had served to bathe the preparation (nervous tissue) was replaced with a freshly prepared solution of equal parts of 110 mM ferric chloride and Ringer's solution. The preparation was allowed to remain in this solution for not less than four hours. At the end of this period, electrophoretic ejection of the ferrocyanide ions was carried out by making the micropipet 4 V negative with respect to an indifferent electrode. The quantity of electricity required to produce 20-μ spots in the processed tissue was $2 \times 10^{-4\text{coul}}$ i.e., 0.2 mA-sec. In a typical experiment with a micropipet resistance of 30 MΩ, a current of 0.13 μA was allowed to flow for 1800 sec. Immediately thereafter the tissue was fixed in 90% alcohol for 15 hours and blocked in paraffin at 50°C. Sections 20 μ thick were made and stained with eosin and mounted in DePeX; the electrode sites appeared as blue spots on a red background.

Tomita et al. (1959) were able to record the electrical activity of excised fish retinas and to mark the sites of the micropipet tips by using a 10% ferricyanide solution in the micropipet. By making the micropipet negative by about 5 to 10 V for 90 to 120 sec with respect to an indifferent electrode on which the retina rested, ferricyanide ions were ejected and marking was accomplished. The electrode site was made visible by gently pouring a small amount of 2% ferrous chloride solution onto the retinal region around the micropipet; the reaction taking place produced a (Turnbull's) blue spot about 20 to 40 μ diam.

Crystal-Violet Marking MacNichol and Svaetichin (1958) were able to mark electrode-tip sites by using a saturated solution of crystal violet in 5% hydrochloric acid to fill their micropipet, which was then advanced into excised retinal cells from fish eyes. The crystal-violet solution was expelled electro-phoretically by passing a current ranging between 1 to 100 nA (nanoamperes) for about 1 minute. Paraffin or methacrylate mounting techniques could not be used; instead they fixed the retinal tissue in "10 percent ammonium molybdate, five-percent glacial acetic acid for 30 minutes, infiltrated with 20 percent gelatin, five percent glycerol just above the melting point of the mixture, and sectioned in the frozen state." They reported that the site of the electrode tip could be identified in sections as thin as 25 μ.

Silver Marking   Oikawa et al. (1958) developed a method for marking the locations of the tips of their 1 to 3 μ micropipets in retinas from carp eyes. Instead of filling their electrodes with $3M$ potassium chloride they employed a saturated solution of silver nitrate. To mark the location of the electrode tip, they applied the positive terminal of a 400-V supply to the micropipet; the negative terminal was connected to the reference electrode—a silver plate on which the eye rested. By a combination of iontophoresis and heating, silver nitrate was expelled from the micropipet into the retina. (The investigators believed that heating was the main reason for ejection of the silver ions because the method could be used equally well with alternating current.) To identify the location of the electrode, the retina was placed between two pieces of liver and fixed with formalin; then serial paraffin sections were made and the tissue was stained with hematoxylin-eosin solution; the electrode site was made visible by treating the tissue sections with sodium thiosulfate solution. Oikawa et al. reported that spots 50×150 μ were obtained when a potential of about 270 V was applied; they stated, however, that controlling the size of the marks by choice of a particular voltage applied for a measured time, (no data given) presented difficulties. To this writer the method developed by Oikawa and colleagues might be improved by the use of a constant-current, rather than a constant-voltage supply.

The procedures just described have all worked satisfactorily in the hands of the investigators. It is obvious, however, that marking the sites of micropipets is more difficult than marking with iron-containing metal electrodes. Nonetheless, the electrophoretic principle is sound and can be used successfully when an ionic species can be ejected and made to produce a visible spot by combining with a radical in the tissue or in the solution used to perfuse the tissue. The size of the spot is dependent on the diameter of the tip of the micropipet and the charge (mA-sec) employed. The ingenious investigator can easily develop a combination ideally suited to his particular needs.

Comparison Between Metal Microelectrodes and Electrolyte-Filled Micropipets

Both metal and micropipet type microelectrodes are used for the measurement of membrane and action potentials. However, it is worthwhile to compare the suitabilities of each for these two quite different tasks. Because the electrical properties of each type of electrode differ substantially, it might be suspected that each is suitable for one particular task; this is indeed so.

The metal microelectrode is characterized by a small-area metal-electrolyte junction which exhibits an impedance that is extremely high at low frequencies and decreases with increasing frequency. This impedance-frequency characteristic makes it difficult to obtain a potential-measuring device with a high enough input resistance to measure resting membrane potentials with a small area metal microelectrode. The impedance-frequency characteristic of an electrolyte-filled micropipet is resistive (i.e., not frequency dependent) in the low-frequency region and has a finite, although high, zero-frequency (dc) resistance. With this type of electrode it is relatively easy to provide a measuring instrument with an input resistance high enough for the accurate measurement of resting membrane potentials.

Both types of electrode can be used to measure action potentials, provided a high input impedance measuring device is employed; if the impedance is not high enough, distortion results. In the case of a metal microelectrode, if the input impedance of the measuring device is too low, the microelectrode behaves like a high-pass filter (i.e., the low-frequency components of the action potential are attenuated and the waveform is electrically differentiated). Even with a high input impedance device connected to an electrolyte-filled micropipet, the distributed capacitance at the tip causes the electrode to behave like a low-pass filter (i.e., the high-frequency components of the action potential are attenuated and the waveform lacks sharpness and detail). Although the metal microelectrode has a distributed tip capacitance, the low high-frequency impedance of the metal-electrolyte junction makes the effect of the tip capacitance negligible. In the case of the micropipet, the tip capacitance is extremely bothersome and, even after all practical steps have been taken to reduce it, further reduction of its effect can only be achieved by the use of a negative-capacity amplifier (see p. 178).

The metal microelectrode and the micropipet can also be compared on the basis of their noise voltages. Because the metal microelectrode has a lower resistance component of its impedance with increasing frequency, its noise voltage is considerably lower than that of the micropipet, which has a uniformly high resistive impedance over a broad frequency spectrum. For this reason, if high-gain recording of action potentials with a wide bandwidth is to be employed, the lower noise of the metal microelectrode is an attractive feature.

For the reasons just given, metal microelectrodes see most service in the measurement of action potentials. Electrolyte-filled micropipets are routinely used to measure resting membrane potentials; with suitable precautions they can measure action potentials. Thus the two types have complementary characteristics and the statement, attributed to Svaetichin by Gesteland et al.

(1959), that metal microelectrodes resemble high-pass filters and micropipets resemble low-pass filters, retains validity.

Multiple Micropipets

In studies of the factors underlying genesis of the membrane and action potential of a cell, it is often necessary to inject substances into the cell or even to pass a current through its enveloping membrane, or to compare the action potentials recorded with intra- and extracellular electrodes. In order to conduct such studies, various investigators have developed multiple micropipets. Some are coaxial (pencil type), some have two lumens (parallel and double-barreled) like a shotgun; others are multibarreled like a Gatling gun. Sketches of these various types appear in Fig. 4-21.

Desirous of constructing a small double micropipet, Tomita (1956, 1965) placed a $3M$ KCl-filled micropipet having a tip diameter on the order of 1 μ inside a seawater-filled micropipet with a tip diameter of about 10 μ; the tip of the inner protruded 12 μ beyond that of the tip of the outer micropipet. Tomita used his coaxial (pencil-type) micropipet to measure simultaneously intra- and extracellular action potentials from cells in the eye of the horseshoe crab. A similar coaxial micropipet was described by Freygang and Frank (1959) and Frank (1959) for recording intra- and extracellular action potentials from single spinal motor neurones in the cat. With the electrode in Fig. 4-21a, they showed that, in their particular recording situation, the action potential recorded extracellularly from a nerve cell resembled the differentiated waveform obtained with the intracellular electrode (see p. 279).

The method of using two micropipets mounted side by side was described by Terzuolo and Araki (1961–1962). Their "parallel-microelectrode" (Fig. 4-21b) was fabricated by starting with two lengths of Pyrex tubing (1.2-mm o.d.), fused together for a short distance using a gas flame. The fused region, whose length determined the spacing of the tips, was placed in a micropipet puller and heat pulled in the conventional manner. The resulting electrode consisted of two quite independent micropipets, each having a tip of 1 μ, separated by about 2.5 μ.

Among the earliest to use the double-barreled micropipet were Coombs et al. (1955); their electrode (Fig. 4-21c) was constructed by placing a glass partition in a capillary tube, sealing one end, and attaching two side tubes. Each side tube communicated with one of the compartments in the capillary, which was heat pulled down to a tip diameter ranging from 0.7 to 2 μ. They placed the tip of their double-barreled micropipet in cat spinal motor neurones and used one barrel for measuring membrane potentials and the other to apply current (from a high-impedance supply) to inject or remove ions from the cell; the reference electrode for measuring membrane potential and the application of current was in the lumbar region of the animal.

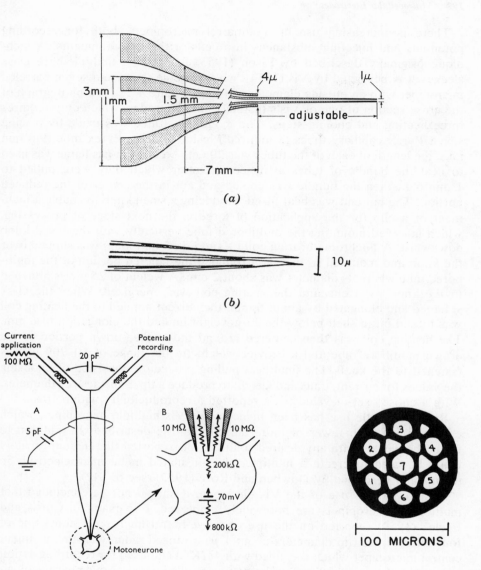

Fig. 4-21. Multiple micropipets: (*a*) coaxial (pencil-type) micropipet; (*b*) parallel micropipet; (*c*) double-barreled micropipet; (*d*) multibarreled (Gatling gun) micropipet. [(*a*) From W. H. Frey-gang and K. Frank, *J. Gen. Physiol.* 1959, **42**:749–760 (By permission). (*b*) From C. A. Terzuolo and T. Araki, *Ann. N.Y. Acad. Sci.* 1961–1962, **94**:547–558 (By permission) © The New York Academy of Sciences. (*c*) From J. S. Coombs, J. C. Eccles, and P. Fatt, *J. Physiol.* 1955, **130**: 291–325 (By permission), (*d*) From V. A. Vis, *Science.* 1954, **120**:152–153 (By permission)]

There is increasing use of multibarrel micropipets, both for recording potentials and injecting substances into cells or their environments. A technique originally described by Elson (1953) for making multiple-bore glass sleeves was perfected by Vis (1954), who was able to make a seven-barreled micropipet with an outside diameter of 100 μ. Figure 4-21d is a photograph of the cross section of one of Vis's (gatling gun) electrodes, which only requires three heating and cooling steps. The model illustrated was made by placing seven Pyrex capillary tubes (3 mm o.d.) inside a larger Pyrex tube (9.6 mm i.d.); the length of each of the tubes was 20 cm. An oxygen-gas flame was used to heat the bundle of tubes at its center, after which they were pulled to 1 mm o.d.; then the bundle was cooled and cut in the center of the reduced portion. The cut end was heat fused, producing a small knob to enable attachment of a clip for the application of force in the next stage of processing, which involved mounting the multibored tube vertically with the fused knob downward. A Nichrome heating coil (3 mm i.d., 5 mm long) was slipped over the knob and mounted so that it surrounded the tapered portion of the multibored tube where its diameter was about 2 mm. A weight of 25 g was attached to the knob by a clip and the heating coil was energized. When the glass softened and elongated by about 6 mm, the current applied to the heating coil was turned off; a shelf below the 25-g weight limited the elongation to 6 mm. The heating coil was then centered around the necked-down portion of the drawn multibore tube and a heavier weight (in the range of 50–900 g) was fastened to the knob. The final heat-pulling procedure consisted of choosing the values for current, time, and weight to produce a tip of the desired diameter. With a chosen set of values, Vis reported a reproducibility within 10%.

Relatively little use has been made of the Vis multiple micropipet, which can be made with fewer or more barrels than he described. In addition to filling the barrels with any desired solution, Vis reported that indium-tin alloy could be used to create a multiple, glass-insulated metal microelectrode in the manner described by Dowben and Rose (1953) (see p. 142).

Despite lack of use of the Vis electrode, it should not be concluded that multibarrel micropipets are not widely employed. For example, Curtis and Eccles (1959) reported on the use of a five-barrel micropipet consisting of four micropipets (tip diameters about 1 μ) arranged radially around a similar central micropipet which was filled with 4MKCl and used for recording action potentials; the surrounding four barrels were filled with various pharmacological agents that were electrophoretically placed in the environment of the cell from which recordings were being made.

Electrical Properties of Multiple Micropipets   When two micropipets are used in the coaxial configuration, or when two or more are placed side by side, there is considerable electrical coupling between them. If the tips are in the same solution, there is a conductive coupling at this site. Because micropipets are usually made of high-quality glass, the conductive coupling through the intervening walls is negligible (Rush et al., 1968). However, there is always appreciable capacitive coupling between the individual micropipets. Although the coupling does not seriously interfere with the measurement of resting membrane potential, it interferes to a serious degree when action potentials are recorded; unless special precautions are taken to exclude the potential changes detected by an adjacent micropipet, the action potential measured cannot be judged to be a faithful representation of the bioelectric event.

Electrically, there are two problems to be solved when using multiple micropipets. The first is to provide adequate compensation for tip capacitance (see p. 178) so that the shortest response time can be attained with each micropipet; the second is to eliminate the signal detected by one micropipet from appearing at the outputs of the amplifiers connected to the other micropipets. The first problem is solved by the use of a negative-capacity (neutralized or positive-feedback) amplifier and the second by injection of a canceling voltage into the amplifiers connected to the adjacent micropipets. The complexity of the circuitry to amplify each signal from a single micropipet in any array increases dramatically with increasing numbers of micropipets. This point and the technique employed are illustrated by analysis of the circuitry required with a double micropipet. The easiest method of establishing the circuitry necessary is to identify the components of the equivalent circuit of a coaxial micropipet; for simplicity, the tips have been placed in a beaker of saline containing a reference electrode (Fig. 4-22a). Neglecting all electrode-electrolyte impedances (which are small with respect to the resistances of the tips) and all electrode-electrolyte and junction potentials (because only changing, i.e., action potentials are of interest), the equivalent circuit appears in Fig. 4-22b results. The principal components of this circuit are the distributed capacitances between each micropipet and the solution C'_{1R}, C'_{2R}, which is in contact with the reference electrode, the interpipet distributed capacitance C'_{12}, the tip resistances R_{t1} and R_{t2}, and the resistance of the electrolytic solution between the tips R_{t12}. When current is passed through one micropipet and voltage is measured by the other, R_{t12} is of considerable importance; however, when only bioelectric events are measured, this value is less significant.

The circuit of Fig. 4-22b can be simplified by collecting the distributed components and synthesizing a lumped circuit, as in Fig. 4-22c, in which the symbols have the same significance. To further illustrate the measurement problem, assume that the inner micropipet (1) detects a bioelectric signal E_1, and the outer micropipet (2) detects another, E_2; both are measured with respect to the reference electrode R. These bioelectric signals have been included in Fig. 4-22c; and inspection reveals that E_1 will present a signal to its amplifying channel, connected between terminals 1 and R. It will also present a capacitively coupled signal to the second amplifying channel connected to terminals

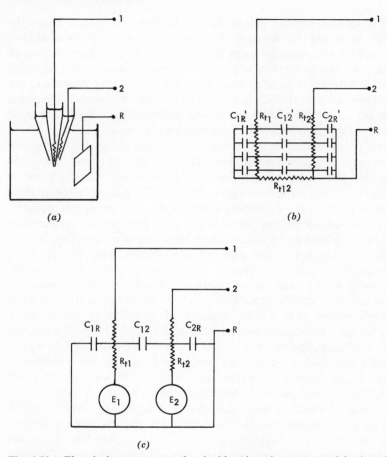

Fig. 4-22. Electrical components of a double micropipet: (*a*) coaxial micropipet and reference electrode in a beaker of saline; (*b*) distributed capacitances and resistances of a coaxial micropipet; (*c*) approximate equivalent circuit for a coaxial micropipet when two bioelectric events E_1 and E_2 are being measured.

2 and R; the converse regarding E_2 is true, as well. In addition, for each amplifier to faithfully reproduce E_1 and E_2, compensation must be provided to cancel most of the tip capacitances C_{1R} and C_{2R} because the multiple micropipet, like the single micropipet (see p. 178), exhibits a considerable distributed tip capacitance. In summary, cancellation for the cross coupling and provision for neutralization of the tip capacitances are the functions demanded of the amplifying system used with a multiple micropipet.

Cross Neutralization

Several ingenious methods have been developed to eliminate the signal detected by one micropipet in an array from appearing at the outputs of the amplifiers connected to the other micropipets. The method of solving the problem can be illustrated by considering the case of the coaxial micropipet. The principle, (often called cross neutralization), is diagrammed in Fig. 4-23 by the use of cathode followers $CF1$, $CF2$ and operational amplifiers A_1, A_2. It should be obvious that source followers and operational amplifiers will accomplish the same goal. The bioelectric event, detected by the inner coaxial micropipet 1, appears between terminals 1 and R, which are connected to the upper cathode follower $CF1$ in Fig. 4-23; the bioelectric event detected by the outer coaxial micropipet 2 appears between terminals 2 and R, which are connected to the lower cathode follower $CF2$. The operational amplifiers provide positive-feedback signals, adjustable by manipulation of potentiometers $+FB1$ and $+FB2$ to neutralize the tip capacitances of each micropipet as identified by C_{1R} and C_{2R} in Fig. 4-22c. As stated previously, because of the interpipet capacitance (C_{12} in Fig. 4-22c) a portion of E_1 will be presented to the input of $CF2$ and a portion of E_2 will be presented to $CF1$. To eliminate the cross-coupled signal due to E_2 from appearing at OUTPUT 1, a fraction of E_2 derived from A_2 is fed into the inverting ($-$) terminal of operational amplifier

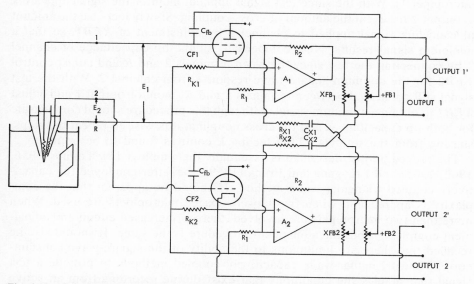

Fig. 4-23. Principle employed to compensate for tip capacitance and to provide cross neutralization for a double micropipet.

1 via the series resistance and capacitance network R_{X1}, C_{X1}. The same procedure is employed to prevent the signal detected by micropipet 1 (i.e., E_1) from appearing at OUTPUT 2. If the operational amplifiers are of the direct-coupled type, slightly larger output signals can be obtained from their outputs; these points are designated OUTPUT 1' and OUTPUT 2' in Fig. 4-23.

Although the principle employed for compensation is simple and straightforward, it is quite difficult to make the correct adjustments to neutralize the capacitances of each micropipet and to eliminate the cross coupling. Two of the methods (the square-wave constant-current and capacitively coupled triangular-wave methods) used for adjusting the amount of positive feedback to compensate for the capacitance at the tip of a single micropipet can be employed to adjust the feedback to compensate for tip capacitance. The same methods can also serve to inject the correct amount of canceling signal to prevent cross-coupled signals from appearing at the outputs of the two recording channels.

In practice, four main steps are necessary to achieve compensation for tip capacitance and to attain cross neutralization. Referring to Fig. 4-23, consider the following procedure: with all feedback FB controls turned to zero, apply the test signal (using either the square-wave constant-current—Fig. 4.18*b* or the triangular-wave capacitively coupled method—Fig. 4-18*c*) to terminals 1 and R and monitor output 1; adjust the $+FB1$ control to obtain the best square-wave response (this procedure provides compensation for the capacitance of micropipet 1). With the same test signal applied, monitor the signal appearing at output 2 to reveal the amount of cross coupling. Slowly increase the amount of canceling signal applied to channel 2 by adjustment of $XFB1$, so that a minimum signal results. Now compensate for the tip capacitance of channel 2 by connecting the test signal between terminals 2 and R and adjust control $+FB2$ for the optimum square-wave response from channel 2. With the test signal still connected between terminals 2 and R, monitor output 1 and adjust $XFB2$ to obtain a minimum signal. Although this procedure should compensate for both tip capacitances and for cross coupling, it is wise to repeat the steps, starting from the settings of the feedback controls found to be optimum.

The procedure just described is essentially that employed by Tomita (1956, 1962, 1965) and Freygang and Frank (1959); the latter employed the capacitively coupled triangular-wave method. It should be apparent that the complexity is appreciable when two closely spaced micropipets are used. When more than two micropipets are involved, there is increased circuit and adjustment complexity, but the adjustment procedure is the same. It should also be recalled that there are limitations to the ability of the square-wave constant-current and triangular-wave capacitively coupled methods to provide a test signal that mimics the conditions that exist during recording from an active cell (see p. 181). However, these circuits and the procedure outlined here are all that is available at present.

5

The Membrane and its Action Potential

THE MEMBRANE

Descriptions of the types and characteristics of electrodes would be sterile without a short discussion of the nature of bioelectric generators that produce the voltages detected by electrodes. Although the underlying processes are complex in origin and function by mechanisms that have not all been assigned their correct roles, much is known about bioelectric phenomena. All have their origins in the membrane potential which results from the ionic gradients that exist across cell membranes. By mechanisms that are not fully understood, but reasonably well described, pass the myriad of substances concerned with cellular function. As the Hokins (1965) put it, "The cell walls are selective gateways for the transport of substances into and out of the cell, and thus they control its form and activities."

Although there are many different types of cells, their membranes have many structural features in common. For example, Bourne (1964) stated

It is of interest that a common structure for the cell membrane exists in cells as diverse as erythocytes, axons of nerve cells, muscle fibers, leucocytes of the blood, yeast cells, algal cells, the cells of higher plants, and the ova of echinoderms, e.g., sea urchins and starfish. It seems likely, however, that there are variations to some extent between the membranes of the various cells; nevertheless, it is highly probable that a general structural pattern does exist for all of them.

Existing structural models are based on x-ray diffraction and electron-microscope studies of the accessible and readily manipulatable membranes referred to by Bourne. It is outside the scope of this chapter to discuss details of the structure of membranes beyond presenting currently acceptable models as an aid in understanding the properties of membranes. The properties of the cell membrane are described as if it were a physical structure, but it must be remembered that it exists because of and for the cell and is maintained by the expenditure of metabolic energy. In most present-day models, the membrane is composed of thin layers of lipid and protein; however, there is considerable evidence that these layers are extremely complex in structure and that they consist of many different kinds of lipid and protein. For additional information, the interested reader is directed to the excellent reviews contained in the books by Bourne (1964), Cole (1968), Chapman (1968), and Dowben (1968).

The most agreed-upon structure for the cell membrane is two layers of lipid material between two layers of protein.* Three such models are presented in Fig. 5-1. The Davson-Danielli model (Fig. 5-1*a*) was derived from marine-egg studies and was reported by Danielli and Davson (1935); Robertson's model (Fig. 5-1*b*) was reported in its present form in 1966 (although it was conceived

---

* It is believed that the protein layers contain minute pores.

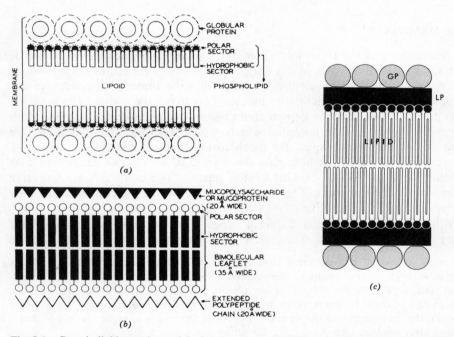

Fig. 5-1. Protein-lipid-protein models for the structure of the cell (plasma) membrane deduced from surface tension, x-ray diffraction and electron-microscope studies of a variety of cells; (*a*) the Davson-Danielli model, derived from studies of marine eggs; (*b*) Robertson's polar model in which the exterior (mucopolysaccharide or protein) is different from the interior (polypeptide) surface; (*c*) Hokins' butter sandwich model consisting of two lipid layers (butter) between two layers of protein (globular layer *GP*) and layered protein (*LP*). [(*a*) and (*b*) From R. M. Dowben, *Biological Membranes,* Boston: Little Brown & Co., 1969, (By permission), (*c*) From L. E. Hokin, and M. H. Hokin, *Sci. Amer.* 1965, **213**:78–96. (By permission) © 1965 by Scientific American, Inc. All rights reserved.]

at an earlier time) and was developed from myelin-sheath studies. The Hokins model (Fig. 5-1c) came from studies on gland cells.

Of the three models, the "butter sandwich" (Hokin, 1965) is perhaps the most popular. There appears to be agreement on the composition of the bulk of the cell membrane—namely, phospholipids, which are fatty substances that contain water-soluble glycerol and phosphate, and water-insoluble fatty acids, which are centrally located in the membrane. Electrically, therefore, the membrane would be expected to behave like a good insulator (dielectric). Because most cell membranes measure less than 100 Å in thickness, their capacitances would be expected to be high in comparison to man-made capacitors. Table 5-1 lists some representative cell-membrane capacitances. Particularly significant is the fact that the capacitance range for living cells (0.5–9 μF/cm² of surface area) is only slightly lower than the values that are known to occur at the interface between a metal electrode and an aqueous electrolyte.

Membrane Dielectric Constant

It is of some interest to speculate on the dielectric constants of cell membranes. For example, if the squid giant axon is chosen, it is possible to calculate its average dielectric constant because the thickness and capacitance of the axon have been measured. Hodgkin (1964) reported that "the total thickness of the axon membrane is about 70 Å; if examined under high power, the membrane can be resolved into two dense lines about 50 Å apart." (Apparently the membrane consists of two layers of protein each 10 Å thick with a lipid layer 50 Å in thickness between them.) A typical value for membrane capacitance is 1 μF/cm². Because the ratio of the axon diameter to membrane thickness is large, the following expression for the capacitance of a parallel-plate capacitor can be employed without serious error.

$$C_{\mu\text{F}} = 0.0885 \times 10^{-6} \frac{A}{d} K$$

Where the capacitance C is in microfarads, the area A and the separation d are in centimeter units, and the dielectric constant K is the desired quantity. Rearrangement of this expression to isolate K gives

$$K = \frac{Cd}{0.0885 \times 10^{-6} A}$$

Entering the values for the axon, $C = 1.0$ μF, $A = 1$ cm², and $d = 70$ Å or 70×10^{-8} cm, which is the axon-wall thickness, the following is obtained for the average dielectric constant:

$$K = \frac{1.0 \times 70 \times 10^{-8}}{0.0885 \times 10^{-6} \times 1} = 7.9$$

Table 5-1 Electrical Properties of Cells

Cell Type	Membrane Capacitance (μF/cm²)	Membrane Resistance (Ω for 1 cm²)	Phase Angle (deg.)	Dielectric Constant	Cytoplasm Resistivity (Ω-cm)	Investigator and Year
Node of Ranvier (frog axon)	3–7	10–15	—	—	—	Tasaki (1955)
Internode	—	$\rho = 10^9$	—	5–10	100	Hodgkin (1951), Stämpfli (1952)
Asterias (marine egg)	0.5	3000	—	—	—	Tyler et al. (1956)
Toad sartorius muscle	5–9	—	—	—	—	Tasaki and Hagiwara (1957)
Frog sartorius muscle	2.6	3000	—	—	—	Falk and Fatt (1964)
Crayfish muscle	3.9	680	—	—	—	Falk and Fatt (1964)
Many different cells	0.5–2.0	1000	—	—	—	Cole (1968)
Sea urchin egg	0.87	—	90	—	—	Cole (1968)
Asterias (marine egg)	1.0	—	90	—	—	Cole and Cole (1936)
Arbacia (eggs)	0.74	—	90	—	—	Cole and Cole (1936)
Yeast	0.6	—	—	—	—	Fricke and Curtis (1934)
Leucocytes	1.0	—	—	—	—	Fricke and Curtis (1935)
E. Coli	0.7	—	—	—	—	Fricke et al. (1956)
Mitochondrium	0.5–1.1	—	—	—	—	Pauly et al. (1960)
Nitella-Valonia	1.0	—	—	—	—	Blinks (1936)
Nitella	0.94	—	80	—	—	Curtis and Cole (1937)
Squid	1.3	—	80	—	—	Curtis and Cole (1938)
Frog eggs	2.0	—	86	—	—	Cole and Guttman (1942)
Squid axon	0.7	—	75–80	—	570	Taylor and Chandler (1962)
Crab (Carcinus leg nerve)	1.1	8000	—	—	90	Hodgkin (1947)
Lobster (Homarus leg nerve)	1.3	23,000	—	—	60	Hodgkin and Rushton (1946)
Frog–sciatic nerve	0.55	?	40?	—	560	Cole and Curtis (1936)
Squid–stellar nerve	1.1	1500	75	—	30	Cole and Marmont (1950)
Cat–sciatic nerve	0.65	?	40?	—	720	Cole and Curtis (1936)
Frog—Sartorius muscle	1.5	~40	70	—	250	Bozler and Cole; Cole and Curtis (1936, 1938)
Red cell—Man	0.8	?∞	90	—	—	Fricke (1931)
Leucocyte—rabbit	1.0	—	80	—	140	Fricke and Curtis (1935)
Red cell—turtle	0.8	?∞	90	—	140	Frick and Curtis (1934)

Table 5-2   Properties of Physical Dielectrics[a]

Type of Material	Dielectric Constant	Resistivity, (Ω-cm)	Dielectric Strength (V/cm)
Distilled water	78	0.5×10^6	—
Oil (transformer)	2.5	—	80,000
Paraffin	2.0–2.5	10^{15}–10^{19}	100,000
Rubber (neoprene)	4.5	8×10^{12}	120,000
Rubber	2.0–3.0	—	120,000–300,000
Rubber (hard)	2.8	2×10^{15}	188,000
Glass (Pyrex)	4.5	—	130,000
Glass (ordinary)	5.5–10	9×10^{13}	60,000–120,000
Quartz (fused)	3.8	$> 10^{19}$	160,000
Polystyrene	2.65	10^{18}	240,000
Barium Titanate	1200	10^{12}–10^{13}	30,000
Epoxy	3.62	$> 3.8 \times 10^7$	162,000
Cellulose acetate	3.2–6.2	10^{10}–10^{12}	100,000–160,000

[a] At room temperature. Data from *Reference Data for Radio Engineers*, 5th ed. H. W. Sams, Inc., Indianapolis, Ind., 1968, and *Handbook of Chemistry and Physics*, 40th Ed. Chemical Rubber Publishing Co., Cleveland, Ohio, 1958. (By permission)

If it is believed that all the membrane capacity is due to the lipid layer 50 Å thick, then by substituting this value into the foregoing expression, the dielectric constant obtained is 5.6. Both values are typical of the materials employed in physical capacitors, for example, oil and plastics. Table 5-2 presents a representative list of values.

Membrane Resistance and Resistivity

Because of the water-insoluble lipid composition of the central portion of cell membranes, it would be expected that their resistivities would be high. This, however, cannot readily be ascertained by inspection of the values presented for membrane resistance, which are the resistances for 1 cm² area of membrane. Measurements of membrane resistance have been made on a variety of cells (e.g., those in Table 5-1). In the investigations, the areas of the membranes measured were small and the values obtained were scaled upward and expressed as the resistance for 1 cm² of membrane area. Sometimes the term "membrane resistivity" is used, but this is an incorrect application of the term resistivity because resistivity is a property of a material and does not depend on how much is present. In the case of the membrane, the measured membrane resistance depends on its thickness in addition to its conducting property. Membrane resistance is merely a convenient term in membranology; the units are ohms for one square centimeter, usually abbreviated as Ω-cm^2. Note that the resistance of an area larger than 1 cm^2 will be less; conductance, the reciprocal of resistance, is proportional to membrane area. An idea of the resistivity of the material constituting the membrane can be estimated by choosing typical values for membrane resistance and thickness. Assuming a membrane resistance of 1000 Ω-cm^2 and a thickness of 100 Å, the resistivity (AR/L) becomes $1 \times 1000/100 \times 10^{-8}$ or 10^9 Ω-cm—a value only slightly below those for good insulators that have resistivity values in the range of 10^{10} to 10^{18} Ω-cm. The resistivity of distilled water is 0.5×10^6 Ω-cm at 18°C. The typical resistivity values for a variety of dielectrics, presented in Table 5-2 are illustrative.

Dielectric Strength   One of the especially interesting areas of membranology involves the strength of the electric field sustained by cell membranes. Electric field strength is defined as the potential difference existing between two points; that is, the field is expressed in terms of volts per centimeter. Insulators are materials known for their ability to sustain a high electric field, which is expressed as the dielectric strength, or the voltage-per-centimeter thickness that an insulator can tolerate without breaking down and becoming a conductor.

The electric field sustained by a cell can be calculated by dividing the membrane potential by membrane thickness. The resting membrane potential of many cells with membranes 50 to 200 Å thick is about 80 mV. These parameters indicate than an average electric field of 40,000 to 160,000 V/cm exists in the membrane. It is interesting to note that this value of dielectric strength compares very favorably with the values for good insulators shown in Table 5-2. For example, the dielectric strengths of plastics, rubber, paraffin, and transformer oils are in the range of 80,000 to 240,000 V/cm. Thus it is clear that the cell membrane qualifies as a dielectric that is not too far from its breakdown voltage.

Cytoplasmic Resistivity

The resistivities of the fluid contents of cells have also been measured (Table 5-1). Like membrane capacitance, the range of variation is not very large, extending from 30 to about 700 Ω-cm. Cole and Curtis (1950) reported that a typical value for the cytoplasmic resistivity of most mammalian cells is 300 Ω-cm.

TRANSPORT PROCESSES

Because the contents of cells are so strikingly different from those of their environments, it is logical to inquire about the mechanisms that account for the difference. Because of the nature of cell membranes it should be recognized that a substance must pass three barriers in order to enter a cell; the first is the interface constituted by the environmental water and the cell membrane, the second is created by the material constituting the membrane, and the third is the interface formed by the membrane and cytoplasm. Every molecule or ion that enters or leaves must negotiate these three barriers. Although the intimate details of the transport processes occurring at these barriers are un-known, it is customary to describe cell (and physical) membranes in terms of their resultant properties, which are termed semipermeable or selectively permeable. A semipermeable membrane allows passage of a solvent but not a solute; a selectively permeable membrane allows passage of one substance more readily than others, although often the substances passed and those rejected have quite similar characteristics. It is important to note that the term "selectively permeable" in no way identifies the mechanism responsible for selective permeability.

Various physical transport phenomena result from the properties of mem-branes. Although cell membranes are physical structures made by living processes, they exhibit many of the transport characteristics of nonliving membranes; they also exhibit at least two transport mechanisms absent in physical membranes. One involves the transport of material into the cell by engulfment (pinocytosis or cell drinking and phagocytosis or cell eating); the other involves the transport of material against a concentration and/or po-tential gradient with the expenditure of metabolic energy (active transport). The details of these processes are discussed subsequently.

As stated previously, the physical properties of membranes make it possible for material to be transported across cell membranes. Furthermore, the move-ment of different kinds of materials contributes to the composition of cellular contents and to the development of the ionic gradient, which expresses itself as the membrane potential. However, movement requires the participation of active transport as well as the phenomena of physical membranes to establish the ionic gradients so characteristically present across cell membranes. The various forces that transport material are now described.

Osmosis

When a semipermeable membrane separates a solvent from a solution of solvent and solute, there will be a movement of solvent through the membrane toward the solute until the chemical potentials of the solvent on both sides of the membrane are equal; the equality of chemical potential is achieved by the development of a hydrostatic pressure. Osmosis is the name given to the process whereby a solvent moves across a semipermeable membrane in response to a concentration gradient.

Since cell membranes are semipermeable, when cells are placed in aqueous solutions that contain a lower concentration of solute, water will move into the cells and cause them to swell and sometimes burst. If cells are placed in an aqueous solution containing a higher concentration of solute, water will move out of the cells, causing them to shrink. If the concentration of solute in the environment is such that the cells neither shrink nor swell, the environmental solution is said to be isotonic; thus hypotonic solutions cause cells to swell and hypertonic solutions cause cells to shrink.

The osmotic behavior of living cells was discovered in 1748 by the Abbée J. A. Nollet, who filled animal bladders with "spirits of wine," closed them, and placed them in pure water. The bladders were seen to swell greatly; some burst. A quantitative study of the phenomenon was made by Pfeffer (1877), who employed nonliving membranes consisting of colloidal films of cupric ferrocyanide; the films were deposited on the interior surfaces of porous (unglazed) earthenware pots by soaking them first in copper sulfate solution and then in potassium ferrocyanide solution. He filled the pots with concentrated solutions of sucrose in water, connected the pots to manometers, and placed them in pure water. He found that the pressure attained, as indicated by the height of the manometer depended on the concentration of the sucrose. Shortly thereafter, van't Hoff (1887) pointed out that, in dilute solutions, the ratio of the osmotic pressure to the concentration of the solute in the solution was a constant. From this evidence he deduced that for nonelectrolytes the osmotic pressure obeyed the ideal gas law ($PV = nRT$).

The driving force in osmosis is the concentration gradient, and the essential requirement for the phenomenon to occur is the ability of the membrane to allow free passage of one component (the solvent) and not the other (the solute). The mechanism by which the solute passes through the membrane may be different for each type of solute and membrane. The action in one case may be sievelike, in which the pore size is bigger than the molecular size of the solvent but smaller than the diameter of the solute molecules. In another case the solvent may be soluble in the membrane although the solute is not. It can be seen that in the case of a semipermeable membrane, the solvent is free to pass through the membrane in both directions but the solute is not.

The physical process by which osmosis occurs is analogous to the equalization of concentration throughout a solution. For example, if sugar crystals are placed in a beaker of water, they will sink to the bottom and start to dissolve. Initially the concentration of the sugar molecules will be high at the bottom. With the passage of time, random movement of all the molecules will result in a uniform concentration throughout the solution and there will be no concentration gradient. With a semipermeable membrane (which does not allow solute molecules to pass but allows free movement of solvent molecules) interposed between a solution containing a high concentration of solute, there exists a concentration gradient across the membrane. Solvent molecules pass through the semipermeable membrane in an attempt to eliminate the concentration gradient; but because the solute molecules cannot leave, solvent molecules continue to enter, the volume of solution on the solute side of the membrane increases, and the pressure rises to a point where the net passage of solvent molecules in one direction is equal to that in the other. At this point, the maximum pressure, or osmotic pressure is attained. In the terms of physical chemistry, at this point the chemical potential on either side of the membrane is the same.

With a membrane appropriate for the solute and solvent, the osmotic pressure P (atm) attained is, as shown by van't Hoff, described by the ideal gas law:

$$P = \frac{n}{V} RT = mRT$$

where n is the number of moles of the solute contained in the volume V (liters) of the solvent, R is the gas constant, m is the molal concentration, and T is the absolute temperature. Note that the value of RT at 25°C is 22.4 liter-atm/mole. Therefore, 1 mole of solute in 1 liter of solvent exerts a pressure of 22.4 atm; this is equivalent to supporting a column of water 760 ft high. Note that the pressure is independent of the nature of the solvent.

Another aspect of osmotic pressure to be noted is the difference between electrolytes and nonelectrolytes. Osmotic pressure is an expression of the number of particles present. In the case of nonelectrolytes, 1 mole contains Avogadro's number (6.02×10^{23} particles, i.e., molecules), however, since the molecules of some substances dissociate into positive and negative ions, the number of particles present is greater—the amount being dependent on the degree of ionization of the particular solute in the solvent. Thus, for a given concentration, an electrolyte would be expected to exhibit a higher osmotic pressure than a nonelectrolyte. To account for the osmotic pressure obtained with electrolytic solutions, the expression for osmotic pressure is modified to include a coefficient G, the cryoscopic coefficient, to give

$$P = G \frac{n}{V} RT$$

The factor G is less than the total number of ions formed from a molecule of electrolyte because it is the activity of the ions, rather than their concentration, which determines the osmotic pressure. Ionic interference reduces the number of ions free to participate. Cryoscopic coefficients, which include ionic activities, are determined by measuring the depression in the freezing

Table 5-3   Cryoscopic Coefficients at Various Molal Concentrations[a]

Electrolyte	0.02	0.05	0.1	0.2	0.5
$MgCl_2$	2.708[b]	2.677	2.658	2.679	2.896
$MgSO_4$	1.393[b]	1.302	1.212	1.125	—
$CaCl_2$	2.673[b]	2.630	2.601	2.573	2.680
LiCl	1.928	1.912	1.895	1.884	1.927
NaCl	1.921	—	1.872	1.843	—
KCl	1.919	1.885	1.857	1.827	1.784
KNO_3	1.904	1.847	1.784	1.698	1.551

[a] Data from Heilbrunn, 1952: *An Outline of General Physiology.* 3rd ed. W. B. Saunders Company, Philadelphia.
[b] 0.025 Molal.

point caused by addition of the substance. Table 5-3 lists the cryoscopic coefficients for several electrolytes for various molal concentrations (0.02–0.5). Note that the range of values extends from a little above 1 to slightly less than 3, but it is always less than the number of ions possible.

That osmosis is probably one of the most important processes for the transport of water molecules through cell membranes can be demonstrated by a variety of simple experiments. Since the cytoplasm is constituted of large molecules that cannot pass outward through the cell membrane, it is possible to manipulate the external environment osmotically by the addition or removal of electrolytes or nonelectrolytes and to predict what will happen to the cell. For example, if sea-urchin eggs are placed in seawater (their natural environment), to which a little distilled water is added, water enters the egg and it swells, the amount increases with increasing dilution of the seawater. If, on the other hand, salt or sucrose is added to the seawater, water leaves the cell and it shrinks. The same phenomenon can be demonstrated with other cells, the red cell being a favorite. The swelling can be taken to the point of rupture of the cell membrane (cytolysis or hemolysis in the case of the red cell); or shrinkage (crenation) can occur with visible prunelike wrinkles.

Partition Coefficient and Molecular Size   Cell membrane permeability studies have revealed that there is a proportional relation between the rate of entry (mass/sec per unit area per unit concentration difference) of a substance into a cell and its partition coefficient, which is defined as the ratio of its solubility in lipid to its solubility in water. A high partition coefficient favors entry into a cell. Nonpolar compounds are fat soluble and tend to have high partition coefficients; polar compounds are water soluble and have low partition coefficients. The higher alcohols and fatty acids are soluble in fats and sparingly soluble in water and have high partition coefficients. The lower alcohols, weak fatty acids, easily ionized compounds, and strong electrolytes have high solubility in water, low solubility in fat, and low partition coefficients; hence they do not readily pass through cell membranes.

Parenthetically it is interesting to note that accurate knowledge of the structure of cell membranes has only recently become available and, in view of the structure, it is not surprising that there is a relation between permeability and partition coefficient. However, long before there was knowledge of the membrane structure, the permeability–partition coefficient relationship was known. For example, Collander (1937), after paying tribute to Overton's work conducted before 1900, wrote "Thus the plasma [cell] membrane seems to act both as a selective solvent and a molecular sieve," and this view still retains validity.

The relation between cell-membrane permeability and partition coefficient has been studied mainly in large plant cells, although a few animal cells have been investigated. Figure 5-2 illustrates the relation between membrane permeability and partition coefficient obtained by Collander (1937), who placed plant algal cells (*Chara ceratophylla*) in different solutions and analyzed the cell contents after various times. Particularly noticeable in this illustration is the positive correlation between membrane permeability and partition coefficient (determined using olive oil as the lipid). In Fig. 5-2 the circle sizes represent the various radii of the molecules; it is quite apparent that the partition coefficient is frequently more important than molecular size as a determinant of permeability.

Although it is clear that the higher the partition coefficient of a substance, the more readily it passes through cell membranes, there must be an upper limit of molecular size for passage through the membrane by a physical process. Molecular configuration is indeed important in determining the effective size of a molecule; but molecular weight provides a rough indicator of molecular size, and membrane permeability studies have been made on this basis. It has been found that for substances having the same partition coefficient, membrane permeability is in general less with increasing molecular weight. Giese (1966) graphically presented pertinent data obtained by Ruhland and Hoffmann. (1925) to demonstrate this point; Fig. 5-3 displays the evidence.

From the foregoing it can be seen that even though a high partition coefficient is an important determinant for passage of a substance through a cell membrane, molecular size cannot be ignored and probably sets an upper limit.

The Donnan Equilibrium   A special type of ionic distribution is expressed by the Donnan equilibrium, and there is no doubt that it is an important contributor to the establishment of the ionic gradient that exists across cell membranes. This interesting type of electrochemical equilibrium resides in a special property of some membranes which are freely permeable to some species of ions but impermeable to others. This situation results in an osmotic pressure gradient and a potential difference, and its sequela was investi-

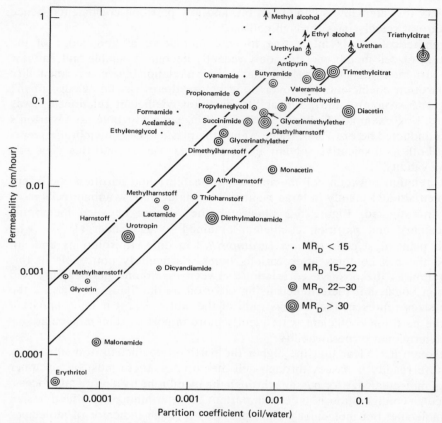

Fig. 5-2. The relation between membrane (*Chara*) permeability and oil-water partition coefficient for a variety of substances. The sizes of the points indicate the relative molecular radii (MR). (From R. Collander, *Trans. Faraday. Soc.* 1937, **33**:985–9990. By permission.)

gated by Donnan and Harris (1911) and Donnan (1925). The underlying thermodynamic theory had been enunciated earlier by Gibbs, therefore the phenomenon is often known as the Gibbs-Donnan equilibrium. The relevance of the Donnan equilibrium to the distribution of some ions across the cell membrane becomes obvious when it is recalled that the cytoplasm contains large charged protein molecules that cannot pass through cell membranes; therefore they can serve as the nondiffusible ions needed to create the Donnan equilibrium.

Donnan and Harris were investigating the osmotic pressure developed by a saline solution of Congo Red* separated from a saline solution by a

---

* A high-molecular-weight dye; a sodium salt of sulfonic acid.

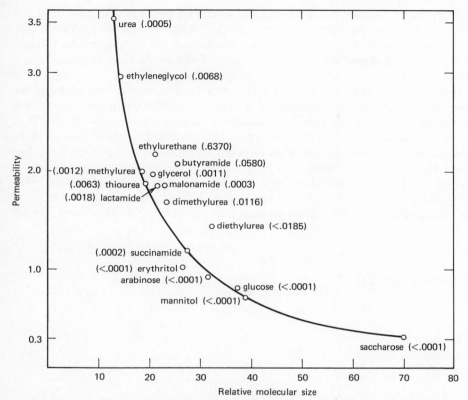

Fig. 5-3. The relation between membrane permeability and molecular size (as measured by refraction of yellow light); the membrane employed is that of *Beggiatoa*, a large sulfur bacterium. Ether-water partition coefficients are given in parentheses. (From A. C. Giese, *Cell Physiology*, 2nd ed. Philadelphia, Pa.: W. B. Saunders Company, 1963. 592 pp. By permission.)

parchment membrane. They found that when equilibrium had been attained, an unexpected ionic gradient resulted because the membrane was permeable to sodium and chloride ions but not to the charged dye molecules. At equilibrium, sodium and chloride were present on both sides of the membrane, but on the dye side the concentration of sodium chloride had increased. Donnan and Harris proceeded to work out the theory to account for this phenomenon. However, it is important to note that the theory only covers the equilibrium state brought about by the permeability property of the membrane; nowhere in the theory does any membrane parameter appear.

The equilibrium state described by Donnan and Harris can be generalized by stating that if a semipermeable membrane separates two electrolytic solutions and one of these contains nondiffusible ions (i.e., ions that cannot pass through the membrane), there will occur an unequal distribution of the diffusible ions. At equilibrium, there must be electrical neutrality on each side of the membrane; this means that the product of the concentration (ideally activity) of the diffusible ions on one side of the membrane is equal to the product of their concentration on the other.

Figure 5-4*a* presents the most frequently cited example to illustrate the Donnan equilibrium; it consists of a semipermeable membrane separating two dilute solutions (so that concentrations can be equated to activities) of sodium chloride (NaCl) and sodium proteinate (NaP). The membrane is freely permeable to sodium and chloride ions but impermeable to protein ions. Because there are no chloride ions on the side of the membrane with the proteinate (side 1), chloride ions will migrate and the electrostatic force will carry along sodium ions. After a short period, the net migration will be stopped and there will be a buildup of sodium ions on side 1. At this point, which

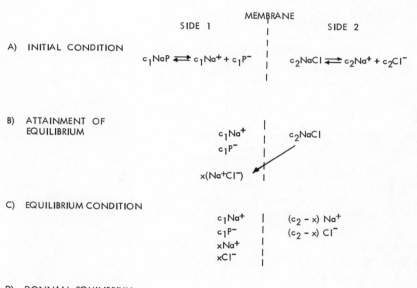

Fig. 5-4. The Donnan equilibrium: (*a*) initial condition; (*b*) attainment of equilibrium; (*c*) equilibrium condition; (*d*) Donnan equilibrium.

establishes the Donnan equilibrium, the migration of the diffusible ions (Na^+ and Cl^-) is equal in both directions. The energies (ΔF) required to transport Na^+ and Cl^- ions across the membrane are

$$\Delta F_{Na} = RT \ln \frac{Na_1}{Na_2}$$

$$\Delta F_{Cl} = RT \ln \frac{Cl_1}{Cl_2}$$

At equilibrium, the total free energy must be zero; therefore

$$RT \ln \frac{Na_1}{Na_2} + RT \ln \frac{Cl_1}{Cl_2} = 0$$

that is

$$RT \ln \frac{Na_1}{Na_2} = -RT \ln \frac{Cl_1}{Cl_2} = RT \ln \frac{Cl_2}{Cl_1} \quad \text{or} \quad \frac{Na_1}{Na_2} = \frac{Cl_2}{Cl_1} = r$$

In this expression, r is designated the differential distribution ratio. The Donnan equilibrium is often written as

$$[Na_1][Cl_1] = [Na_2][Cl_2]$$

and verbalized by stating that the product of the concentrations of the diffusible ions on one side of the membrane is equal to the product of the concentrations of the diffusible ions on the other. The information content of this statement can be illustrated by analysis of the situation in Fig. 5-4a, which represents the initial concentrations of sodium proteinate on side 1 and sodium chloride on side 2 as c_1 and c_2, respectively. During the attainment of equilibrium, a concentration x of Na^+ and Cl^- ions will move from side 2 to 1 because there is no chloride on side 1; proteinate ions cannot move from side 1 to side 2 because the membrane is impermeable to them. At equilibrium (Fig. 5-4c) the concentration of sodium ions on side 1 will be $c_1 Na^+ + xNa^+$ and the concentration of chloride ions will be xCl^-; similarly, on side 2 the concentrations of sodium and chloride ions are $(c_2 - x) Na^+$ and $(c_2 - x) Cl^-$, respectively. Application of the Donnan equilibrium condition is shown in Fig. 5-4c, 5-4d. Solving the equation for x (the concentration of NaCl that moved from side 2 to side 1 to attain equilibrium) gives

$$x = \frac{c_2^2}{c_1 + 2c_2}$$

Note that x is the amount of sodium chloride that moved from side 2 to side 1 to attain equilibrium and the equation indicates that as the concentration c_1 of proteinate (NaP) is increased, less sodium chloride will migrate.

Sollner (1954) presented data illustrating the agreement between predicted and attained equilibria in experiments in which various solutions were separated by a semipermeable (permselective) membrane. In order to balance osmotic pressures, sucrose, a nonelectrolyte, was added. Table 5-4 presents the data obtained and indicates a remarkably close correlation between the predicted and measured concentrations under equilibrium conditions (which required 3–24 hours).

As stated previously, a consequence of the Donnan equilibrium is the existence of an ionic gradient that manifests itself as a potential difference across the membrane. This difference in potential E is given by the Nernst (1889) equation, which was later modified to the following expression:

$$E = \frac{RT}{nF} \ln \frac{A_1}{A_2}$$

where R is the gas constant, T is the absolute temperature, F is the Faraday constant, n is the valence, and A_1 and A_2 are the activities of the ion species on sides 1 and 2 of the membrane. At 20°C for univalent ions, the expression reduces to

$$E = 58.2 \times 10^{-3} \log_{10} \frac{A_1}{A_2}$$

If the solutions on either side of the membrane are dilute, concentrations c_1 and c_2 can be substituted for activities, so that the expression becomes

$$E = 58.2 \times 10^{-3} \log_{10} \frac{c_1}{c_2}$$

Recalling that the concentration ratio for the diffusible ions was designated r, the Donnan membrane potential at 20° C then becomes

$$E = 58.2 \text{ mV} \log_{10} r$$

The Donnan membrane potentials expected for the experiment described in Table 5-4 can be calculated from the foregoing expression. Interestingly enough, although predicted and measured values of concentration are quite close, there have been few studies in which Donnan equilibrium potentials have been measured; an excellent review of some of these was presented by Sollner (1969).

In applying the Donnan equilibrium to single cells, it is important to recognize several complications. The environment of cells contains many monovalent and polyvalent diffusible ions, and the cytoplasm contains many different proteins. Establishment of the equilibrium condition for many different species of ions requires more bookkeeping, as Donnan (1925) himself has shown. Protein molecules carry both positive and negative charges (they are often called zwitterions); therefore, the net effect of cellular proteins in establishing a Donnan equilibrium is difficult to assess. In addition, the charges on

Table 5-4 Gibbs-Donnan Equilibrium across Permselective Collodion Membranes Involving only Strong Inorganic Electrodes"[a] (the Anions are the Nondiffusible Ions)

Ratio of volumes of Solution in Solution out	Solute	Original State (mM/liter)		Equilibrium state				Concentration Ratio In/Out = r	
				Experimental (mM/liter)		Calculated (mM/liter)			
		In	Out	In	Out	In	Out	Experimental	Calculated
1:1	NH_4^+	20.0	10.0	22.4	7.5	22.5	7.5	2.99 ± 0.05	3.00
	K^+	10.0	—	7.4	2.4	7.5	2.5	3.08 ± 0.10	3.00
	Cl^-	30.0	10.0	30.0	10.2	30.0	10.0	2.94 ± 0.05	3.00
	Sucrose	—	33	—	(33)	—	(33)	—	—
1:1	NH_4^+	30.0	—	22.4	7.5	22.5	7.5	2.99 ± 0.05	3.00
	K^+	—	10.0	7.5	2.5	7.5	2.5	3.00 ± 0.10	3.00
	Cl^-	30.0	10.0	29.8	10.1	30.0	10.0	2.95 ± 0.05	3.00
	Sucrose	—	33	—	(33)	—	(33)	—	—
1:10	NH_4^+	50.2	2.51	37.4	3.79	37.5	3.78	9.9 ± 0.3	9.9
	K^+	—	2.56	12.0	1.27	12.7	1.29	9.4 ± 0.4	9.9
	$C_2O_4^-$	25.1	2.54	24.7	2.53	25.1	2.54	9.8 ± 0.2	9.9
	Sucrose	—	39	—	(39)	—	(39)	—	—

[a] From K. Sollner, *Ann. N. Y. Acad. Sci.* 1954, **57**:177–203. By permission. © The New York Academy of Sciences

protein molecules are affected by the pH of their environment, hence the effectiveness of protein in establishing the Donnan equilibrium would be expected to be pH dependent. Moreover, attainment of the Donnan equilibrium requires that there not be a pressure gradient across the semipermeable membrane; the migration of ions can produce an osmotic pressure gradient. In practice, pressures are equalized by the addition of a nonelectrolyte solute. Finally, the Donnan "membrane potential" ignores the existence of two possible surface potentials—one arising at each surface of the membrane. If these potentials were equal and opposite (which is unlikely), the measured and predicted Donnan membrane potentials would agree.

In conclusion, the Donnan equilibrium plays an obvious role in the establishment of the potential across cell membranes. The difficulty that arises is according its contribution the proper place.

Active Transport

It has long been known that living cells have the capability of separating materials in solution; the classical example is the action of a gland in separating substances from the blood stream and deriving from them a particular constituent that is needed to fulfill some key function in the body. Historically, the name applied to all such processes was secretion, derived from the Latin word *secretionem*, which in turn comes from the verb *secernere*—to set aside or distinguish. As time passed, the meaning of the term became broadened to describe the selective movement of substances across cell membranes. To better identify the processes by which the movement of such materials takes place, Rosenberg (1948) employed the term "active transport," which he characterized as a transfer of matter which, for energetic reasons, cannot take place spontaneously. Rosenberg defined active transport as "a transfer of chemical matter from a lower to a higher chemical (in the case of charged components: electrochemical) potential." Because of the high specificity of some cases of active transport, it is believed that the mechanisms must involve chemical combination, therefore the reactions must entail ion formation. Perhaps for this reason Ussing (1949) called attention to some loopholes in Rosenberg's definition and defined active transport as follows:

We shall only speak of active transport when work has to be done to transfer the ion across the membrane, whether this work is used to overcome a potential difference, a concentration difference or a combination of both.

Ussing's definition therefore takes into account the movement of material against a concentration and/or a potential gradient; it does not, of course, provide any clue to the process or processes by which cells can perform this remarkable feat. *That* they do is well documented; *how* they use metabolic energy to perform the task is one of the fascinating unsolved problems in biophysics.

One of the truly outstanding examples of active transport was described in a series of experiments by Chambers and his associates (1932, 1933, 1935), who incubated segments of 9-to-10 day old chick embryo proximal tubules for 18 to 24 hours in a tissue-culture fluid consisting of oxygenated and warmed (39°C) plasma, embryonic fluid, and Tyrode's solution. At the end of the incubation period, they found that the segments of tubule had entrapped the environmental fluid and had become closed cylinders and spheres. When phenol red (or a similar dye) was placed in the environmental culture fluid, the dye was quickly transported by the cells into the lumenal fluid; the movement was observed by microscopic examination (Fig. 5-5a, 5-5b). When metabolism was depressed by cooling to 3 to 6°C, the dye leaked out of the

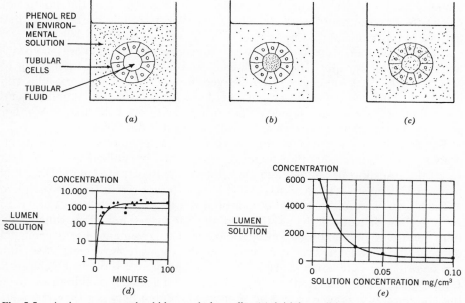

Fig. 5-5.   Active transport by kidney tubular cells: (*a*) initial condition, phenol red in environment; (*b*) movement of phenol red into tubular fluid by active transport; (*c*) leakage of phenol red from tubular fluid by the use of metabolic inhibitors; (*d*) ratio of phenol red concentration inside the tubule lumen to that in the environmental solution, for flounder kidney suspended in the balanced ionic medium of Forster and Taggart (1950) at 28°C. The concentration of phenol red in the solution was 0.01 mg/cm³. Each set of points represents an experiment on a different fish on a different day. (*e*) ratio of maximum phenol red concentration in tubule lumen to that in the environmental solution for varying concentrations of the dye in the medium. [(*d*) and (*e*) from T. T. Puck et al., *J. Cell. Comp. Physiol.* 1952, **40**:73–88. By permission).]

lumen into the environmental tissue-culture fluid; rewarming caused the dye to be transported back into the lumenal fluid.

Chambers et al. carried out other experiments to investigate the phenomenon. They found that transport of dye into the lumen, or retention in the lumen, could be reversibly blocked by metabolic inhibitors (Fig. 5-5c) such as hydrogen sulfide, cyanide, oxygen lack (produced by placing the preparation in an atmosphere of pure nitrogen), and narcotics. Removal of the metabolic blocking agents permitted the cells to concentrate and retain the dye in the lumenal fluid. In a similar series of *in vitro* experiments using flounder tubules in a balanced aerated salt solution maintained at $28 \pm 3°C$, Puck et al. (1952) measured the ratio of the concentration of phenol red in the tubule to that in the environmental fluid. They found that with a low concentration of dye in the environment, the tubular cells quickly concentrated dye in the tubular fluid; within 15 minutes the concentration ratio (lumen/environment) was 500, and in 60 minutes the ratio increased to 2000 (range 1000–4000; Fig. 5-5d). The concentration ratio decreased with an increase in the amount of dye in the environment; however, above a certain level of environmental dye the concentration ratio did not fall below 500 (Fig. 5-5e).

These experiments clearly suggest the presence of a vital, energy-consuming process to transport the dye molecules. In addition, it is interesting to note that the dye molecules were moved through three barriers—the cell membrane between the environmental fluid and cytoplasm, the cytoplasm, and the cell membrane and the lumenal fluid entrapped by the cell cluster. Whether the same or different processes are involved in transit is not known.

The present-day concept of the active transport of ions across cell membranes came largely from the studies of Ussing and his colleagues, who measured the transport of sodium and chloride ions across frog skin. The starting point of Ussing's work was Krogh's (1937) observation that frogs, which live in fresh water, not only conserve their bodily sodium chloride but can take up sodium chloride through their skin when they are placed in a dilute saline solution; such a phenomenon definitely indicates the movement of ions against a concentration gradient. Ussing realized that the separate movement of sodium and chloride ions could be measured by using radioactive sodium and chloride. Ussing's experimental setup employed a 3×3 cm segment of skin taken from the abdomen of a female frog. The skin was mounted between two chambers containing aerated solutions of the desired composition; Ringer's solution was usually placed in the chamber in contact with the inside of the frog skin. The radioactive sodium and chloride ions were added appropriately in the form of NaCl, the concentration of each ion species was measured with a Geiger-Mueller counter, and the ion fluxes were calculated from this measurement. (The flux of an ion species is defined as the number of ions transported per unit time per unit area of membrane.)

In a very thorough investigation, Ussing (1949) found that increasing the concentration of sodium chloride on the outside of the skin resulted in an increase in the influx and outflux of sodium ions; however, the influx always exceeded the outflux, resulting in a net transport of sodium ions. He also found that the Na^+ influx exhibited a considerable dependence on the pH of the inside solution; high pH values increased the influx, and vice versa. The pH of the outside solution had little effect on the Na^+ influx; however, below pH 5, the influx virtually ceased and at this point the Cl^- influx increased. The Cl^- influx was always lower than the Na^+ influx, and there was no constant relationship between the two; Ussing and his colleague Zerahn (1951) later showed that Cl^- was transported passively. The potential difference across the frog skin (inside positive, outside negative) was found to increase with those manipulations, which increased the Na^+ influx. Adrenalin, in very small concentrations in the inside solution, caused an enormous increase in the influx of Na^+. Cyanide, a metabolic poison, dramatically reduced the Na^+ influx without reducing the outflux; the potential difference paralleled the Na^+ influx.

To test his theory that sodium ions are actively transported, Ussing, with Zerahn (1951), applied a counterelectromotive force across the frog skin to reduce the potential to zero; at this point they measured the current and proposed that with zero potential across the frog skin (i.e., with the potential gradient abolished) and equal concentrations of saline on each side of the frog skin (i.e., with the concentration gradient abolished), the passive movement of ions would cease and active transport would continue. Measurement of the charge represented by the flux of the actively transported ion (Na^+) should exactly equal the current required to maintain the skin in the zero potential condition (i.e., short circuited). By adding radioactive sodium (^{24}Na) in the form of NaCl to the outside solution (Ringer's solution bathed both sides), they were able to measure the Na^+ influx and then to apply the current to reduce the skin potential to zero. The Na^+ influx exceeded the total current by about 5%, and by correcting for Na^+ outflux (5%), they found that the short-circuit current was identical with the inward sodium flux, proving the case for active transport of sodium ions. They further confirmed the active transport of Na^+ by repeating the study using agents that increase and decrease the influx of Na^+; in all cases the short-circuit current and the net sodium flux were in excellent agreement.

The metabolic energy required to transport Na^+ across frog skin (and toad bladder) has been measured by a number of investigators. In general they have found that without Na^+ transport there is a "basal" oxygen consumption. When Na^+ is transported, the oxygen used increases directly with the amount of Na^+ transported. For example, from an experiment using frog skin, Zerahn (1956) reported that

The net consumption of oxygen, equal to the oxygen consumed during transport minus that in the absence of transport, bears a constant ratio to the amount of transported sodium, being 1/4–1/5 equivalent for each equivalent of sodium transported.

Using toad bladder, Leaf and Dempsey (1960) reported "on the average some 19 sodium atoms are transported per molecule of oxygen consumed." These figures agree quite well, although they are expressed differently.

Because frog skin is made up of several layers of cells, it is of interest to identify the layer (or layers) responsible for the active transport of sodium. Information on this point has been provided by Ottoson et al. (1952–1953), Engbaek and Hoshiko (1959), and Hoshiko (1961); Ottoson et al. also made electron microscope studies of the frog skin to establish its structure. In this research the potential difference was measured between an external reference electrode and a micropipet (tip diam. ca. 1 μ) as it was advanced through the skin from the outer (epidermal) surface to the inner (corium) side. The results of these investigations (Fig. 5-6) are not in complete agreement, although they are somewhat similar. As the micropipet tip was advanced through the outer layer of dying cells (stratum corneum), all investigators reported that the tip of the micropipet became about 10 to 20 mV negative with respect to the reference electrode. Ottoson et al. found that this potential difference was maintained as the tip traversed the stratum germinativum (the layer that gives rise to the cells that migrate outward under their dead predecessors). When the tip passed the dermoepithelial junction (germinativum–corium border in Fig. 5-6), the potential difference jumped to that of the frog skin—inside positive by about 100 mV. In the studies reported by Engbaek and Hoshiko and Hoshiko, the frog skin potential was reached in two steps. When the tip of the micropipet entered the stratum germinativum, the potential increased to about 50 mV. As the micropipet was advanced further, however, no potential increase occurred until the tip crossed the membrane separating the corium from the germinal layer; at this point the frog skin potential was indicated (the range being 73–145 mV, with the inside positive).

From the studies just described, it is difficult to identify the actual site of the frog skin potential, and hence the region of active transport. That the dermoepithelial (i.e., germinativum–corium) junction is an important site is clearly shown by the investigations cited. Whether there is a transport mechanism operating in the outer border of the germinal layer remains to be demonstrated by future studies. It must be noted, however, that although it is relatively easy to make microelectrode studies such as those just discussed, it is not at all easy to identify the exact location of the tip of the micropipet because of the elasticity of the skin. In addition, the true dimensions of the skin are also difficult to establish because of the variable shrinkage (up to 40%) associated with its histological preparation.

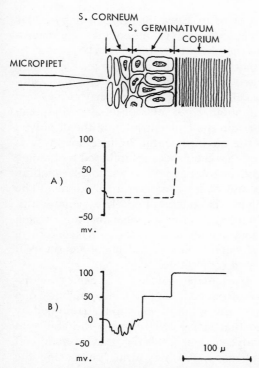

Fig. 5-6. Potential difference measured between an outside reference electrode and a micropipet as it is advanced through frog skin: curve *A* redrawn from Ottoson et al., *Acta Physiol. Scand.* 1953, **29**:611–624; curve *B* redrawn from T. Hoshiko, *Amer. Assn. Adv. Sci.* 1961, **29**:31–47.

Much work has been carried out using the frog skin as a model to study the properties of membranes. An excellent review of the history of the preparation was presented by Ottoson et al. (1952–1953), and a concise summary of some of the experiments carried out was given by Hoshiko (1961). An interesting student experiment in which the active transport of sodium is measured has been designed by Clarkson and Lindemann (1969).

Pinocytosis and Phagocytosis Substances can be accumulated in cells by two similar processes called pinocytosis (cell drinking) and phagocytosis (cell eating). By the former process, many types of cell membranes invaginate or form tiny channels into which the watery environmental fluid enters, soon becoming entrapped by closure of the end of the channel. Thus the cytoplasm contains a tiny pocket (vacuole) of environmental fluid, surrounded by cell membrane. In this manner, environmental fluid, with many dissolved substances that cannot pass through the cell membrane, is now within the cell.

Visual evidence of this process was reported by Edwards (1925) and by Lewis (1931), who made motion pictures of pinocytosis in rat monocytes and macrophages in tissue culture; Lewis named the phenomenon.

Pinocytosis is an active process that can be stimulated (e.g., by certain proteins, salts, and drugs) and inhibited by a variety of environmental substances. It has been estimated that a cell exhibiting this property brings in a volume of environmental fluid equal to about 10 to 30% of its own volume in one hour. It is believed that the nature and the concentration of the dissolved substances are more important than the environmental fluid; temperature is also an important factor. It is theorized that the process is initiated by binding of the environmental solute to the cell membrane. After vacuole formation, the fates of the vacuolar membrane and of its contents are unknown. The vacuole shrinks and the material within it is concentrated. The vacuole ultimately becomes granular in appearance, but the exact mechanism by which materials are exchanged between vacuolar contents and cytoplasm is unknown. An excellent review of the present understanding of the phenomenon as it occurs in the amoeba has been presented by Holter (1959).

Phagocytosis, or the ingestion of particulate matter by a cell, was discovered in 1883 by Metchnikoff, who introduced a splinter into the body of a transparent starfish larva and noted that the splinter was soon surrounded by cells. His subsequent studies showed that many cells possess the ability to engulf material and draw it into the cytoplasm where it is digested. Metchnikoff named cells having this property "phagocytes," (devouring cells), and the process became known as phagocytosis.

Only a limited number of cell types exhibit the phenomenon of phagocytosis, which is of central importance in the human in the defence against disease-producing microorganisms. Although the process may not play an important role in the maintenance of the internal composition of cells, it constitutes another transport mechanism that must be identified. An excellent review of the factors that facilitate and inhibit phagocytosis is due to Mudd et al. (1934).

Ionic Gradient and the Membrane Potential   As a result of the processes just described, there exist ionic gradients across the membranes of living cells, and assignment of the proper role for each process occupies the attention of many biophsicists. However, even without an accurate knowledge of the participation of the various transport processes, it is possible to measure and analyze their end result: concentration gradients and a membrane potential. Some substances are more concentrated within the cell; others exhibit higher concentrations in the environment. Numerous investigators have presented "balance sheets" listing the internal and external concentrations of the various substances important in cellular metabolism. Because of its large size, the squid giant axon is a favorite specimen for study; Table 5-5 identifies many of the substances in the axoplasm and in the environmental fluid. Noticeable in

Table 5-5 Approximate Concentrations of Ions and Other Substances in the Axoplasm of Freshly Isolated Giant Axons and in the External Fluid[a]

Substance	Concentration in Axoplasm	Blood	Seawater[b]	Units
H_2O	865[c]	870[d]	966	g/kg
K	400[e]	20[f]	10[g]	mM/kg H_2O
Na	50[e]	440[h]	460[g]	mM/kg H_2O
Cl	40[i]	560[h]	540[g]	mM/kg H_2O
Ca	0.4[j]	10[h]	10[g]	mM/kg H_2O
Mg	10[k]	54[h]	53[g]	mM/kg H_2O
Isothionate	270[c]	—	—	mM/kg H_2O
Aspartate	75[c]	—	—	mM/kg H_2O
Glutamate	12[c]	—	—	mM/kg H_2O
Succinate + fumarate	17[c]	—	—	mM/kg H_2O
Orthophosphate	2.5–9[l]	—	—	mM/kg H_2O
Adenosinetriphosphate	0.7–1.7[l]	—	—	mM/kg H_2O
Phosphagen	1.8–5.7[l]	—	—	mM/kg H_2O

[a] From A. L. Hodgkin, The Croonian Lecture. "Ionic movements and electrical activity in giant nerve fibers," *Proc. Roy. Soc. (London)* 1958, **B138**:1–37. By permission. References c through l can be found in this paper.

[b] Salinity 3.45%.

[c] Koechlin (1955).

[d] Robertson (1949).

[e] Steinbach and Spiegelman (1943); Keynes and Lewis (1951).

[f] Robertson (1949). See also his Fig. 3, legend.

[g] Webb (1939, 1940).

[h] Robertson (1949).

[i] Steinbach (1941). Value for fresh axoplasm; Koechlin (1955) gives 160 for pooled axoplasm from a large number of squid.

[j] Keynes and Lewis (1956). Koechlin gives 3.65 but regards this as an upper limit.

[k] Koechlin (1955). This is regarded as an upper limit.

[l] Caldwell (1956).

Concentrations in axoplasm have been calculated from the original authors' figures on the basis that axoplasm contains 865 g H_2O/kg axoplasm. For similar data in other fibers see Hodgkin (1951).

the table are the ionic gradients across the cell membrane; potassium ions dominate the cell contents, and sodium and chloride ions dominate the environment. This situation is by no means peculiar to the squid axon; this point is illustrated by Table 5-6, which presents an "ionic balance sheet" for nerve, skeletal, and cardiac muscle.

Since it is unwise to dwell on minute differences in the early stages of knowledge of a subject, it is better to concentrate on major similarities in these data. With this point of view it can be seen that, despite the remarkable variety

Table 5-6 Approximate Concentrations of Potassium, Sodium, and Chloride in Excitable Tissues and in External Fluid (mM/kg H_2O)[a]

a	b	c	d	e	f	g	h	i	j	k	l	m	n	o	p	q
	Potassium[b,c]					Sodium[b,c]					Chloride[b,c]					
	Concentration			Ratio		Concentration			Ratio		Concentration			Ratio		
	Inside	Outside				Inside	Outside				Inside	Outside				
Preparation	$[K]_i$	$[K]_o^R$	$[K]_o^P$	(b/c)	(b/d)	$[Na]_i$	$[Na]_o^R$	$[Na]_o^P$	(g/h)	(g/b)	$[Cl]_i$	$[Cl]_o^R$	$[Cl]_o^P$	(l/m)	(l/n)	Condition of Tissue
Loligo axon	410	10	22	41	19	49	460	440	0.11	0.11	40	540	560	0.074	0.071	Freshly dissected
Loligo axon	360	10	22	36	16	110	460	440	0.24	0.25	83	540	560	0.15	0.15	3 hr in sea water
Loligo axon	360	10	22	36	16	51	460	440	0.11	0.12	80	540	560	0.15	0.14	1½ hr after decapition of animal
Sepia axon	360	10	17	36	21	43	460	450	0.09	0.10	—	540	540	—	—	2 hr after decapitation
Sepia axon	330	10	17	33	19	77	460	450	0.17	0.17	—	540	540	—	—	3 hr in sea water, after decapitation
Carcinus nerve	380	10	12	38	32	—	460	510	—	—	—	540	540	—	—	Whole nerve freshly dissected
Carcinus nerve	230	10	12	23	19	—	460	510	—	—	—	540	540	—	—	30-μ fibers, 10.6 hr crab Ringer
Frog nerve	110 / 170	2.5	2.6	44 / 68	42 / 65	37	120	110	0.31	0.34	—	120	77	—	—	Freshly dissected
Frog sartorius muscle	125	2.5	2.6	50	48	15	120	110	0.12	0.14	1.2	120	77	0.01	0.016	Freshly dissected
Frog sartorius muscle	115	2.5	2.6	46	44	26	120	110	0.22	0.24	11	120	77	0.092	0.14	2 hr in Ringer's containing 2.5 m
Rat cardiac muscle	140	2.7	4.0	52	35	13	150	150	0.087	0.087	—	140	120	—	—	Freshly dissected
Dog skeletal muscle	140	2.7	4.0	48	35	12	150	150	0.08	0.08	—	140	120	—	—	Freshly dissected

[a] From A. L. Hodgkin, The ionic basis of electrical activity in nerve and muscle. *Biol. Rev.* 1950–1951, **25-26**: 399–409. (Cambridge Philo. Soc.) By permission.
[b] Concentrations (mM/kg H_2O) in fiber or solution have been rounded to two significant figures.
[c] $[\ ]_i$ gives concentration inside fiber; $[\ ]_o^R$ gives concentration in physiological solution such Ringer's fluid or seawater; $[\ ]_o^P$ gives concentrations in plasma or dialyzed blood.

of substances within and without cells, proteins and potassium are concentrated inside cells and sodium and chloride dominate the extracellular fluid.

In any physical system, the work done in separating charged particules is described in terms of potential and is measured in volts. In the case of a living cell, the dynamic state of "equilibrium" is revealed by the ionic gradients, which in turn express their presence by the existence of a "resting" membrane potential (mV).

In practice, the membrane potential is measured between an electrode (which is small with respect to the cell size) advanced into the cell and a larger one placed on the surface of the cell; the latter is sometimes in the electrolytic environment of the cell. When the small (intracellular) electrode penetrates the cell membrane, the membrane potential suddenly appears between the electrode terminals. In order to know the true membrane potential it is necessary to apply corrections for any liquid-junction and electrode-electrolyte potentials (see p. 9). The values for the resting membrane potentials for a large variety of cells have been measured by many investigators; Table 5-7 presents a representative list.

The present day concept of the factors determining the resting membrane and action potentials derive largely from studies of the nerves and muscles of squid, crabs, and frogs. Despite the ionic gradients that are known to exist across the membranes of these experimental subjects, their resting membrane potentials are determined largely by only one of the species of ions present. In a series of carefully executed experiments in which the ionic composition of the environment was varied, Curtis and Cole (1942) showed that the resting membrane potential of the squid axon varies logarithmically with the external concentration of potassium $[K_o]$; a tenfold increase in $[K_o]$ decreased the resting membrane potential by 50 mV (i.e., $- 50mV/10[K_o]$). Increasing $[K_o]$ to 18 times the normal value in seawater decreased the resting membrane potential to zero, and increasing $[K_o]$ fortyfold reversed the membrane potential by 15 mV. Similar results were obtained in frog sartorius muscle $(-44mV/10[K_o])$ by Ling and Gerard (1950); in sepia axon $(-50 mV/10[K_o])$ by Hodgkin and Keynes (1950); in frog sciatic nerve $(-50 mV/10[K_o])$ by Huxley and Stämpfli (1951); in sepia nerve $(-50 mV/10[K_o])$ by Hodgkin and Keynes (1955); in frog sartorius $(-52.3 mV/10[K_o])$ by Adrian (1956); in protozoa $(ca.-35mV/10[K_0])$ by Hisada (1957); in guinea pig gut $(-33 mV/10[K_o])$ by Holman (1958); and in frog cardiac muscle $(-40mV/10[K_o])$ by Suttgau and Niedergerke (1958). A selection of these findings is presented in Fig. 5-7.

From these data, which are in surprisingly close agreement considering the wide variety of experimental subjects, it is concluded that the resting membrane potential is largely (although not entirely) a potassium potential. The restriction "not entirely" is applied because the resting membrane potential

Table 5-7   Resting Membrane Potentials

Type of Cell	Size (μ)	Temperature (°C)	Type of Environment	Membrane Potential (mV)	Investigators and Year
Nerve					
Squid giant axon	500	20	Seawater	45	Hodgkin and Huxley (1939)
Frog myelinated axon	12–15	37	Ringer's solution	71	Huxley and Stämpfli (1951)
Chick embryo spinal ganglion	30–40	Body	Tissue culture fluid	50–65	Crain (1956)
Cat spinal motoneuron	70	Body	In vivo	70	Coombs et al. (1955)
Cat pyramidal cell	70		In vivo	55	Phillips (1955)
Rabbit sup. cervical ganglion	—	36–38	Physiological solution	65–75	Eccles (1955)
Frog neuro-myal junction	—	22–23	Ringer's solution	90	Nastuk (1953)
Toad symp. B neuron	35	Room	Ringer's solution	65	Nishi et al. (1965)
Muscle (Skeletal)					
Frog sartorius	80	20–22	Ringer's solution	97.6	Ling and Gerard (1949)
Rat	—	Body	In vivo	99.8 ± 0.19	Bennett et al. (1953)
Barnacle	1500	Room	Barnacle saline	70–8	Hagiwara et al. (1964)
Muscle (Cardiac)					
Frog ventricle	30	12–15	Ringer's solution	62 (50–90)	Woodbury et al. (1950)
Tortoise ventricle	30–80	Room	In vivo	56 (50–63)	Sano et al. (1956)
Chick embryo auricle	—	38.5	In vivo	29.2	Fingl et al. (1952)
Chick embryo ventricle	—	38.5	In vivo	39.3	Fingl et al. (1952)
Rat auricle	—	30	In vitro	62	Hollander and Webb (1955)
Rabbit auricle	—	37	In vitro	78	West (1955)
Rabbit auricle (pacemaker)	—	37	In vitro	56	West (1955)
Dog ventricle	16	Body	In vivo	80 (65–95)	Hoffman and Suckling (1952)
Dog auricle	10	Body	In vivo	85 (66–94)	Hoffman and Suckling (1952)
Dog papillary muscle	16	38	In vitro	85 (70–95)	Hoffman and Suckling (1953)
Dog Purkinje	25–35	37	Krebs' solution	71 (60–82)	Coraboeuf and Weidmann (1949)
Muscle (Smooth)					
Guinea pig intestine	—	36	Physiological solution	51.5	Holman (1958)
Guinea pig uterus (pregnant)	—	Body	In vivo	32.6	Woodbury and McIntyre
Protozoa					
Noctiluca scintillans	300	Room	Seawater	45	Hisada (1957)

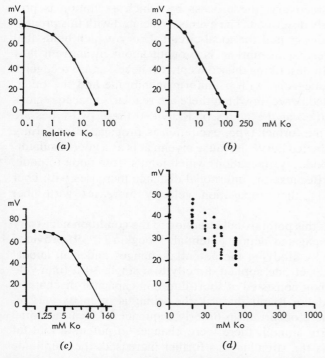

Fig. 5-7. The relation between resting membrane potential (mV) and the concentration of potassium in the environmental fluid (mM K_o): (*a*) squid giant axon; (*b*) frog sartorius muscle; (*c*) frog myelinated nerve; (*d*) protozoa. [(*a*) Redrawn from H. J. Curtis and K. S. Cole, *J. Cell. Comp. Physiol.* 1942, **19**:135–144; (*b*) Redrawn from G. Ling and R. W. Gerard, *Nature,* 1950, **165:** 113–114. (*c*) redrawn from A. F. Huxley and R. Stampfli *Journ. Physiol.* 1951, **112**:496–508. (*d*) Redrawn from M. Hisada, *J. Cell. Comp. Physiol.* 1957, **50**:57–71.]

does not coincide exactly with the value predicted by the Nernst equation (i.e., 58 mV per tenfold ratio of potassium concentration at 20°C). Dependence of the membrane potential on extracellular potassium concentration implies that the resting membrane is permeable to potassium and sparingly permeable to sodium and other ions. Dynamic permeability studies soon confirmed the correctness of this view.

THE PROPAGATION OF EXCITATION

Long before there was accurate knowledge of the magnitude and time course of the excursion in membrane potential and the ion fluxes that underlie

it during excitation and recovery, the process by which excitation is propagated had been accurately described. The events associated with this process
are embraced in what is designated the local-circuit theory, which holds that
stimulation of adjacent resting membrane is brought about by current flow
in the fluid environment. In describing this concept, it is necessary to recognize
two types of irritable tissue—one with a uniform membrane and the other,
characterized by myelinated nerve, in which there are regularly spaced regions
where myelin is thinner (i.e., at the nodes of Ranvier); these two types are
sketched in Fig. 5-8. In the former type, excitation is propagated uniformly
and is slower than in myelinated nerve. Because myelin acts as a good insulator,
excitation occurs at the nodes, propagation, which jumps from node to node,
is described as saltatory. Because the internodal distance increases with fiber
diameter (Rushton, 1951), the propagation velocity increases with fiber
diameter.

It is perhaps pertinent at this point to call attention to the condition necessary
for the initiation of a propagated action potential. Hodgkin (1939) provided
the essential information by studying the potential changes under an anodal
and cathodal stimulating electrode applied directly to a single 30-μ (unmyeli-
nated) crab axon; the stimuli consisted of short-duration capacitor discharges.
In this investigation Hodgkin found that, as the stimulus intensity was increased incrementally, there occurred an orderly sequence of local potential
changes; with low-intensity stimuli, there were changes in potential that far
outlasted the stimulus. As the stimulus was further increased, the amplitude
of the local response increased; but in addition, that occurring under the
cathode became longer in duration than that under the anode. After a certain
level of stimulus had been reached, a propagated action potential arose at the

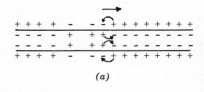

(a)

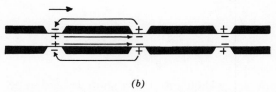

(b)

Fig. 5-8.   Local-circuit theory: (*a*) uniform membrane; (*b*) myelinated membrane.

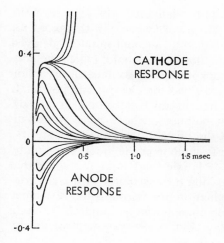

Fig. 5-9. Local potential changes under the cathode and anode with increasing stimulus intensity. Note that under the cathode, when the stimulus intensity reduced the local potential to about 0.38 of the amplitude of an action potential, excitation occurred; excitation did not occur under the anode with increasing stimulus intensity. (Redrawn from A. L. Hodgkin, *J. Physiol.* 1939, **126:**87–121.)

cathode. Hodgkin's data, (Fig. 5-9) illustrate that with low-intensity stimuli there were brief local responses in the form of polarization changes occurring under each electrode. Under the cathode, as the stimulus intensity was increased, the amplitude and duration of the local response increased. Finally, when a critical stimulus level was reached (producing a local hypopolarization of about 38% of the 40-mV amplitude of a propagated action potential), depolarization became regenerative and a propagated action potential developed. Under the anode, the local response to each stimulus merely increased in amplitude with the intensity of the stimulus; the duration of the local response was essentially the same for each stimulus and was always shorter in duration than for a cathodal stimulus. This study clearly illustrates that the genesis of a propagated action potential merely requires a small local hypopolarization, reaching a critical value that is much less than the membrane potential, for the membrane to undergo its cyclic permeability changes that result in the propagation of excitation.

Because an excited (depolarized) membrane is electronegative to a resting area, and because conducting fluid constitutes the environment, charges will flow from the resting (+) region to the active (−) region. Inside the cell, charges will flow in the opposite direction. This movement of charge constitutes an ionic current that hypopolarizes the resting region to such an extent that it becomes active and a traveling wave of depolarization is created whose velocity is characteristic for the particular type of membrane.

Validation of the local-circuit theory was presented by Hodgkin (1939), who showed that current does indeed flow in the electrolytic environment of an active membrane. Using frog sciatic nerve he created a block by cold and then by pressure and verified that threshold stimuli delivered to one end of the nerve did not pass the blocked region. However, when the threshold for stimulation beyond the block was measured with an appropriately delayed stimulus, there occurred an 80 to 90% reduction in threshold, indicating considerable hypopolarization distal to the blocked region. By measuring action potentials just proximal to the blocked region and the electrotonic potentials at different distances beyond, Hodgkin then found electrotonic potentials, having the same waveform as the action potential, whose peak values diminished with increasing distance beyond the blocked region (Fig. 5-10). He also proved that excitation could be made to jump over the blocked region by bridging it with a metallic conductor. In 1949 Huxley and Stampfli showed that the propagation could be blocked by increasing the resistance of the environment between the nodes of Ranvier. A reduction in environmental resistance restored propagation. The pattern of action current and potential distribution around an active area in frog sciatic nerve (Fig. 6-19) was mapped out by Lorente de Nó (1947).

An interesting corollary to the local-circuit theory was presented by Katz and Schmitt (1940), who demonstrated a lowering of threshold in crab nerve when the propagated wave of excitation passed in an adjacent nerve. They

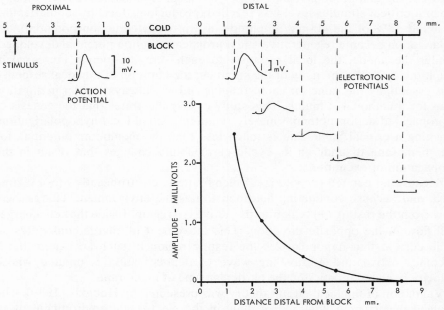

Fig. 5-10. Electrotonic potentials beyond a region of block in a nerve. (Redrawn from A. L. Hodgkin, *J. Physiol.* 1939, **90**:183–210).

also showed that when excitation in the two nerve fibers was advancing in step, the local-circuit currents interfered and the propagation velocity was slowed. If one impulse preceded the other, it accelerated the propagation velocity of the slower one and tended to produce synchronication of the impulses. A theoretical analysis of such excitability changes in an adjacent fiber has been presented by Clark and Plonsey (1970).

Cable Analog

To illustrate how electrotonic current spreads in advance of a propagated wave of excitation, recourse is made to a simplified version of cable theory; the application derives from likening a long irritable cell to a conductor surrounded by an insulator located in a conducting environment. Figure 5-11a diagrams an axon in which unit length Δx represents a typical segment having distributed electrical components. Figure 5-11b illustrates the spatial arrangement of these components, which are the resistances of the axoplasm R_a and membrane R_m; the resistance of the membrane is shunted by the membrane capacitance C_m. To better illustrate the cable analog, the membrane potential has been omitted. Figure 5-11c depicts the situation for the application of a steady potential E_0 of low amplitude; in this simplified case it is permissible to neglect the membrane capacitance to illustrate an important passive component—the length or space constant, sometimes called the characteristic length. A solution of the differential equation for this situation was presented by Hodgkin (1937), who showed that the potential E in advance of the steady potential decays exponentially with distance x and has the form $E_0/e^{x/\lambda}$, where λ is the length constant which is the distance for the potential to decay to 37% of E_0 (Fig. 5-11d). The length constant λ is equal to $\sqrt{R_m/(R_e + R_a)}$.

Despite the obvious oversimplification of cable theory, this model nonetheless provides useful information. For example, a decrease in the length constant would be expected to accompany an increase in the resistance of the environment. A sufficient decrease in the length constant could block propagation, and a study by Huxley and Stampfli (1949) showed that this did occur. It would therefore appear that propagation of excitation requires a length constant that is in excess of some critical value in each type of irritable tissue. Typical values for length constants are 2 mm in kid Purkinje fibers (Weidmann, 1952) 2.5 mm in 100-μ frog skeletal muscle fibers (Ruch and Patton, 1965), 6.5 mm in squid giant axon (Cole, 1968), and 2.0 mm in frog sciatic nerve (Hodgkin, 1937).

Cable theory has been applied quite extensively to predict some of the properties of irritable tissues using measured values for R_e, R_m, R_a, and C_m; the methods employed are well described by Taylor (1963) and Cole (1968). In these equations an important membrane descriptor appears; namely, the membrane time constant T_m, which is defined as the product of membrane

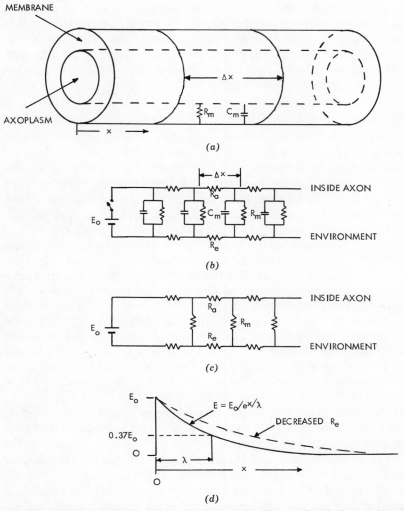

Fig. 5-11. The cable analog of irritable tissue: (*a*) uniform cylindrical axon; (*b*) simplified cable analog; (*c*) simplified equivalent for a steady potential; (*d*) potential decrement along cable.

resistance R_m and capacitance C_m. A value often noted for membrane resistance (expressed as the resistance for 1 cm² of membrane area) is 1000 Ω-cm²; the corresponding value for membrane capacitance is 1 μF/cm², with a range from 0.1 to 10. These values provide a range of 0.1 to 10 msec for the membrane time constant.

A discussion of propagation that focuses only on the length constant is informative but has limited value because it ignores the dynamic aspects of propagation. For excitation to be propagated, adjacent membrane must be hypopolarized by the sudden depolarization of the region stimulated. This means that there must occur a sufficient reduction in potential to which the membrane capacitance C_m is charged. Only with a sufficient reduction (i.e., generally between one-third and one-half) in this voltage will the dynamic membrane permeability changes occur and progress rapidly to depolarization; such depolarization can be likened to a voltage source E_0 applied to the cable analog (Fig. 5-11b). In order for the potential of C_m to be altered, current must flow through the axoplasm R_a and the environment R_e. The time required to reduce the potential of C_m depends on the resistance of the environment plus that of the axoplasm and the membrane capacitance. A reduction in this time will increase the propagation velocity. Gasser and Erlanger's discovery (1927) that propagation velocity in nerve increases with fiber diameter can be explained by these facts.

Hodgkin (1964) pointed out that the resistance of the axoplasm R_a varies inversely with the diameter squared, the resistance of the environment R_e is low with respect to R_a, and the capacitance of the membrane C_m varies directly as the diameter; therefore, the time required to change the charge on the membrane capacitance will decrease as fiber diameter is increased. Consequently, the propagation velocity would be expected to increase with fiber diameter. Although prediction of the actual relation between propagation velocity and fiber diameter is more complex than is indicated by this simple line of reasoning, the fact remains that propagation velocity is almost linearly proportional to fiber diameter. Rushton (1951) and Hodgkin (1954) have investigated the subject theoretically, demonstrating that a linear relationship it expected for nerve fibers larger than about 4 μ; for smaller fibers, the relationship for propagation velocity would be expected to vary as the square root of fiber diameter.

The rate of propagation of excitation is a quantity that is characteristic of each type of irritable tissue; perhaps the most extensively investigated has been nerve. As stated earlier, Gasser and Erlanger (1927) showed that in a variety of nerve fibers, propagation velocity V (in m/sec) is proportional to fiber diameter D (μ) and to temperature. In an investigation relating to the type of proportionality that exists, Hursh (1939) showed with data obtained from the cat and kitten that the velocity was equal to six times the fiber diameter (Fig. 5-12a). Tasaki (1953) summarized his data obtained from individual bullfrog sciatic nerve (Fig. 5-12b), according to which the velocity was equal to 2.05D. It should be noted that the relationship reported by Hursh (i.e., $V = 6.0D$), was obtained on warm-blooded animals (37°C) and that reported by Tasaki (i.e., $V = 2.05D$), was obtained on cold-blooded animals (24°C). Table 5-8 presents a few representative values for propagation velocity.

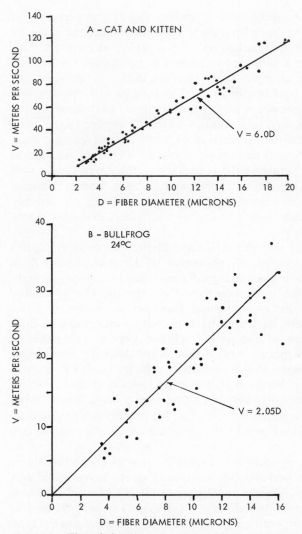

Fig. 5-12. The relation between propagation velocity V and fiber diameter D: curve A redrawn from J. B. Hursh, *J. Physiol.* 1939, **127**:131–139 (by permission); curve B redrawn from I. Tasaki, *Nervous Transmission,* Springfield, Ill., 1953: Courtesy of publisher Charles C. Thomas, (by permission).

Table 5-8 Propagation Velocities

Tissue	Diameter D (μ)	Temperature (°C)	Velocity (m/Sec)	Investigators and Year
Nerve				
Squid axon	500	20	25	Katz (1948)
Crab axon	30	20	4.5	Katz (1948)
Frog axon	D	24	2.05D	Tasaki (1953)
Cat axon	D	Body	6.0D	Hursh (1939)
Muscle (skeletal)				
Frog sartorius	60	20	1.6	Katz (1948)
Muscle (cardiac)				
Dog atrium	—	Body	0.8	Hoffman and Cranefield (1960)
Rabbit atrium	—	Body	0.4–0.6	Hoffman and Cranefield (1960)
Rabbit A-V node	—	Body	0.02–0.05	Hoffman and Cranefield (1960)
Rabbit His bundle	—	Body	0.8–1.0	Hoffman and Cranefield (1960)
Ox Purkinje fiber	—	13	1.2	Hoffman and Cranefield (1960)
Ox Purkinje fiber	—	37	4.2	Hoffman and Cranefield (1960)
Dog Purkinje fiber	—	Body	2.0	Hoffman and Cranefield (1960)
Muscle (smooth)				
Chick amnion	—	38–40	0.029	Prosser and Raferty (1956)

THE ACTION POTENTIAL

As shown by Hodgkin (1939); (Fig. 5-9), hypopolarization of the membrane beyond a critical level sets off an explosive-like series of regenerative changes in membrane permeability which give rises to a propagated action potential. It had long been known (Bernstein, 1912) that the permeability change during activity was accompanied by an outward migration of potassium ions. Ion movement is current flow and the measurement by Cole and Curtis (1939) of a fortyfold decrease in membrane impedance, which occurred virtually coincident with the action potential (Fig. 5-13), offered conclusive proof of the dynamic nature of the permeability change. Almost simultaneously, Hodgkin and Huxley (1939) and Curtis and Cole (1940, 1942) discovered that the peak amplitude of the action potential exceeded the resting membrane potential. This event forced consideration of two phenonena: a process consisting of more than simple depolarization (which would only provide an action potential equal to the potassium-controlled resting membrane potential) and the entry of sodium (or other cations) as a source of the observed potential. Although the following explanation of the genesis of the action potential was derived from the squid giant axon, it is believed that a very similar sequence of events occurs in other irritable cells.

Curtis and Cole (1942) noted that the overshoot in membrane potential varied considerably with little or no change in resting membrane potential;

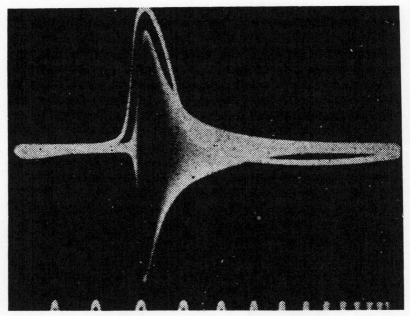

Fig. 5-13. The relation between the action potential and decrease in membrane impedance shown by the imbalance voltage of a 20-KHz impedance bridge; time marks 1 msec. (From K. S. Cole, and H. J. Curtis, *J. Gen. Physiol.* 1938–1939, **22**:649–670. By permission.)

this led then to postulate a separate mechanism for the production of each. Confirmation came from a study by Hodgkin and Katz (1949), who showed that the height of the action potential depends on the external sodium concentration. This in turn indicates that, during activity, the membrane becomes permeable to sodium; a change in the environmental concentration of sodium was without significant effect on the resting membrane potential. The data of Hodgkin and Katz (Fig. 5-14) employ the spike height in normal seawater as the reference.

A number of different investigators have validated this observation for other irritable cells (see reviews by Hodgkin, 1964; Katz, 1966). Thus the two events of major importance during the action potential are permeability changes and ion movements; sodium enters and potassium leaves the cell. Ion-flux studies carried out with radioactive tracers by Keynes (1948, 1950, 1951) and Keynes and Lewis (1951) showed that for one impulse, a squid giant axon takes up 3 to 4 picomoles (pM) of sodium ion per square centimeter of membrane surface and loses the same amount of potassium. Hodgkin (1964) stated that the entry of this amount of sodium was equivalent to an inward movement of

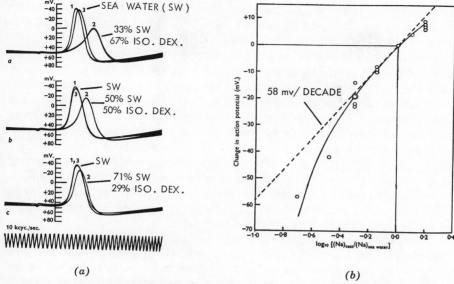

Fig. 5-14. The relation between the amplitude of the action potential and the environmental concentration of sodium: (*a*) action potentials with different environmental concentrations of sodium—SW = seawater, ISO. DEX. = isotonic dextrose; (*b*) change in amplitude of action potential produced by changing external sodium concentration. (From A. L. Hodgkin, and B. Katz, *J. Physiol.* 1949, **108**:37–77. By permission.)

20,000 ions per square micron of membrane and was more than enough to account for the action potential. He also reported that a squid giant axon had the capability of conducting 100,000 impulses in the presence of metabolic blockade of the processes that establish the ionic gradients. This of course means that the axon is capable of operating for some time on its stored ions. It is important to note, however, that these interesting and fundamental studies merely documented the translocation of ions; they provided no data on the temporal realtion between the entry of sodium and the exit of potassium; a solution to this problem came from the voltage-clamp studies.

Voltage-Clamp Studies

Identification of the temporal translocation of sodium and potassium ions that occurs during an action potential lasting only a few milliseconds, was truly a remarkable feat accomplished by Hodgkin and his colleagues; however, the basic information could only come from application of the voltage-clamp method described by Cole (1949) and Hodgkin et al. (1949). The principle of the method is similar to that used by Ussing (1949), who measured the steady current necessary to reduce the potential across frog skin to zero and thereby prevent the active transport of sodium ions. Whereas Ussing's method employed steady current, the voltage-clamp technique involves a sudden shift and maintenance of transmembrane potential at a chosen value (i.e., clamping) and measuring the time course of current flow, which reflects the time course of ion movements. Then by varying the ionic composition of the extracellular fluid, Hodgkin and Huxley (1951, 1952, 1953) separated the ionic current flowing through the membrane into the components carried by sodium and potassium and thereby calculated the temporal variation in membrane permeability to these ions.

The principle underlying the voltage clamp is easily described, for it merely involves the application of a feedback amplifier. Accurate practical application, however, requires special care and leads to considerable electronic complexity; a good account of the precautions to be taken was presented by Moore and Cole (1963). The simplest form of the voltage-clamp circuit appears in Fig. 5-15a, in which the output V_o of a differential amplifier (of gain A) delivers current I to an intracellular electrode (1) through a resistor R. The magnitude of R includes all the resistance between the output terminal of the amplifier and the intracellular electrode. The extracellular electrode (2) is connected to the potential E, to which the membrane is to be clamped. The membrane in turn is connected to the noninverting input (+) terminal of the differential amplifier. The membrane potential V_m, which constitutes the feedback information, is obtained from the intracellular electrode (l) and is fed into the inverting (−) terminal of the differential amplifier. The behavior of the voltage-clamp circuit can be understood by referring to Fig. 5-15a, in which the impedances of the electrodes and electrolytes have been omitted.

The output V_o of the differential amplifier is equal to the gain a multiplied by the input, which is the difference between the clamping voltage E and the membrane potential V_m; therefore, $V_o = A(E - V_m)$. The output voltage V_o is also equal to $IR + V_m$; equating these two and manipulating the equation leads to

$$A(E - V_m) = IR + V_m$$
$$E \frac{A}{1 + A} = I \frac{R}{1 + A} + V_m$$

This expression describes the voltages and currents in the circuit in Fig. 5-15b and indicates that the membrane potential V_m will be clamped to E if the A gain of the differential amplifier is large with respect to unity [i.e., $A/(1 + A)$ is close to 1.0]. In addition, the clamping voltage E is almost directly connected to the membrane because the value of R is diminished by $1 + A$. (It is important to note that practical application of the voltage-clamp principle requires recognition of several impedances and the use of a separate pair of electrodes to monitor the transmembrane potential V_m during clamping.)

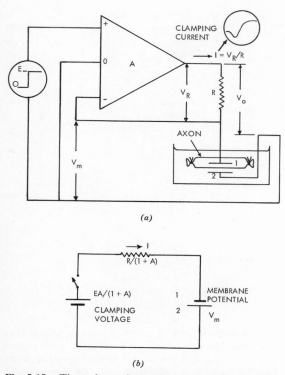

(a)

(b)

Fig. 5-15. The voltage-clamp circuit: *(a)* voltage clamp principle; *(b)* equivalent circuit.

Figure 5-16, derived from the Croonian Lecture presented by Hodgkin (1958), illustrates the type of current flow obtained when the membrane potential of the squid giant axon was suddenly raised from -50 to $+15$ mV and the environmental fluid (seawater) contained a normal amount (460 mM) of sodium ion and when the sodium was replaced by the inert cation choline (i.e., the environment was sodium free). Immediately on application of the (de-polarizing) clamping voltage (Fig. 5-16*a*), there occurred a small surge in current lasting about 20 μ sec (Fig. 5-16*b*); this reflects changing the charge on the membrane capacitance. The magnitude of this initial physical transient is proportional to the clamping voltage; because of its short duration, it is not easily identified in most voltage-clamp records. The direction and magnitude of the subsequent ionic current that flows during maintenance of a fixed membrane potential depends on the magnitude of the clamping voltage and the concentration of the ions in the external environment. By comparing Fig. 5-16*b* and 5-16*c*, it can be seen that the initial inward current transient requires the presence of external sodium. It was later shown that the initial rise in cur-

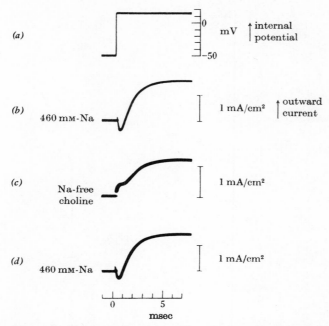

(a)

0 mV ↑ internal | potential

-50

(b) 460 mᴍ-Na

1 mA/cm² ↑ outward | current

(c) Na-free choline

1 mA/cm²

(d) 460 mᴍ-Na

1 mA/cm²

0   5
msec

Fig. 5-16. Membrane currents associated with depolarization of 65 mV in presence and absence of external sodium ions: (*a*) change in membrane potential is shown in A; (*b*), (*c*), (*d*) present the membrane current density; temperature, 11°C, outward current and internal potential shown upward. (From A. L. Hodgkin, *Proc. Roy. Soc.* (*London*). 1958, **B148**:1–37. By permission.)

rent in the sodium-free environment (Fig. 5-16*c*), is due to an outward movement of sodium ions from the axoplasm. Note also that the contour of the later rise in current is relatively unaffected by changes in the environmental sodium concentration. When the concentration of sodium was kept at a normal value (460 mM) and clamping was carried out using different voltages, a family of current-time curves, such as those in Fig. 5-17, was obtained. It is interesting to note that when the membrane was clamped to 117 mV, the current remained constant for about 300 μ sec, indicating the existence of an equilibrium potential at which one component of the membrane current is zero. To explain this Hodgkin, used the electrical model (Fig. 5-18), which is now believed to adequately represent the components of nerve membrane that have been derived both from the measured cable properties and studies featuring manipulation of ionic concentrations to identify the components underlying the membrane potential.

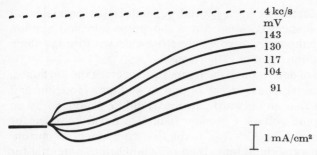

Fig. 5-17. Membrane currents associated with incremental depolarizations; squid axon in sea water (460 mM Na) at 3.5°C, outward current upward. The displacement of the membrane potential from its resting value is shown at the right. (From A. L. Hodgkin, *Proc. Roy. Soc.* (*London*). 1958, **B148**:1–37. By permission.)

In the electrical model of the nerve membrane (Fig. 5-18), R_a and R_e are the resistances of the axoplasm and the environmental fluid; because R_e is so much lower than R_a, the former is often represented as a conductor without resistance. The membrane capacitance is C_m, and at rest, the membrane resistance is high enough to be ignored. The three voltage sources E_K, E_{Na}, and E_X reflect the potentials due to the ionic gradients; E_K is the internal potassium-controlled membrane potential, E_{Na} is the external sodium-dependent potential, which determines the amount of overshoot in the action potential, and E_X is a small potential included to account for the contribution of other ions that change little during activity. The resistance R_K, R_{Na}, R_X (or better,

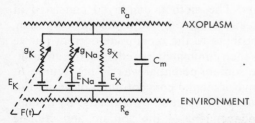

Fig. 5-18. Electrical model of the nerve membrane: E_K, E_{Na}, and E_X are the ion-dependent *potentials,* g_K, g_{Na}, and g_X are the membrane conductances (which are time dependent), C_m is the membrane capacitance, and R_a and R_e are the equivalent resistances of the axoplasm and environment. (Redrawn from A. L. Hodgkin, and A. F. Huxley, *J. Physiol.* 1952, **117**:500–544.)

their reciprocals, the conductances g_K, g_{Na}, and g_X) reflect the permeability of the membrane to the respective ions. Since the potassium and sodium conductances are related temporally, arrows have been drawn through them and coupled by a time-function symbol $F(t)$.

Returning to the problem of determining the temporal aspects of the ion fluxes, to prove that the initial membrane current is due to sodium, Hodgkin and Huxley (1952) pointed out that an outward current would be produced by an increase in g_{Na} for a clamping voltage E greater than E_{Na}, inward current for E less than E_{Na}, and no current for $E = E_{Na}$. If the first component of current were due to the movement of sodium ions, then the equilibrium potential for which there is no sodium current should be given by the Nernst equation for the sodium gradient ($[Na]_o/[Na]_i$). Hodgkin and Huxley (1952) tested this relationship in a study in which the external sodium concentration $[Na]_o$ was varied and voltage clamping was carried out. They found that the Nernst equation held to within 1 mV over a tenfold range of $[Na]_o$, indicating that the initial current is due to a sudden increase in membrane permeability to sodium.

The later phase of membrane current (Figs. 5-16 and 5-17) was revealed to be independent of environmental sodium concentration and shown by Hodgkin and Huxley (1952) to be due to the efflux of potassium; they later (1953) offered additional evidence on this point by measuring the efflux of radioactive potassium (which had been placed within the axon) with varying amounts of membrane depolarization. The efflux of potassium for a given area of membrane in a given time agreed with the total charge passing simultaneously through the membrane. Additional studies showed that the delayed component of membrane current varied with external potassium concentration, although not exactly as predicted by the Nernst equation.

Having established that the initial membrane current was due to the influx of sodium ions and that the delayed current was due to the efflux of potassium ions, Hodgkin and Huxley (1952) set about separating the total membrane current into its temporal components. The method employed consisted of voltage clamping the squid axon (in seawater) to 56 mV; under this condition, both sodium and potassium ions contribute to the membrane current (curve A, Fig. 5-19). The sodium current was then eliminated by reducing the environmental sodium concentration until the sodium potential was 56 mV (curve B, Fig. 5-19), which occurred with an environmental concentration of 46 mM of sodium; inert choline was used to balance the solution osmotically. By assuming that the potassium current is independent of the sodium and choline concentrations, it was possible to obtain the sodium current by subtracting the potassium current from the total current; the result appears as curve C in Fig. 5-19.

From these temporal variations in sodium and potassium currents, Hodgkin

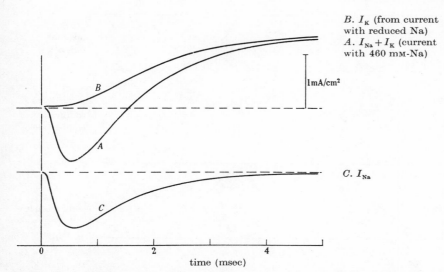

B. I_K (from current with reduced Na)

A. $I_{Na} + I_K$ (current with 460 mM-Na)

1mA/cm²

C. I_{Na}

time (msec)

Fig. 5-19. Separation of the total membrane current (curve *A*) into the components carried by potassium (curve *B*) and sodium (curve *C*). (From A. L. Hodgkin, *Proc. Roy. Soc. (London)*. 1958, **B148**:1–37. By permission.)

and Huxley (1952) calculated the temporal changes in permeability of the membrane to these ions. Permeability was expressed as conductance and the data they obtained are presented in Fig. 5-20. These interesting curves show that, during maintained depolarization, sodium permeability rises quickly, then decays, and, at about the peak of the sodium permeability, the permeability to potassium starts to increase and is sustained as long as depolarization is maintained.

The sequence of events just described relates to the ion fluxes during voltage clamping of the squid axon. Hodgkin and Huxley (1952) then used these fundamental data to calculate the sodium and potassium conductances that occur during the action potential. By making certain assumptions and using probability theory (see Hodgkin, 1958), they developed a series of empirical equations for predicting (*a*) the form, amplitude, and velocity of propagation of the action potential; (*b*) the amplitude and time course of the conductance changes; and (*c*) the quantities of sodium and potassium ions that were transported across the membrane during the action potential. Figure 5-21 presents the results of their calculations for the dynamic events that accompany propagation of the action potential in the squid giant axon, which is believed to behave like many other irritable tissues.

Hodgkin (1964) reported that, for a single propagated impulse in the squid axon, the translocation of sodium and potassium ions (3–4 pM/cm² of mem-

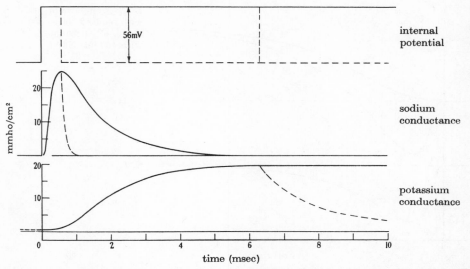

Fig. 5-20. Changes in sodium and potassium permeability (expressed as conductance) associated with a membrane depolarization of 56 mV. The continuous curves show the changes in conductance when the depolarization was maintained; the broken curves show the effect of repolarizing the membrane after 0.6 or 6.3 msec. (From A. L. Hodgkin, *Proc. Roy. Soc.* (*London*) 1958, **B148**:1–37. By permission.)

brane surface area) represents more than enough charge to alter the charge on the membrane capacitance and to account for the amplitude of the action potential. The loss in potassium is small and amounts to about one-millionth of the intracellular potassium. Similarly, the influx of sodium is small, and for this reason impulses can be propagated for a long period. However, for the axon to continue to function, the sodium must be expelled and potassium must re-enter. In studies in which radioactive sodium is placed intracellularly, sodium is expelled by active transport (sodium pump); potassium re-enters because the resting membrane is permeable to it, but the detailed nature of the processes by which cells accumulate potassium and expel sodium are unknown. There is, however, some evidence that the sodium pump and the potassium transport mechanisms are linked (Hodgkin 1964)].

Perhaps there are no better words to summarize the dynamic events that underlie the action potential than those of Hodgkin (1958):

At rest, although the potassium conductance is small, it is much greater than the sodium conductance so the membrane potential is fairly close to the equilibrium potential of the potassium ion. As the impulse advances, the membrane just ahead of the active region becomes depolarized by electric currents flowing in a local circuit through the axoplasm and external fluid. Under the influence of the change in membrane

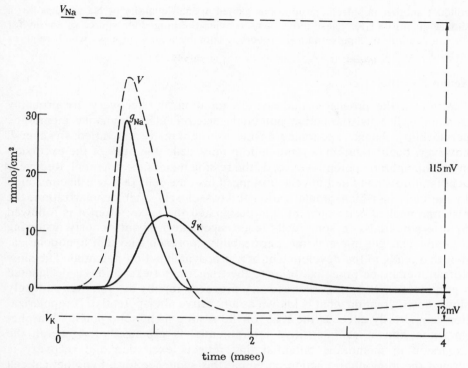

Fig. 5-21. Dynamic changes in membrane conductance (g_{Na}, g_K) occurring during the action potential V in the squid giant axon; V_K is the potassium-dependent membrane potential and V_{Na} is the sodium potential. (From A. L. Hodgkin, *Proc. Roy. Soc.* (*London*). 1958, **B148**:1–37. By permission.)

potential the sodium conductance rises and sodium ions enter the fiber; this inward current makes the inside of the fibre positive (by carrying charge through the dielectric) and provides the current required to depolarize the resting membrane ahead of the active region. At the crest of the spike, the slow changes which result from depolarization begin to take effect. The sodium conductance declines and the potassium conductance rises so that the rate at which potassium ions leave the fibre exceeds the rate at which sodium ions enter; this makes the potential swing towards the equilibrium potential of the potassium ion. As the potential approaches the resting level, any sodium conductance which has not been inactivated is cut off and this may accelerate the rate of repolarization. The slow effects of depolarization, namely raised potassium conductance and inactivation of the sodium carrying system, persist for a few milliseconds after the spike and give rise to the refractory period. Under some, but not all, conditions the potential and conductance may return to their resting value in an oscillatory manner. About 10 msec after a spike the membrane is back in its original state and can conduct another impulse. The fibre has gained a small quantity of Na and has lost a similar quantity of K. These movements provide the immediate source of energy for the conduction of impulse and are reversed later by a slow process which requires metabolic energy.

Action Potentials

Although the processes underlying excitation and recovery are probably similar in all cells, the action potentials that reflect the dynamic membrane permeability changes and ionic translocations are many and varied. In general, however, depolarization is rapid and, in most cells, the peak of the excursion in transmembrane potential exceeds the resting membrane potential. Recovery occurs more slowly and is very prolonged in some cells (e.g., cardiac muscle). In the cells in which repolarization follows closely after depolarization, the resulting peaked action potential is designated the spike; often it is followed by afterpotentials. In some cells (e.g., pacemaker), spontaneously occurring rhythmic changes in membrane permeability approach a level of hypopolarization that leads to the development of a propagated action potential. The slow drift in membrane potential, prior to the triggering of excitation, is designated a prepotential. In some cells recovery proceeds rapidly at first, then suddenly slows down creating what is known as a negative after-potential. If repolarization temporarily exceeds the level of the resting membrane potential, a phenomenon known as a positive afterpotential is produced. In general, the excursion in membrane potential that reflects excitation and recovery is termed the monophasic action potential; this sequence for a hypothetical cell having all these potentials is represented in Fig. 5-22.

The action potential that started serious investigation of the processes underlying excitation and recovery is that of the giant axon of the squid *Loligo*. The unusually large size of certain axons in the squid (1 mm diam. and almost

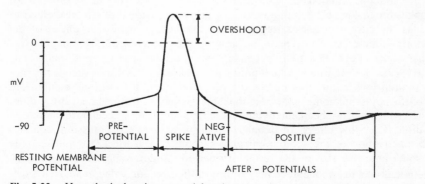

Fig. 5-22. Hypothetical action potential and commonly used nomenclature.

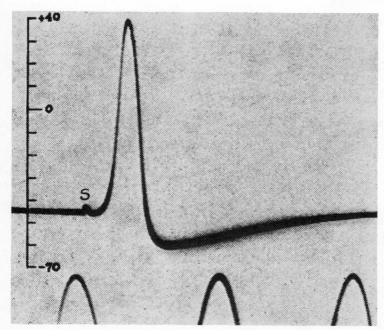

Fig. 5-23. Action potential of the giant axon of the squid *Loligo*; ordinate in millivolts, time marks 2 msec apart, S = stimulus. (From A. L. Hodgkin and A. F. Huxley, *Nature.* 1939, **144**:711. By permission.)

as long as the body, which may be 1– to 2 ft) was noted by Young (1936), who called it to the attention of physiologists. (Incidentally the giant axons innervate the muscles that forcibly expel a jet of water from the mantle cavity and allow the squid to swim backward.)

Very soon after the properties these giant axons were known, they were used to determine the true temporal excursion in membrane potential during excitation and recovery. Recorded in 1939 by Hodgkin and Huxley (Fig. 5-23), the transmembrane action potential proved that the spike height (85 mV) exceeded the resting membrane potential (−45 mV) and immediately brought forward the notion that excitation consisted of more than simple depolarization. During the recovery phase, in the original record for which the axon was in a water environment, a positive after-potential followed the spike. In subsequent studies Hodgkin (1958) used much smaller micropipets, obtaining a resting membrane potential of −65 mV and an overshoot of 45 mV. Hodgkin also showed that the action potentials recorded *in vivo* and seawater were virtually the same except for a slightly smaller positive afterpotential in the former case.

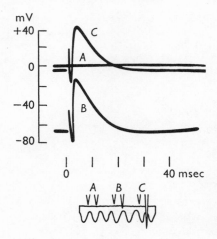

Fig. 5-24. The resting membrane and action potentials of the electric organ of *Raia clavata* and the location of the micropipets (*A, B, C*) used to obtain the recordings; record *B* was obtained with transmembrane electrodes and *C* is an extracellular recording. (From L. G. Brock et al. *J. Physiol*. 1953, **122**:4P–5P. By permission.)

Electric fish are capable of suddenly developing hundreds of volts, which sends an appreciable current through the environment and shocks any organism in it. The high voltage is produced by the discharge of a large number of specialized cells called electroplates which are modified myoneural junctions, each capable of producing millivolt action potentials. Using micropipets, Brock et al. (1953) recorded the resting membrane and action potentials of single electroplates of the fish *Raia clavata*. Their recordings (Fig. 5-24), obtained by the application of single stimuli, revealed a resting membrane potential of −65 mV and a spike height of 50 mV. Interestingly enough, the membrane potential did not diminish to zero during activity. Brock et al. reported that in other electric fish they encountered only small overshoots, amounting to a few millivolts.

The action potential of a single frog skeletal muscle fiber and its relation to the twitch is shown in Fig. 5-25. In this preparation, the resting membrane potential *C* was −92 mV, and at the peak of the spike the potential difference

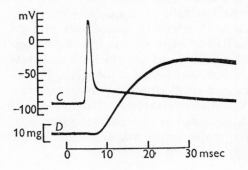

Fig. 5-25. Action potential (*C*) of a single skeletal muscle fiber in the frog, and *D*, the tension developed. (From A. L. Hodgkin and P. Horowicz, *J. Physiol*. 1957, **136**:17P–18P. By permission.)

was +30 mV. During recovery a negative afterpotential is clearly visible. Careful inspection of the twitch record *D* at the peak of the action potential reveals the small precontractile relaxation known as the latency relaxation.

Called to the attention of physiologists only recently by Hoyle and Smyth (1963) are the giant adductor and depressor muscle fibers of the barnacle *Balanus nubilus Darwin*. Each muscle is made up of 25 to 40 fibers about 4 cm long, ranging from 0.5 to 2mm diam. Because of their large size and the ease with which they can be removed, they are becoming popular specimens for investigating the relation between the excitatory and contractile processes. Like other skeletal muscle, the giant barnacle muscle is innervated and contracts; but here the similarity ends. Each muscle fiber is multiply innervated, and excitation produces only local hypopolarization and contraction; propagated action potentials are not ordinarily developed. However, spike formation can be induced by decreasing the concentration of intracellular calcium by using a calcium binding agent (e.g., EDTA); Fig. 5-26 illustrates such a procedure. Increasing the stimulus intensity (Fig. 5-26C1) increases the level of hypopolarization. In preparations in which the internal calcium concentration was reduced, Hagiwara et al. (1964) showed that spikes of increasing height (rising from a resting membrane potential of −60 mV) were obtained by increasing the concentration of calcium in the extracellular fluid (Fig. 5-26d). Hoyle and Smyth had previously demonstrated that the force of the twitch was related to the amount of hypopolarization.

The resting membrane and action potentials of frog and human cardiac muscle are shown in Fig. 5-27a,b; the former shows the manner in which the transmembrane potential is encountered as a micropipet is advanced into a single cell. When the micropipet tip rested on the surface of a frog ventricular cell, the familiar *QRS-T* complex was recorded; when the tip of the micropipet was pushed through the cell membrane, the resting membrane potential (−55 mV) suddenly appeared. As the ventricle continued to beat, the monophasic action potential typical of cardiac muscle was recorded. The outstanding characteristics of the monophasic action potential are a rapid depolarization, an overshoot, and then a prolonged period of depolarization before complete recovery occurs; there is no short-duration spike as seen in skeletal muscle and nerve. Figure 5-27b illustrates that the electrical activity of the human ventricles is essentially the same as in the frog. This remarkable recording was obtained by Woodbury et al. (1957), who applied a sterilized floating micropipet (Fig. 4-9) to a human ventricle that had been exposed for cardiac surgery. Despite the difficulty in holding the electrode by hand and despite movement of the ventricles, the monophasic action potentials are distinct and resemble those obtained from the ventricles of other species.

Although there is considerable similarity among the waveforms recorded from the different parts of the heart, important individual differences exist.

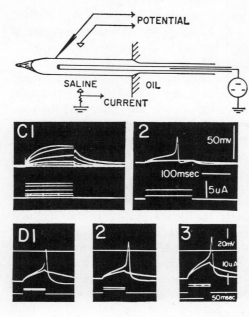

Fig. 5-26. Transmembrane potential and current changes in the giant barnacle muscle in response to square-wave stimuli. The graded response to an increase in stimulus intensity is shown in *C*1; local spike formations produced by first decreasing the intracellular concentration of calcium and then varying the extracellular calcium concentration (20, 84, 338 mM), appear in *D*1, 2, 3. (From S. Hagiwara et al. *Science.* 1964, **143**:1446–1448. By permission.) Copyright 1964 by the American Association for the Advancement of Science.

For example, the prepotential exhibited by the pacemaker (sinoatrial node) decays to a level that results in a spontaneous depolarization, followed by recovery. Atrial muscle, the atrioventricular node, the bundle of His, the Purkinje fibers, and the ventricular myocardium all have their own special types of action potential. An excellent review of these has been presented by Hoffman and Cranefield (1960).

One of the tissues of considerable interest to those concerned with the mechanisms underlying excitation and contraction in smooth muscle is the *Tenia coli* of the guinea pig. Specimens in the form of sheets several centimeters in length can be easily stripped from the intestine. Such preparations are easily mounted in irrigating solutions, and their thinness permits rapid transport of material across the individual cell membranes. Figure 5-28 illustrates the resting membrane and action potential of *Tenia coli*. During activity the membrane potential is seen to change from −46 to +10 mV. The recovery phase exhibits both positive and negative after-potentials.

Membrane and action potentials are associated with plant as well as animal cells. In fact, the large size of many plant cells makes them ideal specimens for the study of membrane phenomena. According to Osterhout (1931), specimens of such fresh-water algae as *Nitella* and *Chara* can be obtained in lengths up to 6 in.; the marine algae *Valonia* and *Halicystis* are single cells up to 2 in. long, consisting of a delicate layer of protoplasm not more than 10 μ thick

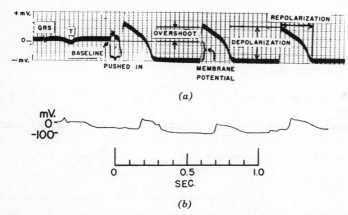

(a)

(b)

Fig. 5-27. The action potential of cardiac muscle: (*a*) the resting membrane and action potential of frog ventricular muscle revealed by pushing in the tip of a micropipet. [(*a*) From H. Hecht, *Ann. N.Y. Acad. Sci.* 1956–1957, **65**:7 (By permission © The New York Academy of Sciences.); (*b*) the resting membrane and action potential of human ventricular muscle recorded with a hand-held floating micropipet. (*b*) From J. W. Woodbury et al. *Circ. Res.* 1957, **5**:179 (By permission).]

covering a central vacuole filled with sap. Outside the protoplasm is a cellulose membrane which is so permeable that its presence can usually be ignored; its mechanical properties permit insertion of glass micropipets.

Resting membrane and action potentials of various algal cells have been reported by Osterhout and Hill (1940), Walker (1955), Findlay (1959), Radenovitch (1968), and Findlay et al. (1969); in Findlay's study, the dependence of the resting membrane potential on ionic gradients was investigated in some detail.

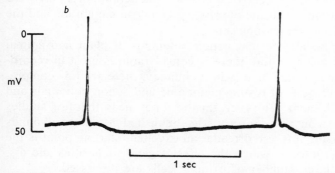

Fig. 5-28. Resting membrane and action potentials of single, smooth muscle fiber of guinea pig gut. (From M. Holman, *J. Physiol.* 1958, **141**:466. By permission.)

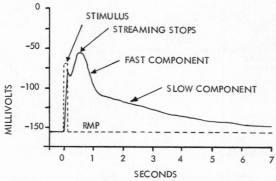

Fig. 5-29. The resting membrane potential (RMP) and action potential of *Nitella*. (Redrawn from G. P. Findlay, *Aust. J. Biol. Sci.* 1959, **12:**412–426.)

The action potentials of a typical algal cell starts from a resting membrane potential that is usually slightly larger than that found in most animal cells. Depolarization and repolarization occupy seconds, rather than milliseconds as in animal cells. At the peak of depolarization, the membrane potential seldom reverses and usually does not reach zero. For example, in *Nitella* immersed in White's culture solution, Findlay (1969) reported a resting membrane potential of about −155 mV (Fig. 5-29). When stimulated to produce an action potential at the peak of depolarization (when protoplasmic streaming was seen to cease), the transmembrane potential amounted to about −50 mV. Recovery occurred in two phases; one was rapid, lasting about 0.5 sec; the other occupied a much longer time. The total duration of a typical action potential in *Nitella* is about 15 sec. Radenovitch et al. (1968) recorded similar action potentials, which were made to occur rhythmically by the application of a drop of 0.1 *M* potassium chloride to one end of the cell. In this study the action potentials were measured with one micropipet in the enveloping cellulose membrane and the membrane and the other in the cytoplasm.

Because the resting membrane and action potentials of plant and animal cells are expressive of cell function, there is considerable interest in recording them. Probably because this field is in its infancy, there are few comprehensive reviews or tabulations of resting membrane and action potentials and the ions associated with them. However, this does not mean that few studies have been carried out; on the contrary, the published literature abounds with such reports. For the interested reader an excellent starting point is the review by Grundfest (1966), in which the physiological similarities and dissimilarities between a large number of irritable cells are discussed.

6

The Measurement of Action Potentials with Extracellular Electrodes

In the previous chapter it was shown that the fundamental bioelectric event consists of a propagated wave of depolarization, reverse polarization, and repolarization. This phenomenon manifests itself across the membrane of an excitable cell, and the resulting potential excursion is easily recorded with transmembrane electrodes. However, a signal that reflects the same event can be recorded with appropriately placed extracellular electrodes. Because such electrodes are extensively used in the practical recording of bioelectric events, this chapter analyzes the waveforms to be expected with different types of electrodes located on or near various types of irritable cells.

Bipolar Recording

The simplest situation of measurement with extracellular electrodes is idealized in Fig. 6-1, which shows two electrodes (A,B) on the surface of a long strip of isolated irritable tissue. It is assumed that the electrodes are small and widely separated in relation to the amount of tissue occupied by excitation. In addition, it will be assumed that the fundamental bioelectric event consists of depolarization, reverse polarization, and repolarization, giving a simple monophasic action potential without pre- or after-potentials. For simplicity in explanation, charges appear only on the upper surface of the tissue where the electrodes are located. It must be realized, however, that the same charge distribution exists over the whole surface of the irritable tissue. When the tissue is inactive, (Fig. 6-1-1), both electrodes are in regions of equal positivity and the potential difference seen by the potential indicator is zero. Assume now that the tissue has been excited to the left of electrode A; when the wave of excitation reaches the region under electrode A, it becomes negative with respect to electrode B and the indicator rises (Fig. 6-1-2). As the

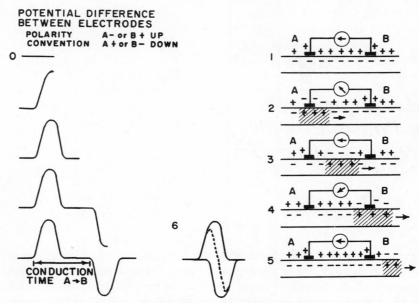

Fig. 6-1.   The measurement of action potentials with electrodes placed on the surface of isolated irritable tissue.

wave of excitation passes onward toward electrode B and occupies the region between the two electrodes, the region under A is recovered and that under B has not yet become excited. The potential indicator registers no voltage under these conditions (Fig. 6-1-3), and the first (upward) phase of the monophasic action potential is thus complete. As the wave of excitation occupies the region under electrode B, it becomes negative with respect to A and hence the potential indicator falls (Fig. 6-1-4). Recovery occurs as the wave of excitation passes B, the membrane potential is re-established, the potential indicator reads zero, and the downward phase of the action potential is complete. In this measurement situation, the time between onset of the action potentials is determined by the velocity of propagation in the tissue and the spacing of the electrodes. If the interelectrode distance is reduced, the two monophasic action potentials will be closer to each other. If the temporal relations are such that excitation occurs under electrode B before recovery is complete under A, a smaller action potential will result (Fig. 6-1-6).

The situation just described applies equally to an isolated single strip or bundle of irritable tissues having the same propagation velocity. If the tissue consists of a bundle of fibers having different velocities of propagation, then the waves of excitation will arrive under each electrode at different times; hence the waveform displayed by the recording instrument will be complex. It must also be recognized that the activity of the tissues closest to the recording electrodes will contribute the most to the recorded potential.

Experimentally it is possible to provide verification for the preceding explanation for the waveform of potential recorded with two electrodes on the surface of an isolated strip of irritable tissue. The frog sartorius muscle, which consists of a bundle of very similar muscle fibers running parallel for the whole length of the muscle, constitutes one of the best preparations for such an experiment. Therefore (in the curarized muscle) the application of a stimulus to one end of the muscle will cause a wave of excitation to travel along each fiber at the same rate, reaching the end of the muscle at the same time. By recording the response with two widely separated electrodes, the diphasic action potential in Fig. 6-1-5 can be obtained; a typical result appears in record 1-5 of Fig. 6-2. When the electrode spacing is reduced so that the monophasic action potentials overlap (i.e., excitation of the distal electrode occurs before recovery at the proximal electrode), the action potential (shown in record 1-2 of Fig. 6-2 is that predicted by the preceding analysis and shown by the dashed curve in Fig. 6-1-6.

The diphasic action potential recorded in the manner just described permits determination of the direction of the spread of excitation. Even with closely spaced electrodes, the direction of the initial deflection of the potential indicator provides this information if its deflection is known in terms of the

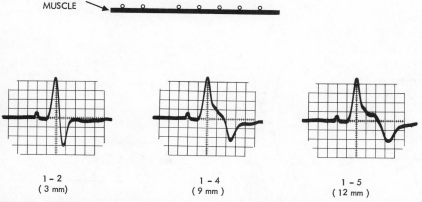

Fig. 6-2. The effect of electrode spacing on the action potential of isolated frog sartorius muscle at room temperature: S = stimulus 0.5 msec; propagation velocity (calculated from record 1-5) = $12 \times 10^{-3}/6 \times 10^{-3} = 2$ m/sec; time scale-2 msec. (Courtesy of Dr. L. E. Geddes)

polarity applied to its terminals. For example, in Fig. 6-1-1, the polarity convention chosen was such that when electrode *A* was negative to electrode *B*, the indicator of the potential-measuring instrument rose; therefore, when excitation traveled from *A* to *B*, the first phase of the action potential was upward. If the tissue were excited at its opposite end (i.e., beyond *B*), electrode *B* would become negative first and the initial deflection of the potential indicator would be downward. These two situations are diagrammed in Fig. 6-3.

Thus far, discussion of the meaning of the polarity of the potential difference between the electrodes has been devoted to the case of the spread of excitation being in the same direction as a line joining the electrodes. The importance of orientation of the electrodes with respect to the direction of excitation and recovery can be demonstrated by placing the electrodes opposite each other on the tissue and causing a wave of excitation to be propagated (Fig. 6-4). Regardless of the direction of propagation of excitation, if everything is symmetrical, depolarization and repolarization will occur simultaneously under each electrode, and the potential indicator will not be deflected as excitation and recovery pass.

In some tissue, notably cardiac muscle, excitation occupies all the tissue before recovery occurs under either electrode. In many instances recovery does not travel in the same direction as excitation. Therefore, the action potentials recorded from a pair of electrodes on the surface of such tissue are expected to be different from those previously discussed.

Figure 6-5 diagrams a strip of isolated irritable tissue in which excitation occupies all of the tissue before recovery occurs under either electrode.

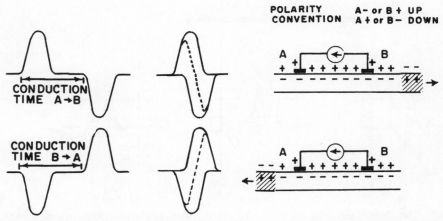

Fig. 6-3.   The relation between direction of propagation of excitation and the recorded action potential with electrodes on an isolated strip of irritable tissue.

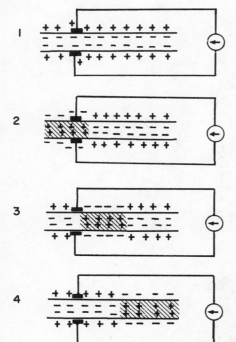

Fig. 6-4. Excitation and recovery propagated at right angles to the axis of a pair of electrodes on an isolated strip of irritable tissue.

Assume that the tissue has been stimulated to the left of electrode *A* (Fig. 6-5-1) and that excitation advances and occupies the region under electrode *A* (Fig. 6-5-2), making this electrode negative with respect to electrode *B*; with the polarity convention adopted, the potential indicator rises. Excitation continues to advance and ultimately occupies the region under electrode *B* (Fig. 6-5-3). Because in this type of tissue recovery will not have occurred under electrode *A* and because both electrodes are now over active tissue, the indicator shows no potential difference, and the first upward phase of the action potential will be completed as shown in Fig. 6-5-3. If the strip of irritable tissue is uniform, recovery will follow in the same direction as excitation, occurring first under electrode *A*. Under this condition, electrode *B* is negative with respect to *A* and the potential indicator falls (Fig. 6-5-4). When recovery occurs under electrode *B*, the potential indicator reads zero and the second (downward) phase of the action potential is completed as shown in Fig. 6-5-5.

In the sequence of events just described, the two monophasic action potentials have a special meaning. The peak of the first upward monophasic action potential indicates excitation under electrode *A*; the end of this action potential indicates that the whole tissue is active. The beginning of the downward wave

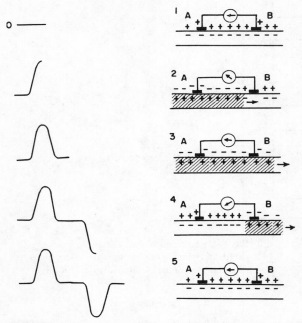

POTENTIAL DIFFERENCE
BETWEEN ELECTRODES
POLARITY   A − or B + UP
CONVENTION   A + or B − DOWN

Fig. 6-5. Action potentials recorded from the surface of isolated tissue which becomes depolarized under both electrodes and with recovery following in the same direction as excitation.

indicates recovery is starting under electrode A and recovery under this electrode becomes complete when the peak of the downward action potential is reached. Completion of the downward action potential indicates full recovery of the tissue.

If the strip of irritable tissue is not uniform, or if there exists a metabolic gradient, the sequence of events just described will not be the same. If, when all of the tissue is active, (as in Fig. 6-6-3), recovery proceeds in the direction opposite that of excitation, the second phase of the action potential will be different. In Fig. 6-6-4, recovery occurring first under electrode B results in electrode A being negative with respect to B. Thus the potential indicator will rise and the second phase of the action potential will be upward (i.e., in the same direction as the first). As the tissue recovers under electrode A, the second (upward) phase of the action potential is completed (Fig. 6-6-5).

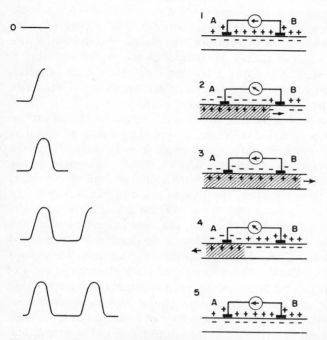

Fig. 6-6. Action potentials recorded from the surface of isolated irritable tissue that becomes depolarized under both electrodes and recovery occurs in the direction opposite to excitation.

In the case just analyzed, the peak of the first upward phase described excitation under electrode *A*. At the end of the first monophasic action potential, when the indicator read zero, the whole tissue was active. The beginning of the second upward phase indicated the start of recovery under electrode *B*; total recovery occurred when the second upward monophasic action potential was completed. This analysis can be generalized by stating that, in tissue which is totally occupied by excitation before recovery occurs anywhere, if the two phases of the action potential are in the opposite direction, excitation and recovery travel in the same direction. If the two phases are in the same direction, excitation and recovery travel in opposite directions. The former situation is often found in the heart of a cold-blooded animal and in homogeneous tissue; the latter is characteristic of the mammalian ventricles.

The Injury and Action Potential

Perhaps one of the most surprising observations associated with the measurement of action potentials with extracellular electrodes, applied directly to irritable tissue, is the appearance of waveforms that resemble, to a remarkable degree, those obtained with transmembrane electrodes. The papers presenting such records usually state that one electrode was placed on uninjured tissue and the other was over injured tissue. Deviation of the waveform from a monophasic action potential is explained by incorrect placement—that is, one electrode was not on totally injured tissue. Such an explanation needs clarification in terms of the membrane-potential concept.

It has been shown in this chapter that if two electrodes are placed on the surface of a uniform strip of irritable tissue, a diphasic action potential is recorded when the tissue responds to a stimulus (Fig. 6-1). The first phase reflects excitation and recovery under the first electrode encountered; the second indicates occurrence of the same event under the second electrode. If the two electrodes are close together, the phases are closer temporally. In the previous chapter it was shown that if one of the surface electrodes is advanced through the membrane into the cell, the membrane potential will appear. If the cell is excited, the monophasic action potential will be recorded rising from, and returning to, the resting membrane potential. These two situations represent two boundary conditions (i.e., both electrodes are extracellular), which give rise to the idealized diphasic action potential; when one electrode is extracellular and the other is intracellular, the idealized monophasic action potential results. Consider now a strip of irritable tissue, injured at one end (i.e., depolarized) by crushing at *B* as in Fig. 6-7-1. The membrane potential is not fully maintained all the way to the site of injury.

Practical studies on this point have been carried out by Graham and Gerard (1946), who used frog sartorius muscle and explored the potential along the

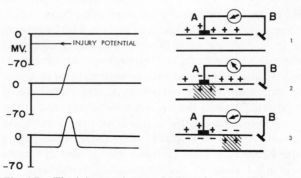

Fig. 6-7. The injury and monophasic action potential.

membrane with transmembrane electrodes up to and within the site of injury. They found that the potential between the exploring electrode (*B* in Fig. 6-8) and the intracellular electrode *A* was equal to the membrane potential to within 5 mm of the site of injury. As electrode *B* was moved toward the cut end, the potential decreased; at 2 mm from the site of injury, the potential was 25% of the membrane potential. In another series of experiments the same investigators placed one electrode on the intact surface of a muscle cell and another in the region of injury, comparing the potential difference so measured with the resting membrane potential. The injury potential was 30 to 39% of the membrane potential.

It is obvious, therefore, that at the site of injury the spatial distribution of membrane potential, whatever it may be, causes current to flow through the fluid environment. Thus in the fluid there will be established a potential field. Consequently, the potential measured between an electrode inside the cell and one at the site of injury will depend on the local conditions at the site of injury and the position of the electrode in the fluid environment. As shown by Graham and Gerard, when this potential—the injury potential—is measured under optimum conditions, it may amount to slightly more than one-third of the membrane potential. The same type of information was developed by Woodbury et al. (1951), who demonstrated (Fig. 4-8) that if the diameter of an intracellular electrode was too large with respect to the size of a cell, the potential measured was considerably less than the membrane potential and approximated 30% of the true membrane potential. From these studies it is clear that a typical injury potential may be about one-third of the membrane potential.

The foregoing situation carries with it an important implication when an action potential is measured with one electrode on the surface of an irritable tissue and the other in an area of injury. In Fig. 6-7-1, for example, suppose that before excitation, the resting membrane potential is −70 mV, that elec-

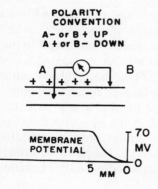

Fig. 6-8. Membrane potential in the vicinity of injury. (Redrawn from data presented by Graham and Gerard, 1946.)

trode A is on the intact surface of the irritable tissue, and that electrode B is in the site of injury. Under this condition the potential difference between the electrodes may be 35% of the membrane potential and amount to about -25 mV. Now if the tissue is stimulated to the left of electrode A (Fig. 6-7-2), when excitation reaches this electrode the potential difference measured between the electrodes will be the algebraic sum of the potentials at the two electrodes. For example, assume that the membrane depolarizes and reverse polarizes to $+20$ mV; the potential difference was -25 mV just before depolarization and $+65$ mV at the peak of reverse polarization. It will then return to -25 mV when the wave of excitation passes the surface electrode. This sequence, (Fig. 6-7-2, 6-7-3) illustrates that a fair representation of the waveform of the transmembrane action potential can be obtained by injuring the tissue under one electrode. Important to note that, although the magnitude of the reverse polarization of the membrane amounted to only 20 mV, in the record it showed up as a much larger potential of $+65$ mV. This situation probably serves to explain the considerable reverse potential observed by Bernstein (1871) when he measured the nerve action potential with the rheotome (see Hoff and Geddes, 1957).

There is another point to consider when the action potential is measured with one electrode on an intact membrane and the other in a region of injury. Before excitation there will be a standing potential difference (the injury potential), whose magnitude will depend primarily on the location of the electrode at the site of injury. If electrode B in Fig. 6-9 is over the injured area, an appreciable percentage of the membrane potential may be detected; if it is moved a short distance from the site of injury and is over excitable tissue, the steady (injury) potential difference between the electrodes will be less. Now if the tissue is excited and excitation and recovery passes under the surface electrode (A in Fig. 6-9), the usual monophasic action potential will occur, superimposed on a baseline of the injury potential (Fig. 6-9-2, 6-9-3). If the strip of irritable tissue is long with respect to the time of propagation of the impulse and the amount of tissue occupied by excitation is small with respect to the interelectrode distance, excitation and recovery will take place under the first electrode before it enters the region of electrode B, which is near the area of injury. Electrode B may also be close to uninjured tissue, and therefore detect not only the injury potential but also an attenuated action potential as it advances toward the area of injury. Thus the resulting action potential measured between the two electrodes will be diphasic, consisting of a large monophasic action potential superimposed on the injury potential, followed by a smaller monophasic action potential in the opposite direction reflecting what electrode B detects from the depolarization and repolarization of normal tissue near the site of injury (Fig. 6-9-4, 6-9-5). If the two electrodes are moved closer together, or if the region of the tissue occupied by excitation

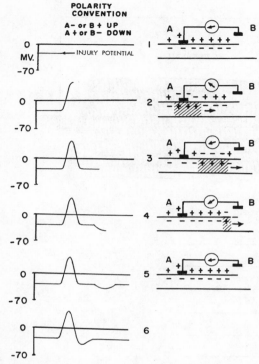

Fig. 6-9. Action potential near a site of injury.

is large with respect to the interelectrode distance, the smaller downward phase of the action potential will be moved closer to the upward phase as in Fig. 6-9-6. This type of waveform is often recorded when a needle electrode inserted into active tissue is paired with another electrode on uninjured tissue.

Measurement from Multiple Irritable Tissues   The foregoing has analyzed the situations involving the potential expected from electrodes on the surface of a strip of isolated irritable tissue. The same reasoning can be applied to predict the anticipated potential from electrodes on a bundle of isolated irritable tissues. In particular, this line of reasoning has value in explaining the action potentials recorded from the surface of a nerve trunk and the effect of injury on the action potentials recorded from myocardial tissue. Although in some instances the analysis is better performed by use of the dipole concept (see next section), the following discussion permits visualization of the electrophysiological principles underlying the action potentials detected by local electrodes.

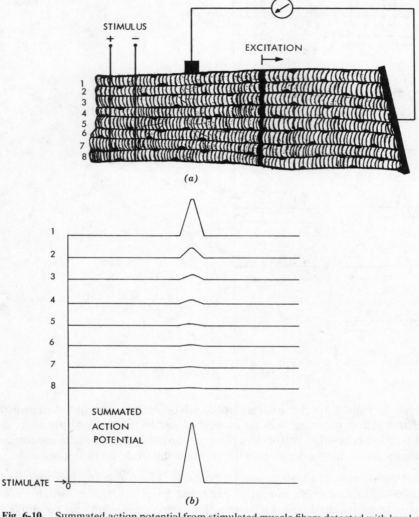

Fig. 6-10. Summated action potential from stimulated muscle fibers detected with local electrodes.

Consider a bundle of irritable fibers having the same propagation velocity. On the surface of the bundle is placed one electrode, and the other electrode is located at the cut (injured) end (Fig. 6-10). In the absence of excitation there will be a standing potential difference (the injury potential) between the electrodes. Now if all the fibers are stimulated simultaneously at the end opposite the cut, all the propagated excitations will pass by the surface electrode at the same time. Ideally, the surface electrode will preferentially detect the action potentials in fiber 1, which is immediately under it. The action potentials in the more distant underlying fibers (2–8) will also be detected, but the more distant fibers will contribute less to the voltage detected by the surface electrode. In accordance with Fig. 6-10b, the resulting action potential will be a combination of all the action potentials of the local and distant fibers. Because all fibers were chosen to be identical, the action potential will be a smooth monophasic wave; no action potentials will be detected at the site of injury.

If now the individual fibers are not excited simultaneously, as for example in skeletal muscle by nerve stimulation, the action potentials of the individual fibers will not pass under the surface electrode synchronously. Therefore, the potential measured between the electrodes will reflect this situation and the action potential recorded, although still unidirectional, will be polyphasic; its form, of course, will reflect the temporal pattern of excitation and the spatial distribution and velocities of propagation of the various fibers (Fig. 6-11).

The situation just discussed is by no means uncommon in the routine measurement of bioelectric events with local extracellular electrodes. For example, in nerve trunks there is a spatial distribution of fibers of differing diameters, and the velocities of propagation are related to fiber diameters— the lager fibers propagate excitation much more rapidly than the smaller ones. Even if all the fibers were stimulated simultaneously, the further down the nerve, the larger will be the time separation between the action potentials of the rapidly and slowly propagating fibers. Hence many sequentially appearing action potentials will be detected by a surface electrode. In fact, this is how the differences in nerve propagation velocity were discovered by Erlanger and Gasser (1937); the apparatus employed in their Nobel Prize winning study, and some sample oscillograms, appear in Fig. 6-12. In this study, which was carried out well before the availability of electrodes to measure transmembrane potentials, the investigators employed injury to good advantage to obtain unit activity. In addition to proving that the propagation velocity in nerve is related to fiber diameter, Erlanger and Gasser showed that the waveform of the action potential recorded by a surface electrode placed on a mixed nerve trunk, in which all of the axons are stimulated simultaneously, will depend on the propagation velocities and the distance from the point of

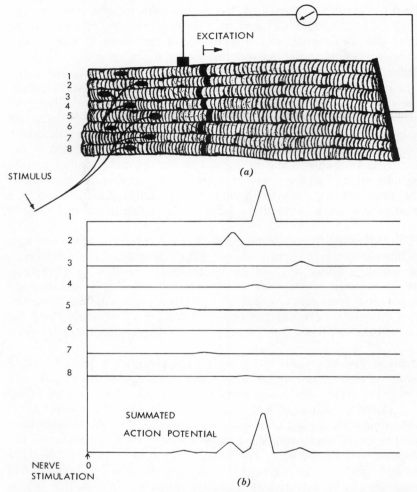

Fig. 6-11. Summated action potential in muscle due to nerve stimulation detected with local electrodes.

stimulation to the active (surface) electrode, which will detect the action potentials of the fibers below it; those in the more distant fibers will contribute less to the recorded action potential.

If the action potentials of a mixed nerve are recorded with a pair of surface electrodes during physiological activation of its neurones (or receptors), the action potentials recorded will reflect the asynchrony of activation of the axons, the differences in their propagation velocities, and the electrode separ-

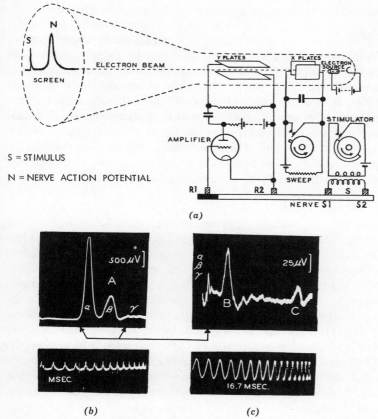

S = STIMULUS

N = NERVE ACTION POTENTIAL

(a)

(b)   *(c)*

Fig. 6-12. The action potentials of a nerve trunk containing a population of fibers having different diameters and therefore different propagation velocities: *(a)* recording method; *(b)* action potentials from the fastest propagating fibers ($A_{\alpha,\beta,\gamma}$), *(c)* action potentials *B* and *C* from the fibers with slower propagation velocity. In Fig. 6-12*a*, one electrode *R*1 is placed on an injured area and the other *R*2 on the surface of the nerve. The nerve fibers are stimulated by electrodes *S*1 and *S*2 and, at the same time, the oscilloscope sweep begins. With this arrangement, the stimulus *S*, which appears at the beginning of the sweep, is followed by the action potentials of the various fibers. (From J. Erlanger and H. S. Gasser, *Electrical Signs of Nervous Activity*. University of Pennsylvania Press, 1937.) By permission.

ation. Action potentials occur with a similar asynchrony when the activity of skeletal muscle is recorded. This situation is diagrammed in Fig. 6-11, which shows skeletal muscle in which there is a spatial distribution of motor end plates. If all the axons were excited simultaneously by a single stimulus, all the muscle fibers would not be excited simultaneously. An electrode near the end of the muscle will detect the action potentials of the individual fibers as they arrive at different times because of the different distances from the end plates. Hence the action potential recorded will by polyphasic. In an actual situation in which the motor neurones are activated physiologically, simultaneous excitation does not occur. Therefore, there will be an added asynchrony to the arrival of the action potentials under the muscle electrode and the electrical activity will consist of a train of action potentials.

The Interference Theory

The foregoing discussion considered primarily the case of electrodes on the surface of isolated active tissue and in regions of injury. Clearly, if both electrodes are on the surface of a bundle of fibers or group of cells, the potential measured will reflect the temporal relationship of arrival of excitation under each electrode and the distances of the individual fibers from each electrode. Although the situation is more complex when electrodes are distant from the irritable tissue (see next section), the principle of algebraic temporal summation is nonetheless useful. This concept, often called the interference theory, seems to have originated with Burdon Sanderson (1879), who used it to explain the genesis of the *QRS* and *T* waves of the ECG from the monophasic action potentials recorded by each electrode. Suppose that a pair of electrodes is placed on a bundle of identical uninjured fibers that are excited asynchronously, or on a bundle of dissimilar fibers excited synchronously. The interference theory predicts that the action potential appearing between the electrodes will be polyphasic and complex, consisting of upward and downward waves whose amplitudes and temporal sequence are dependent on the electrode geometry and characteristics of the particular type of irritable tissue.

The interference theory has value in understanding genesis of some electrocardiographic waveforms; as will be shown later, it is particularly handy in explaining the contribution of injury to the ECG. As stated previously, Burdon Sanderson (1879) pioneered use of the interference theory by using injury under one electrode to reveal the waveform of the action potential of cardiac muscle. The true form of this action potential was first recorded with transmembrane electrodes much later by Coraboeuf and Weidmann (1949). Sanderson showed that the addition of two temporally displaced monophasic action potentials recorded from the ventricle of a frog gave rise to the *R* and *T* waves. Use of the interference theory for the same purpose has since received considerable support from Lewis (1925) and Hoff et al. (1941). Although the dipole concept (see next section) is perhaps a better way of viewing the genesis of some of the electrocardiographic waveforms, particularly when recorded with a "monopolar" electrode, the interference theory is still helpful and may be applied to the situation in which a pair of electrodes are placed on the surface of cardiac muscle.

Assume that a pair of electrodes is placed on the surface of intact cardiac muscle (Fig. 6-13) and that excitation and recovery of each of the cardiac muscle fibers will contribute a potential to each electrode—the further away, the less will be the contribution. The sketches in Fig. 6-13 show the amount of potential contributed by fibers at different depths (1–6) to electrodes *A* and *B*. Because active tissue is electronegative to inactive tissue and because active tissue under electrode *A* will move the potential indicator in one direction and active tissue under electrode *B* will cause the potential indicator to move in the opposite direction, the contributions of potential to the active fibers under electrode *B* are drawn inverted.

The interference theory states that the potential difference recorded between terminals *A* and *B* is the algebraic sum of the temporal development of voltages provided by the active fibers under each electrode. Summation of all these potentials appears in the bottom graph of Fig. 6-13, which diagrams genesis of the *R* and *T* waves of the electrogram of simple ventricular myocardium. Parenthetically it should be noted that, should recovery be caused to occur earlier under electrode *B* than *A*, the duration of the monophasic action potential under *B* will be less and the *T* wave will be upward; Fig. 2-32 illustrates this point.

If some of the myocardial fibers under electrode *B* are now injured, as by ischemia, the electrical activity detected by electrode *B* will be altered. Figure 6-14 shows tissue injury under electrode *B* at the level of the fibers corresponding to depth 2; therefore, although there will be no excursion in membrane potential in this region, there will be a standing injury potential. Spread of excitation over the myocardial fibers under electrode *A* will thus produce normal monophasic action potentials (Sum *A*, Fig. 6-14). Excitation passing under electrode *B* will produce monophasic action potentials in the uninjured fibers and nothing but a standing injury potential from the area of injury (Fig. 6-7). Therefore, the temporal summation of action potentials under electrode *B* will be less (Sum *B*, Fig. 6-14), and the potential indicator will be presented with the sum of the action potentials detected by electrode *A* (Sum *A*), the sum detected by electrode *B* (Sum *B*), and the standing injury potential. The temporal sum of these three components (lower graph in Fig. 6-14), reveals that the *R* wave starts at the level of the injury potential and rises and falls, reaching a plateau of zero potential when all the tissue is depolarized; this is the *S-T* segment. As recovery starts, the *T* wave is described and ends at the level of the injury potential. The elevation in the *S-T* segment, which is in

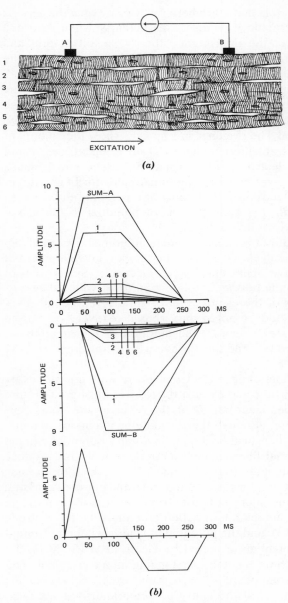

(a)

(b)

Fig. 6-13. The potentials from electrodes placed on the surface of cardiac muscle.

268

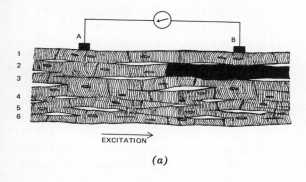

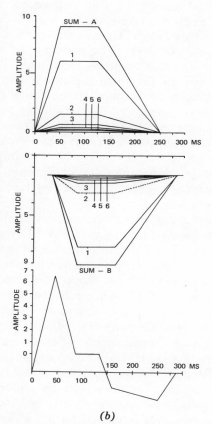

Fig. 6-14. Action potentials of injured cardiac muscle idealized by use of the interference theory.

reality a depression of the diastolic baseline, is the cardinal sign of injury to the ventricular myocardium. Whether it appears as an *S-T* segment elevation or depression depends, of course, on the proximity of the injury to one electrode or the other.

In summary, it can be seen that when electrodes are placed on irritable tissue, the potential measured reflects the nature of the excitatory and recovery process in the individual tissues, the manner in which the active tissues are excited and are spatially located with respect to the electrodes and the presence or absence of an injury potential. Whether the action potential will have symmetrical upward and downward components will depend on whether one electrode is located in an area of injury or not and the sequence of recovery.

The Dipole Concept   In the practical measurement of a bioelectric event it is often impossible to place both extracellular electrodes directly on the irritable tissue; one may be nearby and the other at a considerable distance, constituting a reference or "indifferent" electrode. The principal difference between this method of measurement and that featuring electrodes directly in contact with the irritable tissue is that the potentials measured reflect the flow of current in the conducting environment surrounding the active region of the irritable tissue. This situation is best analyzed by considering activity (depolarization and reverse polarization) and recovery (repolarization) as traveling dipoles. This concept was first enunciated for nerve by Bernstein's pupil Hermann (1879); it was later extended by Craib (1927), Wilson et al. (1933), and Macleod (1938) to include cardiac muscle. Verification of its applicability to human electrocardiography has been presented by Hecht and Woodbury (1950).

If a source of potential (i.e., a dipole) is embedded in a uniform conducting medium of infinite extent (i.e., a volume conductor), current will flow and a potential field will be established. Figure 6-15a illustrates the manner in which the potential field is distributed. The isopotential lines (of which there is an infinite number) describe the potential measured by a "monopolar" electrode located anywhere in the environment of the dipole when referred to another electrode in a region of zero potential (i.e., at an infinite distance—or on the zero isopotential line passing midway between the poles of the dipole). Suppose now that a monopolar electrode starts from a remote point and is moved along a line ($d = 1$, Fig. 6-15a) parallel to the dipole axis (the line joining its positive and negative poles); the isopotential lines are encountered in an orderly sequence and the potential will first increase, then fall to zero (when the electrode is over the midpoint of the dipole), then reverse polarity and increase magnitude, and then decrease as the electrode is moved further away as in Fig. 6-15b. It should be noted that the same sequence will be measured if the electrode is fixed and the dipole moves. If the procedure were repeated by moving the monopolar electrode along another line parallel to the dipole

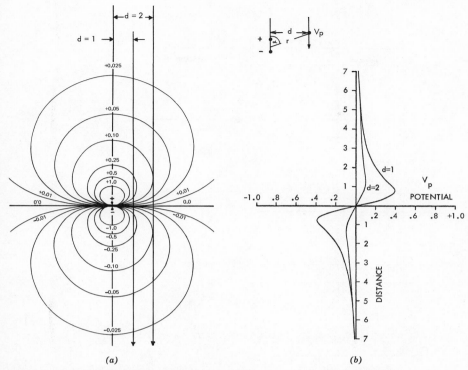

(a)

(b)

Fig. 6-15. The dipole and its field of potential: (*a*) potential distribution; (*b*) potential encountered by exploring electrode moving along lines (d = 1, d = 2) parallel to the dipole axis.

axis but more distant ($d = 2$), the same sequence of events would occur, but the magnitude of the excursion in voltage would be less (Fig. 6-15*b*, d = 2).

Application of the dipole concept is illustrated in Fig. 6-16; in which *a* shows a long strip of irritable tissue at rest. In a method of recording often termed "monopolar," the potential V_p at a nearby point *P* would be measured with respect to a truly indifferent electrode. Since, as previously noted, a truly indifferent electrode is one at an infinite distance in the conducting environment, the potential will be essentially zero. If the tissue is stimulated, as in Fig. 6-16*b*, the active region (which is negative to the resting region) will cause current to flow in the conducting environment and to establish a potential field. Because the boundary between the active and inactive regions is characterized by charges of opposite sign, the wavefront of excitation can be equated with a dipole with its positive pole facing the direction of propagation of excitation, in Fig. 6-16*f*. If the active region occupies a large segment of the irritable tissue, then the potential changes appearing at the point *P* are those

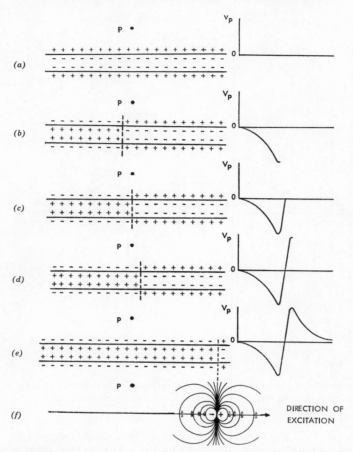

Fig. 6-16. Extracellular potential variations at point *P* reflecting excitation represented by a dipole traveling with its positive pole facing the direction of propagation of excitation.

represented by the dipole accompanied by its potential field as it moves by (Fig. 6-16*b*-6-16*e*). The potential difference appearing between a nearby electrode and a distant reference electrode is clearly diphasic (positive followed by negative) as the wave of excitation passes the nearby electrode. Whether the polarity chosen for the indicator presents a down-up or an up-down sequence depends on the convention adopted.

The same line of reasoning can be applied to the process of recovery diagrammed in Fig. 6-17. Because the active area is negative to inactive tissue, recovery can be equated to a dipole with its negative pole facing the direction of advancing recovery. Therefore, passage of recovery by the nearby measuring electrode will produce a negative-positive variation in potential; this sequence of events appears in Fig. 6-17*b* through 6-17*e*, using the same polarity convention as that in Fig. 6-16.

From the foregoing it can be seen that when excitation passes a nearby monopolar electrode a diphasic (positive-negative) potential change is recorded (Fig. 6-16*e*). If recovery travels in the same direction as excitation, a negative-positive diphasic potential change is recorded (Fig. 6-17*e*). If the active region is narrow, the time between excitation and recovery will be short; consequently

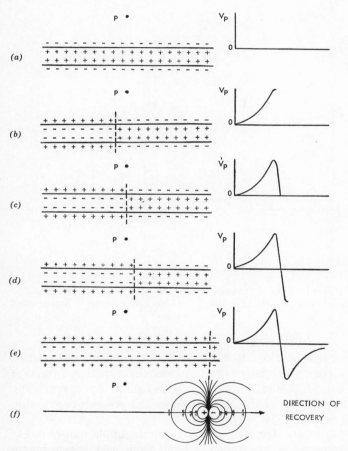

Fig. 6-17. Extracellular potential variations at point *P* reflecting recovery represented by a dipole traveling with its negative pole facing the direction of propagation of recovery.

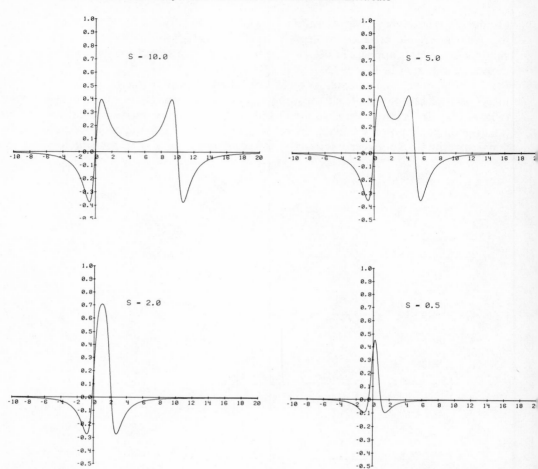

Fig. 6-18. Action potentials recorded by a monopolar electrode in the environment of a strip of irritable tissue in a volume conductor when the length *S* of the active region is decreased (S is in arbitrary units).

the two diphasic waves will be close to each other and may indeed overlap, resulting in a complex positive-negative-positive waveform to signal passage of excitation and recovery. Figure 6-18 clarifies this point by showing the effect of decreasing the width *S* of the active region; the polarity convention chosen is the same as that of Figs. 6-16 and 6-17.

Lorente de Nó (1947) nicely demonstrated the double dipole nature of excitation and recovery in nerve by placing an isolated bullfrog sciatic nerve on a piece of Ringer's-soaked blotting paper measuring 30 × 20 cm.

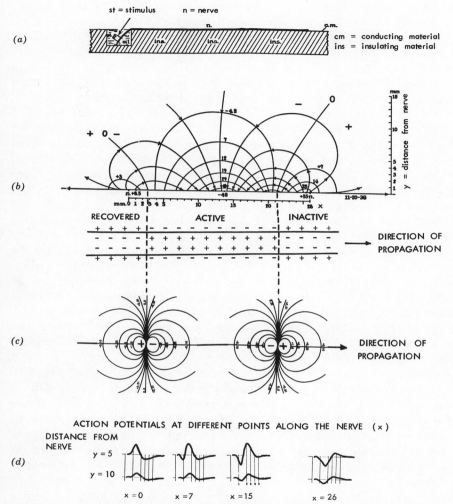

Fig. 6-19. The field pattern surrounding an active region of nerve on a conducting plane and its relation to the dipole concept and the action potentials recorded from different points on the conducting plane. (Redrawn after R. Lorente de Nó, *A Study of Nerve Physiology.* New York: Rockefeller Institute, 1947. Part 2, Chapter 16.

Figure 6-19*a* shows how the blotting paper was supported by an insulating block that had a small recess filled with mineral oil for accommodating the end of the nerve and the stimulating electrodes. By stimulating the nerve it was possible to map the potential field with a monopolar microelectrode (tip radius 20 μ) at the point near the stimulating electrodes where the nerve entered the blotting paper ($x = 0$), near the middle, and at the end of the nerve ($x = 26$). In Fig. 6-19*b*, where the lines of equal voltage and current appear, the latter are identified by arrows denoting the direction of current flow. The inactive and active regions are identified by the row of positive and negative charges on a strip of irritable tissue below the field pattern; the equivalent dipoles of excitation and recovery, accompanied by their idealized potential fields, are shown in Fig. 6-19*c*. Figure 6-19*d* gives the action potentials recorded with a monopolar electrode located at different distances ($x = 0, 7, 15$, and 26 mm) along the nerve and at different distances from the nerve ($y = 5$, and y 10 mm). Two facts are apparent from Fig. 6-19*d*. First, with the exploring electrode 5 mm from the nerve and located in the central region ($x = 7$ and $x = 15$ mm), the action potentials are triphasic, as would be expected from the movement of an idealized dipole of excitation followed closely by one of recovery (as in Fig. 6-18). In addition, the action potentials recorded from the ends of the nerve ($x = 0$ and $x = 26$ mm) are distorted because the potential field has been distorted by environmental inhomogeneities and injury. Second, with the exploring electrode located at the same points but more remote and along a line 10 mm parallel to the nerve ($y = 10$), the action potentials have a similar waveform but are reduced in amplitude.

Lorente de Nó showed that the dipole concept can be substantiated *in vivo* by femoral exposure of a branch of the sciatic nerve of a frog by stimulating it antidromically and recording action potentials with a metal microelectrode placed at different sites on the adjacent muscle. Figure 6-20, which shows the recording he obtained, illustrates the two facts predicted by theory: passage of the wave of excitation and recovery gives a triphasic action potential, and the recorded amplitude diminishes with increasing distance from the irritable tissue (nerve).

An excellent example of the applicability of the dipole concept to human electrocardiography was presented by Hecht and Woodbury (1950). These investigators used a monopolar esophageal electrode to record the action potential that reflects excitation of the atria (Fig. 6-21*a*); they compared this potential with that obtained by moving a dipole past a local monopolar electrode in a volume conductor (Fig. 6-21*b*), signaling approaching positivity by an upward deflection of the potential indicator. The similarity of the waveforms for these two situations is quite striking. Hecht and Woodbury pointed out that the equivalent dipole of excitation is in reality a band of dipoles (Fig. 6-21*c*) in which there is a spacing between the poles that represents the transition region between active (−) and resting (+) tissue.

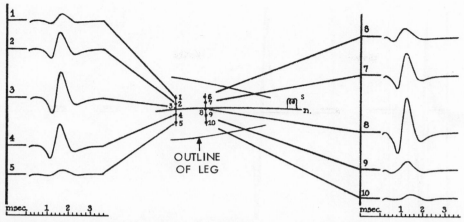

Fig. 6-20. Extracellular action potentials recorded *in situ* from the stimulated (*s*) bullfrog sciatic nerve (*n*) on the right side of the animal. The numbers on the recordings in the vicinity of the nerve identify the locations of the monopolar metal microelectrode (tip radius 20 μ); the "indifferent" (ground) electrode was placed on the left leg. (Redrawn from R. Lorente de Nó. *A Study of Nerve Physiology*, New York: Rockefeller Institute, Part 2, Chapter 16, Fig. 9.

In the preceding discussion of the dipole theory, excitation and recovery have been represented as traveling dipoles. In a practical situation, when recordings are made from a mass of tissue that becomes active, excitation and recovery are characterized by a considerable spatial distribution of dipoles. It is also important to realize that, although these two processes are closely related, they are quite different, and each has its own temporal and spatial characteristics. In general, depolarization is rapid and the transition between active and inactive tissue occupies only a short distance; therefore, the waveform representing excitation usually conforms quite well to that predicted by a traveling dipole. Recovery is a much slower process, however, and it is unevenly distributed over a greater amount of tissue, which is why the waveform representing recovery is usually smaller in amplitude and longer in duration. Macleod (1938) was perhaps the first to point out this difference in studies using the dipole theory to explain the recovery (*T*) wave of an ECG that was recorded with an electrode pair consisting of one active and one "indifferent" (reference) electrode. His statement, which described application of the dipole concept to cardiac muscle, applies equally well to all irritable tissue and elegantly indicates the important points to be considered in the practical application of the dipole concept. He wrote (1938)

Muscle does not become active instantaneously. The active process spreads with a given velocity so that one length of muscle will be coming active, another will be fully active, and a third will be regressing from the active state. The lengths which are in

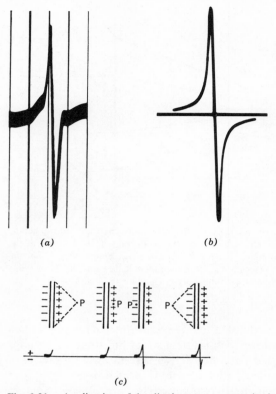

(a)   (b)

(c)

Fig. 6-21. Application of the dipole concept to excitation of the human atria. (*a*) Atrial excitation electrogram illustrating the action potential recorded by an esophageal monopolar electrode; (*b*) advancing dipole waveform, showing the potential predicted by moving a dipole (in a volume conductor) past a local monopolar electrode; (*c*) genesis of potential wave produced by an advancing dipole, clarifying the manner in which excitation advances—note movement of a band of dipoles with the spacing between the + and − poles representing the transition between inactive (+) and active (−) tissue. (From L. A. Woodbury, *Circulation*, 1950, **2**:37–47. By permission.)

transition are the distances over which the potential difference which exists between resting and active muscle must be distributed. It is possible to represent the potential difference either by a chain of doublets [dipoles] distributed along the transitional region or by a single positive and a single negative pole located at its beginning and end, respectively. Conversely the length of the doublet chain or the distance apart of the positive and negative poles measures the length of the transitional region.

Thus, in a practical situation, the distances between the poles of the dipoles of excitation and recovery are expected to be different and hence the contours of the potentials recorded due to passage are also expected to be different.

The foregoing discussion of use of the dipole concept to predict the potential recorded with a monopolar electrode is obviously very greatly simplified and caution should be exercised in extrapolating it to all *in vivo* situations. An idea of the possible complexity can be appreciated by considering what might happen if both electrodes are in the environment of the active tissue (i.e., one electrode not in a region of zero potential). Add to this the fact that the *in vivo* environmental conducting medium does not extend to infinity in all directions and is constituted by inhomogeneous tissue. It should then be apparent that a relatively complex waveform, reflecting excitation and recovery, can be detected by extracellular electrodes. Although accurate prediction of the waveform is impossible in many practical circumstances, the previous discussion can serve as a guide to what can be expected in a given situation.

Transmembrane and Extracellular Potentials It would be very convenient if there were an easy way to relate the action potential detected by an external monopolar electrode (i.e., one paired with an indifferent electrode) to the transmembrane potential. Unfortunately, in practical situations of measurement, no simple and constant relationship can be derived because of environmental inhomogeneities of various kinds. However, a relationship does exist, and it can be demonstrated in an idealized situation.

When an irritable tissue in a volume conductor becomes active, there is a current flow in the environment and a potential field is established (see Fig. 6-19). A nearby monopolar electrode detects the potential due to the flow of current through the resistance of the environmental material. The cause of the current is of course the active region of the membrane, which experiences an excursion in potential. It can be shown by the use of field theory (Lorente de Nó., 1947; Clark and Plonsey, 1968; Plonsey, 1969) and by use of the cable analog (Huxley and Stampfli, 1949; Tasaki, 1959; Clark and Plonsey, 1966) that the membrane current does not have the same waveform as the excursion in transmembrane potential. The mathematical analyses presented in the papers cited, which considered the case of a cylindrical irritable tissue located in a uniform volume conductor, showed that the membrane current is proportional to the second derivative of the transmembrane potential.

A modified form of the cable analog (Fig. 6-22) for a long, cylindrical, irritable cell can be used as a model to illustrate that the external action potential detected by a nearby monopolar electrode in the environmental volume conductor is proportional to the second derivative of the transmembrane action potential. In this analysis it is convenient to designate the environment as a resistance having a value r_1 Ω/unit length; similarly, the resistance per unit length of the cytoplasm is designated r_2. During activity there is a

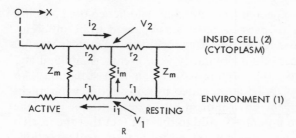

i_m = MEMBRANE CURRENT

Z_m = MEMBRANE IMPEDANCE

V_m = MEMBRANE POTENTIAL

V_m = $V_1 - V_2$

Fig. 6-22. Cable analog for a long cylindrical cell in a uniform environmental volume conductor carrying current due to activity.

current flow in the environment i_1, in the cytoplasm i_2, and through the membrane i_m. In Fig. 6-22 the currents are identified, along with the coordinate system in which x increases to the right. Along the cell there is a decrement in current within and without, and this decrement reflects the current i_m flowing through the membrane. Because of the current flow, at any point there are potentials developed; at a point outside the cell, a potential V_1 will exist and within the cell a potential V_2 will exist.

Because the membrane current i_m is the decrement in the cytoplasmic and environmental current

$$i_m = \frac{\partial i_2}{\partial x} \qquad \text{and} \qquad i_m = \frac{-\partial i_1}{\partial x}$$

Cytoplasmic and environmental potential gradients exist because there is current flow, therefore:

$$\frac{\partial V_2}{\partial x} = i_2 r_2 \qquad \text{and} \qquad \frac{\partial V_1}{\partial x} = i_1 r_1$$

from which

$$\frac{\partial^2 V_2}{\partial x^2} = \frac{r_2 \partial i_2}{\partial x} \qquad \text{and} \qquad \frac{\partial^2 V_1}{\partial x^2} = \frac{r_1 \partial i_1}{\partial x}$$

Now

$$r_2 i_m = \frac{r_2 \partial i_2}{\partial x} \qquad \text{and} \qquad r_1 i_m = \frac{-r_1 \partial i_1}{\partial x}$$

Therefore

$$r_2 i_m = \frac{\partial^2 V_2}{\partial x^2} \quad \text{and} \quad r_1 i_m = \frac{-\partial^2 V_1}{\partial x^2}$$

Because the transmembrane potential V_m is the difference between the potential outside V_2 and inside V_1 the cell,

$$V_m = V_2 - V_1$$

Therefore

$$\frac{\partial^2 V_m}{\partial x^2} = \frac{\partial^2 V_2}{\partial x^2} - \frac{\partial^2 V_1}{\partial x^2} = r_2 i_m + r_1 i_m = i_m (r_1 + r_2)$$

Now because the excursion in membrane potential is a wave that is propagated with a constant velocity u and without decrement, it can be represented by

$$V_m = f\left(t - \frac{x}{u}\right)$$

This expression satisfies the wave equation

$$\frac{\partial^2 V_m}{\partial x^2} = \frac{1}{u^2} \frac{\partial^2 V_m}{\partial t^2}$$

Therefore, the expression for the membrane potential can be transformed from the distance (x) coordinate to the time domain t; making this substitution yields

$$\frac{1}{u^2} \frac{\partial^2 V_m}{\partial t^2} = i_m (r_1 + r_2)$$

Therefore

$$i_m = \frac{1}{u^2(r_1 + r_2)} \frac{\partial^2 V_m}{\partial t^2}$$

The foregoing analysis, which is based on the cable analog, shows that the membrane current is proportional to the second derivative of the transmembrane potential with respect to time. Experimental verification was presented by Tasaki (1959), who recorded simultaneously the membrane current and the transmembrane action potential of the squid giant axon (Fig. 6-23). The membrane current i_m was detected by forcing it to flow through a low value of resistance r, connected to a small central pool of seawater 2 mm wide; on either side of this pool, and insulated from it, were two other pools containing electrodes joined together and connected to the other side of the resistor. Therefore, the potential difference appearing across r was proportional

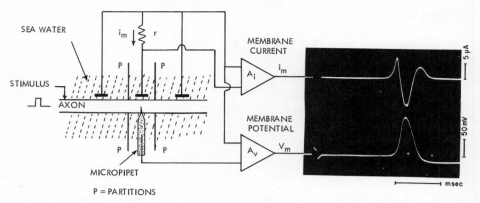

Fig. 6-23. Simultaneously recorded membrane current i_m and transmembrane potential V_m. (From I. Tasaki, *Handbook of Physiology and Neurophysiology,* Vol. I, J. Field, H. W. Magoun, and V. E. Hall, Eds., Washington, D.C.: American Physiological Society, 1959. By permission.)

to the membrane current flowing during activity that was initiated by the application of a stimulus (square wave) to one end of the nerve. The transmembrane potential of the central segment of the nerve was measured by inserting a micropipet into the axon. The voltage appearing across r and that detected by the micropipet were applied to two amplifiers A_i and A_v, whose outputs are shown in Fig. 6-23. This display nicely illustrates that, although the transmembrane potential is a monophasic wave, the membrane current has an entirely different waveform, and is, in fact, decidedly triphasic.

The cable theory just presented predicts that the membrane current varies as the second derivative of the transmembrane potential; the study carried out by Tasaki permits examination of this relationship. Figure 6-24*a* illustrates the excursion in transmembrane potential as recorded by Tasaki and Fig. 6-24*b* shows the second derivative of the membrane potential; Fig. 6-24*c* presents the transmembrane current recorded by Tasaki (Fig. 6-24*a* and 6-24*c* were directly from Fig. 6-23). Comparison of the second derivative of the transmembrane potential *b* with the membrane current *c* reveals that they have the same general contour. The difference is probably due to experimental factors. It must be recalled that in the theoretical derivation, electrode size and cell dimensions were not considered; potentials and currents were said to exist at points. In the experiment, neither the axon nor the electrode pair was infinitely small; nor did the volume conductor environment extend to infinity in all directions. However, despite these necessary practical limitations, there is a reasonable similarity between the waveform of the membrane current and the second derivative of transmembrane potential.

Because the waveform of the membrane current is proportional to the second derivative of the transmembrane potential, the potential detected by a local monopolar electrode ought also to be proportional to the second derivative of the transmembrane potential. To illustrate this point, a specimen (2 × 1 mm) of dog Purkinje fiber was placed in a 3-ml beaker of oxygenated Krebs-Ringer solution and connected to a tiny bipolar stimulating electrode that was connected to a stimulator having an isolated output circuit. A silver-silver chloride reference electrode was placed in the solution about 15 mm distant, and the

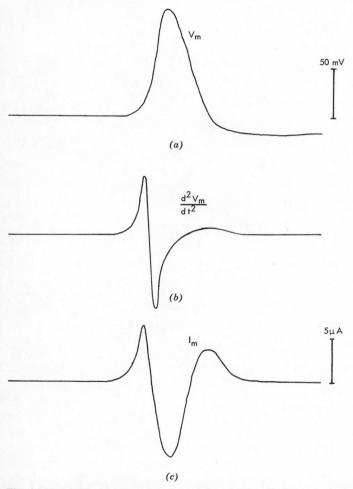

Fig. 6-24. (*a*) The transmembrane potential of the squid axon; (*b*) second derivative of the transmembrane potential; (*c*) membrane current. (Fig. 6-24*a* and 6-24*c* redrawn from Fig. 6-23.)

potential developed in the solution when the specimen was stimulated was measured with a 1-μ micropipet filled with $3M$ potassium chloride. Single stimuli were delivered as the tip of the micropipet was brought toward the specimen from a distance of about 3 mm (A in Fig. 6-25) and continuing until the tip of the micropipet penetrated the membrane of a Purkinje fiber F. An orderly sequence of increase in amplitude was obtained with almost no change in the waveform ($A-E$) until a cell membrane was penetrated, where-

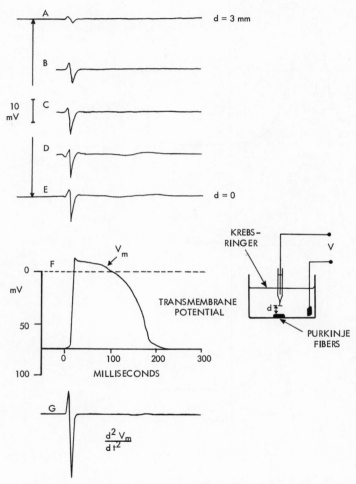

Fig. 6-25. The relation between extracellular potentials $A-E$, transmembrane potential F, and the second derivative of the transmembrane potential G. (Data courtesy of Drs. D. Riopel and R. Vick, 1970.)

upon the transmembrane potential excursion was recorded (F), revealing a quite different waveform with a much larger excursion in potential. The similarity of the second derivative of the transmembrane potential (G in Fig. 6-25) to the extracellularly recorded action potentials $A-E$ is striking.

Under ideal conditions (i.e., a single cell in an infinite and homogeneous volume conductor environment), the evidence just presented indicates that the membrane current is proportional to the second derivative of the trans-membrane potential. Furthermore, since the membrane current flows through the environment, then the potential detected by a local monopolar electrode is expected to be proportional to the second derivative of the membrane potential. As the monopolar electrode is moved more distant, the waveform is the same, but the amplitude is diminished.

SUMMARY

In this chapter the measurement of bioelectric events has been discussed from two viewpoints; one considered one or both electrodes located directly on isolated irritable tissue, the other dealt with measurement with the tissue embedded in a volume conductor. In the latter case, one electrode was nearby and the other (the reference electrode) was in a region of zero potential, achieved by locating the reference electrode at a great distance from the active tissue. Although in both cases the potential measured was derived from depolarization and repolarization of cell membranes, the waveforms measured are quite different. With both electrodes on isolated irritable tissue, the inter-ference theory accounts quite well for the waveforms recorded; Fig. 6-2 offers excellent supportive evidence. If now the region under one electrode is in-jured, the excursion in potential recorded during activity resembles that measured with transmembrane electrodes; however the amplitude is con-siderably less and the action potential rises from and returns to the injury potential.

When irritable tissue is surrounded by a volume conductor, the potential appearing between a local monopolar electrode and its distant reference electrode does not resemble the excursion in transmembrane potential of the irritable cells; rather, it is related to the membrane current, which varies as the second derivative of the transmembrane potential. The dipole concept represents another way of viewing the potential detected by a local mono-polar electrode—that is, excitation can be equated to a dipole traveling with its positive pole facing the direction of propagation; recovery is represented by a dipole moving with its negative pole facing the direction of propagation recovery.

References

Adey, W. R., C. W. Dunlop, and C. E. Hendrix. Hippocampal slow waves. *A.M.A. Arch. Neurol.* 1960, **3**:74–90.

Adey, W. R., D. O. Walter, and C. E. Hendrix. Computer techniques in correlation and spectral analyses of cerebral slow waves during discriminative behavior. *Exp. Neurol.* 1961, **3**:501–524.

Adey, W. R., W. D. Winters, and M. R. Delucchi. EEG in simulated stresses of space flight with special reference to problems of vibration. *EEG Clin. Neurophysiol.* 1963, **15**:305–320.

Adrian, R. H. The effect of internal and external potassium concentration on the membrane potential of frog muscle. *J. Physiol.* 1956, **133**:631–658.

Adrian, E. D., and D. W. Bronk. The discharge of impulses in motor nerve fibers. Part II. *J. Physiol.* 1929, **67**:119–151.

Adrian, E. D., and G. Moruzzi. Impulses in the pyramidal tract. *J. Physiol.* 1939, **97**:153–199.

Afanasiev, A. S. Influence of the solvent on the electromotive force of silver-silver halide cells. *J. Amer. Chem. Soc.* 1930, **52**:3477–3483.

Agin, D. P. Electrochemical properties of glass microelectrodes. In *Glass Microelectrodes,* M. Lavallée, O. F. Schanne, and N. C. Hebert, Eds. New York, John Wiley & Sons, 1969, pp. 62–75. 446 pp.

Agin, D., and D. Holtzman. Glass microelectrodes: origin and elimination of tip potentials. *Nature.* 1966, **211**:1194–1195.

Alberts, W. W., E. W. Wright, B. Feinstein, and G. von Bonin. Experimental radiofrequency brain lesion size as a function of physical parameters. *J. Neurosurg.* 1966, **25**:421–423.

Alexander, J. T., and W. L. Nastuk. An instrument for the production of microelectrodes used in physiological studies. *Rev. Sci. Instr.* 1953, **24**:528–531.

Allen, A. C. *The Skin.* New York: Grune and Stratton, 1967. 1182 pp.

Almasi, J. J., and O. H. Schmitt. The dependence of skin-thru-electrode impedance on individual variations, skin preparation and body location. *Proc. 21st Ann. Conf. Eng. Med. Biol.* 1968, **10**:13A2.

Almasi, J. J., and O. H. Schmitt. Systematic and random variations of ECG electrode system impedance. *Ann. N.Y. Acad. Sci.* 1970, **170** (Art. 2) :509–519.

A.M.A. Counc. on Physical Medicine. Minimum requirements for acceptable electrocardiographs. *JAMA* 1947, **134**:455. (For additional data see also *JAMA,* 1950, 143, 654–655 and *IEEE Trans. on Bio-Med. Engng.* 1967, *BME*-**14**:60–68.)

Amatneik, E. Measurement of bioelectric potentials with microelectrodes and neutralized input capacity amplifiers. *IRE Trans. Med. Electron.* 1958, **PGME-10:**3–14.

Anand, B. K., S. Dua, and K. Shoenberg. Hypothalamic control of food intake in cats and monkeys. *J. Physiol.* 1955, **127:**143–152.

Andrews, H. L. A new electrode for recording bioelectric potentials. *Amer. Heart J.* 1939, **17:**599–601.

Aronow, S. The use of radio-frequency power in making lesions in the brain. *J. Neurosurg.* 1960, **17:**431–438.

d'Arsonval, A. Electrodes impolarisables et excitateur electrique. *C. R. Soc. Biol.* 1886, **38:**228–229.

Artz, W. Silicone rubber electrodes for long-term patient monitoring. *Bio-Med. Eng.* 1970, **5:**300–301.

Asa, M. M., A. H. Crews, E. L. Rothfield, E. S. Lewis, R. I. Zucker, and A. Bernstein. High fidelity radioelectrocardiography. *Amer. J. Cardiol.* 1964, **14:**530–532.

Atkins, A. R. Measuring heart rate of an active athlete. *Electron. Eng.* 1961, **33:**457.

Barnett, A. The basic factors involved in proposed electrical methods for measuring thyroid function. *West. J. Surg. Obstet. Gynecol.* 1937, **45:**540–554.

Basmajian, J. V., and G. Stecko. A new bipolar electrode for electromyography. *J. Appl. Physiol.* 1962, **17:**849.

Baudoin, A., H. Fischgold, and J. Lerique. Une nouvelle electrode liquide. *C. R. Soc. Biol.* 1938, **127:**1221–1222.

Bell, G. H., J. A. C. Knox, and A. J. Small. Electrocardiograph electrolytes. *Brit. Heart J.* 1939, **1:**229–236.

Bennett, A. L., A. L. Dunn, and A. R. McIntyre. The normal membrane resting potential of mammalian skeletal muscle measured *in vivo. J. Cell. Comp. Physiol.* 1953, **42:**343–357.

Benedetti-Pichler, A. A., and J. R. Rachele. Limits of identification of simple confirmatory tests. *Ind. Eng. Chem. Anal. Ed.* 1940, **12:**233–241.

Bergey, G. E., R. D. Squires, and W. C. Sipple. Electrocardiogram recording with pasteless electrodes. *IEEE Trans. Bio-Med. Eng.* 1971, **BME 18:**206–211.

Bernstein, J. *Elektrobiologie.* Braunschweig: Druck und Vérlag von Friedr. Virweg & Sohn, 1912. 215 pp.

Berson, A. S., and H. V. Pipberger. Electrode to skin impedance problems in ECG recordings. *Proc. 7th Int. Conf. Med. Biol. Eng. 1967,* Stockholm, Sweden. p. 415.

Blank, I. H., and T. G. Finesinger. Electrical resistance of the skin. *Arch. Neurol. Psychiat.* 1946, **54:**544–557.

Block, M. T. The electrical and biological properties of tungsten microelectrodes. *Med. Biol. Eng.* 1968, **6:**517–525.

Boter, J., A. den Hertog, and J. Kuiper. Disturbance-free skin electrodes for persons during exercise. *Med. Biol. Eng.* 1966, **4:**91–95.

Bourne, G. H. *Division of Labour in Cells.* New York: Academic Press, 1964. 248 pp.

Brown, A. S. A type of silver chloride electrode suitable for use in dilute solutions. *J. Amer. Chem. Soc.* 1934, **56:**646–647.

Brock, L. G., R. M. Eccles, and R. D. Keynes. The discharge of individual electroplates in *Raia clavata. J. Physiol.* 1953, **122:**4P–5P.

Brown, C. W., and F. M. Henry. The central nervous mechanism for emotional responses II. *Proc. Natl. Acad. Sci. (U.S.)* 1934, **20:**310–315.

Bultitude, K. H. A technique for marking the site of recording with capillary microelectrodes. *J. Microsc. Sci.* 1958, **99:**61–63.

Burch, G. E., and N. P. DePasquale. *A History of Electrocardiography*. Chicago: Year Book Publishers, 1954. 309 pp.

Burdon-Sanderson, J., and F. J. M. Page. On the time-relations of the excitatory process in the ventricles of the heart of the frog. *J. Physiol.* 1879, **2:**384–435.

Bures, J., M. Petran, and J. Zachar. *Electrophysiological Methods in Biological Research*. New York: Academic Press, 1962.

Burns, D. C., and P. D. Gollnick. An inexpensive floating-mesh electrode for EKG recording during exercise. *J. Appl. Physiol.* 1966, **21:**1889–1891.

Burns, R. C. Study of skin impedance. *Electronics.* 1950, **23:**190–196.

Burr, H. S., and A. Mauro. Millivoltmeters. *Yale J. Biol. Med.* 1948–1949, **21:**249–253.

Burr, H. S. In *Medical Physics*. Glasser, O. Ed. Chicago: 1944, Vol. 2, 1950, Chicago: Year Book Publishers. 1227 pp.

Byzov, A. L., and V. L. Chemystov. Machine for making microelectrodes. *Biophysics.* 1961, **6:**79–82.

Caldwell, P. C., and A. C. Downing. The preparation of capillary microelectrodes. *J. Physiol.* 1955, **128:**31P.

Carbery, W. J., W. E. Tolles, and A. H. Freiman. A system for monitoring the ECG under dynamic conditions. *Aerosp. Med.* 1960, **31:**131–137.

Carmody, W. R. A study of the silver chloride electrode. *J. Amer. Chem. Soc.* 1929, **51:**2901–2904.

Carmody, W. R. Studies in the measurement of electromotive force in dilute aqueous solutions. II. The silver chloride electrode. *J. Amer. Chem. Soc.* 1932, **54:**188–192.

Carmody, W. R. Studies in the measurement of electromotive force in dilute aqueous solutions. I. A study of the lead electrode. *J. Amer. Chem. Soc.* 1929, **51:**2905–2909.

Castellan, G. W. *Physical Chemistry*. Reading, Mass.: Addison-Wesley Publishing Company, 1964. 717 pp.

Carpenter, M. B., and J. R. Whittier. Study of methods for producing experimental lesions of the central nervous system with special reference to stereotaxic technique. *J. Comp. Neurol.* 1952, **97:**73–131.

Carpenter, M. B., J. R. Whittier, and F. A. Mettler. Analysis of choreoid hyperkinesia in the rhesus monkey. *J. Comp. Neurol.* 1950, **92:**293–331.

Chambers, R., and G. Cameron. Intracellular hydrion concentration studies. *J. Cell. Comp. Physiol.* 1932, **2:**99–103.

Chambers, R., and R. T. Kempton. Indications of function of the chick mesonephrons in tissue cultures with phenol red. *J. Cell. Comp. Physiol.* 1933, **3:**131–167.

Chambers, R., L. Beck, and M. Belkin. Secretion in tissue cultures. *J. Cell Comp. Physiol.* 1935, **6:**425–439.

Champan, D. *Biological Membranes*. New York: Academic Press, 1968. 438 pp.

Chartrain, G. E., H. W. Dodge, M. C. Petersen, and R. G. Bickford. A multielectrode lead for intracerebral recording. *EEG Clin. Neurophysiol.* 1959, **11:**165–169.

Chowdhury, T. K. Fabrication of extremely fine glass micropipette electrodes. *J. Sci. Instr.* 1969, **2:**1087–1090.

Churney, L., R. Ashman, and E. Byer. Electrogram of turtle heart strip immersed in a volume conductor. *Amer. J. Physiol.* 1948, **154:**141–148.

Clark, L. C., and T. J. Lacey. An improved skin electrode. *J. Lab. Clin. Med.* 1950, **35**:786–787.

Clark, J., and R. Plonsey. A mathematical evaluation of the core conductor model. *Biophys. J.* 1966, **6**:95–112.

Clark, J. W., and R. Plonsey. A mathematical study of nerve fiber interaction. *Biophys. J.* 1970, **10**:937–956.

Clarkson, T. W., and B. Lindemann. Experiments on Na transport of frog skin epithelium. In *Laboratory Techniques in Membrane Biophysics,* H. Passow and R. Stampfli, Eds. Berlin: Springer-Verlag, 1969. 201 pp.

Clendenning, W. E., and R. Auerbach. Traumatic calcium deposition in the skin. *Arch. Dermatol.* 1964, **89**:360–363.

Cohn, A. E. A new electrode for use in clinical electrocardiography. *Arch. Int. Med.* 1920, **26**:105–113.

Cole, K. S. Personal communication (1967).

Cole, K. S. *Membranes, Ions and Impulses.* Berkeley, Calif.: University of California Press, 1968. 569 pp.

Cole, K. S. Dynamic electrical characteristics of the squid axon membrane. *Arch. Sci. Physiol.* 1949, **3**:253–258.

Cole, K. S., and U. Kishimoto. Platinized silver chloride electrode. *Science.* 1962, **136**:381–382.

Collander, R. The permeability of plant protoplasts to nonelectrolytes. *Trans. Faraday Soc.* 1937, **33**:985–990.

Collias, J. C., and E. E. Manuelidis. Histopathological changes produced by implanted electrodes in cat brains. *J. Neurosurg.* 1957, **14**:302–328.

Coombs, J. S., J. C. Eccles, and P. Fatt. The electrical properties of the motoneurone membrane. *J. Physiol.* 1955, **130**:291–325.

Coombs, J. S., J. C. Eccles, and P. Fatt. The specific ionic conductances and the ionic movements across the motoneuronal membrane that produce the inhibitory postsynaptic potential. *J. Physiol.* 1955, **130**:326–373.

Cooper, R. Electrodes. *Amer. J. EEG Technol.* 1963, **3**:91–101.

Cooper, R. Storage of silver chloride electrodes. *EEG. Clin. Neurophysiol.* 1956, **8**:692.

Coraboeuf, E., and S. Weidmann. Potentiel de repos et potentiels d'action de muscle cardiaque measurée a l'aide d'électrodes internes. *C. R. Sci. Soc. Biol.* 1949, **143**:1329–1331.

Craib, W. H. A study of the electrical field surrounding active heart muscle. *Heart,* 1927, **14**:71–109.

Crain, S. M. Resting and action potentials of cultured chick embryo spinal ganglion cells. *J. Comp. Neurol.* 1956, **104**:285–329.

Curtis, H. J., and K. S. Cole. Membrane resting and action potentials from the squid giant axon. *J. Cell. Comp. Physiol.* 1942, **19**:135–144.

Curtis, H. J., and K. S. Cole. Transverse electric impedance of the squid giant axon. *J. Gen. Physiol.* 1938, **21**:757–765.

Curtis, D. R., and R. Eccles. The excitation of Renshaw cells by pharmacological agents applied electrophoretically. *J. Physiol.* 1958, **141**:435–445.

Danielli, J. F., and H. Davson. A contribution to the theory of permeability of thin films. *J. Cell. Comp. Physiol.* 1935, **5**:495–508.

Davis, D. A., and W. E. Thornton. *Radiotelemetry in Anesthesia and Surgery.* International Anesthesiology Clinics, Boston, Mass.: Little, Brown & Co., 1965. 586 pp.

Dawson, H., and M. G. Eggleton. *Starling and Lovatt Evans' Principles of Human Physiology*. Philadelphia, Pa. Lea and Febiger, 1962. 1879 pp.

Day, J., and M. Lippitt. A long-term electrode system for electrocardiography and impedance pneumography. *Psychophysiology*. 1964, **1**:174–182. U.S. Patent #3,420,223.

Day, J. L. Review of NASA-MSC electroencephalogram and electrocardiogram electrode systems including application techniques. *NASA Tech. Note*. 1968, TND-4398, 15 pp. U.S. Patent #3,420,223.

de Bethune, A. J. *Electrode Potentials: Temperature Coefficients*. In *The Encyclopedia of Electrochemistry*. C. A. Hampel, Ed. New York: Reinhold Publishing Corp., 1964. 1206 pp.

del Castillo, J., and B. Katz. Local activity at a depolarized nerve-muscle junction. *J. Physiol.* 1955, **128**:396–411.

Delgado, J. M. R. Electrodes for extracellular recording and stimulation. In *Physical Techniques in Biological Research*, W. L. Nastuk, Ed. New York: Academic Press, 1964, Chapter 3. 460 pp.

Delgado, J. M. R. Evaluation of permanent implantation of electrodes within the brain. *EEG Clin. Neurophysiol.* 1955, **7**:637–644.

Delgado, J. M. R. Permanent implantation of multilead electrodes in the brain. *Yale J. Biol. Med.* 1952, **24**:351–358.

De Wane, H. J., and W. J. Hamer. Electrochemical data. XII. *Natl. Bur. Std. (U.S.) Rept.* 9979. U.S. Dept. of Commerce, N.B.S., Washington, D.C.: Jan. 15, 1969.

Dieckmann, G., E. Gabriel, and R. Hassler. Size, form and structural peculiarities of experimental brain lesions obtained by thermocontrolled radiofrequency. *Confin. Neurol.* 1965, **26**:134–142.

Dodge, H. W., C. Petersen, C. W. Sem-Jacobsen, G. P. Sayre, and R. G. Bickford. The paucity of demonstrable brain damage following intracerebral electrography: report of a case. *Proc. Staff Meet. Mayo Clinic.* 1955, **30**:215–221.

Donnan, F. G. The theory of membrane equilibria. *Chem. Rev.* 1925, **1**:73–90.

Donnan, F. G., and A. B. Harris. Osmotic pressure and conductivity of aqueous solutions of congo red and reversible membrane equilibria. *J. Chem. Soc.* 1911, **99**:1554–1577.

Dowben, R. M., and J. E. Rose. A metal-filled microelectrode. *Science*. 1953, **118**:22–24.

Dowben, R. M. *Biological Membranes*. Boston: Little Brown & Co., 1969. 503 pp.

DuBois, D. A machine for pulling glass micropipets and needles. *Science*. 1931, **73**:344–345.

Eccles, J. C. *The Physiology of Nerve Cells*. Baltimore: Johns Hopkins Press, 1957. 270 pp.

Eccles, R. M. Intracellular potentials recorded from a mammalian sympathetic ganglion. *J. Physiol.* 1955, **130**:572–584.

Edelberg, R. Personal communication (1963).

Edelberg, R. "Electrical Properties of the Skin." In *Methods in Psychophysiology*, C. Brown, Ed., Baltimore: The Williams and Wilkins Company, 1967, Chapter I. 502 pp.

Edelberg, R., and N. R. Burch. Skin resistance and galvanic skin response. *Arch. Gen. Psychiatr.* 1962, **7**:163–169.

Edwards, J. G. Formation of food-cups in amoeba induced by chemicals. *Biol. Bull.* 1925, **48**:236–239.

Einthoven, W. *Die Aktionsstrome des Herzens. Handbuch Der Normalen und Pathologischen Physiologie*. Berlin: Julian Springer, 1928, pp. 785–862.

Einthoven, W. Weiteres über das Elektrokardiogramm. *Arch. ges. Physiol.* 1908. **122**:517–584.

Einthoven, W. Die galvanometrische Registrirung des menschlichen Elektrokardiogramm, zugleich eine Beurtheilung der Anwendung des Capillar-Elektrometers in der Physiologie. *Arch. ges. Physiol.* 1903, **99:**472–480.

Elson, L. Multi-bore glass sleeves for insulation. *J. Sci. Instr.* 1953, **30:**140.

Engbaek, L., and T. Hoshiko. Electrical potential gradients through frog skin. *Acta Physiol. Scand.* 1957, **39:**348–355.

Ettisch, G. H., and T. Peterfi. Zur Methodik der Elektrometric der Zelle. *Arch. ges. Physiol.* 1925, **208:**454–466.

Faraday, M. Experimental researches in electricity. *Phil. Trans. Royal Soc. London* 1839, 1–12 (Item 23).

Fascenelli, F. W., C. Cordova, D. G. Simons, J. Johnson, L. Pratt, and L. E. Lamb. Biomedical monitoring during dynamic stress testing. I. *Aerosp. Med.* 1966, **37:**911–922.

Feates, F. S., D. J. G. Ives, and J. H. Pryor. Alternating current bridge for measurement of electrolytic conductance. *J. Electrochem. Soc.* 1956, **103:**580–585.

Findlay, G. P., A. B. Hope, and E. J. Williams. Ionic relations of marine algae. *Austr. J. Biol. Sci.* 1969, **22:**1163–1178.

Findlay, G. P. Studies of action potentials in the vacuole and cytoplasm of *Nitella*. *Austr. J. Biol. Sci.* 1959, **12:**412–426.

Fingl, E., L. A. Woodbury, and H. Hecht. Effects of innervation and drugs upon direct membrane potentials of the embryonic chick myocardium. *J. Pharmacol. Exp. Ther.* 1952, **104:**103–114.

Fischer, G., G. P. Sayre, and R. G. Bickford. Histologic changes in the cat's brain after introduction of metallic and plastic coated wire used in electroencephalography. *Proc. Staff Meet. Mayo Clinic.* 1957, **32:**14–22.

Fischer-Williams, M., and R. A. Cooper. Depth recording from the human brain in epilepsy. *EEG Clin. Neurophysiol.* 1963, **15:**568–587.

Fischmann, E. J., R. N. Selye, and L. R. Crutcher. Clinical trial of a balsa-lithium electrode for conventional electrocardiography. *Amer. J. Cardiol.* 1962, **10:**846–851.

Forbes, A., S. Cobb, and McK. Cattell. An electrocardiogram and an electromyogram in an elephant. *Amer. J. Physiol.* 1921, **55:**385–389.

Forbes, T. W. An improved electrode for the measurement of potentials on the human body. *J. Lab. Clin. Med.* 1934, **19:**1234–1238.

Frank, K. Basic mechanisms of synaptic transmission in the central nervous system. *IRE Trans. Med. Electron.* 1959, **ME-6:**85–88.

Frank, K., and M. C. Becker. "Microelectrodes for Recording and Stimulation." In *Physical Techniques in Biological Research,* VA., W. L. Nastuk, Ed., New York: Academic Press, 1964, Chapter 2. 460 pp.

Frank, K., and M. G. F. Fuotes. Potentials recorded from the spinal cord with microelectrodes. *J. Physiol.* 1955, **130:**625–654.

Frank, K., and M. G. F. Fuortes. Unitary activity of spinal interneurones of cats. *J. Physiol.* 1956, **131:**424–435.

Freygang, W. H. An analysis of extracellular potentials from single neurons in the lateral geniculate nucleus of the cat. *J. Gen. Physiol.* 1958, **41:**143–564.

Freygang, W. H., and K. Frank. Extracellular potentials from single spinal motoneurones. *J. Gen. Physiol.* 1959, **42:**749–760.

Fricke, H. The theory of electrolyte polarization. *Phil. Mag.* 1932, **14:**310–318.

Fricke, H. The electric conductivity and capacity of disperse systems. *Physics.* 1931, **1:**106–115.

Fricke, H., and H. J. Curtis. Electric impedance of suspensions of leucocytes. *Nature*. 1935, **135**:436.

Fricke, H., and H. J. Curtis. Electric impedance of yeast cells. *Nature,* 1934, **134**:102–103.

Fricke, H., and H. J. Curtis. Specific resistance of the interior of the red blood corpuscle. *Nature,* 1934, **133**:651.

Fricke, H., H. P. Schwan, K. Li, and V. Bryson. A dielectric study of the low-conductance surface membrane in E-coli. *Nature,* 1956, **177**:134–135.

Gasser, H. S., and J. Erlanger. The role played by the sizes of the constituent fibers of a nerve trunk in determining the form of its action potential. *Amer. J. Physiol.* 1927, **80**:522–547.

Geddes, L. A. Cortical electrodes. *EEG Clin. Neurophysiol.* 1949, **1**:523. Illustrated on cover of *Sci. Amer.* 1948, **179,** No. 4.

Geddes, L. A., and L. E. Baker. The relationship between input impedance and electrode area in recording the ECG. *Med. Biol. Eng.* 1966, **4**:439–450.

Geddes, L. A., and H. E. Hoff. The discovery of bioelectricity and current electricity (The Galvani–Yolta controversy) *IEEE Spectrum* 1971. **8**:38–46.

Geddes, L. A., and L. E. Baker. Chlorided silver electrodes. *Med. Res. Eng.* 1967, **6**:33–34.

Geddes, L. A., L. E. Baker, and M. McGoodwin. The relationship between electrode area and amplifier input impedance in recording muscle action potentials. *Med. Biol. Eng.* 1967, **5**:561–569.

Geddes, L. A., L. E. Baker, and A. G. Moore. The use of liquid-junction electrodes in recording the human electrocardiogram (ECG). *J. Electrocardiol.* 1968, **1**:51–56.

Geddes, L. A., L. E. Baker, and A. G. Moore. Optimum electrolytic chloriding of silver electrodes. *Med. Biol. Eng.* 1969, **7**:49–56.

Geddes, L. A., C. P. Da Costa, and G. Wise. The impedance of stainless-steel electrodes. *Med. Biol. Eng.* 1971, **9**:511–521.

Geddes, L. A., J. D. McCrady, H. E. Hoff, and A. Moore. Electrodes for large animals. *Southwest. Vet.* 1964, **18**:56–57.

Geddes, L. A., M. Partridge, and H. E. Hoff. An EKG lead for exercising subjects. *J. Appl. Physiol.* 1960, **15**:311–312.

Geddes, L. A., J. Rossborough, H. Garner, J. Amend, A. G. Moore, and M. Szabuniewicz. Obtaining electrocardiograms on animals without skin preparation or penetration. *Vet. Med.* 1970, **12**:1163–1168.

Geddes, L. A., H. E. Hoff, and F. Cohen. The electrocardiogram of the elephant. *Southwest. Vet.* 1967, **20**:211–216.

Gelfan, S. A non-polarizable micro-electrode. *Proc. Soc. Exp. Biol. Med.* 1925, **23**:308–309.

Gerstein, G. L., and W. A. Clark. Simultaneous studies of firing patterns in several neurones. *Science.* 1964, **193**:1325–1327.

Gesteland, R. C., B. Howland, J. Y. Lettvin, and W. H. Pitts. Comments on microelectrodes. *Proc. IRE.* 1958, **47**:1856–1862.

Gibson, T. C., W. E. Thornton, W. P. Algary, and E. Craige. Telecardiography and the use of simple computers. *N. Engl. J. Med.* 1962, **267**:1218–1224.

Giese, A. C. *Cell Physiology,* 2nd ed. (reprinted 1966), Philadelphia: W. B. Saunders Company, 1952. 592 pp.

Giffin, R. R., and C. Susskind. EEG skin "burn": electrical or chemical. *Med. Res. Eng.* 1967, **6**:32–34.

Goldstein, A. G., W. Sloboda, and J. B. Jennings. Spontaneous electrical activity of three types of silver EEG electrodes. *Psychophysiol. Newsl.* 1962, **8:**10–16.

Gouy, M. Sur la constitution de la charge électrique à la surface d'un électrolyte. *J. Phys.* (*Paris*) 1910, **9:**457–468.

Graham, J., and R. W. Gerard. Membrane potentials and excitation of impaled single muscle fibers. *J. Cell. Comp. Physiol.* 1946, **28:**99–117.

Grahame, D. C. Mathematical theory of the faradic admittance. *J. Electrochem. Soc.* 1952, **99:**370C–385C.

Gray, J. A. B., and G. Svaetichin. Electrical properties of platinum-tipped microelectrodes in Ringer's solution. *Acta Physiol. Scand.* 1951–1952, **24:**278–284.

Green, J. D. A simple microelectrode for recording from the central nervous system. *Nature.* 1958, **182:**962.

Greenwald, D. U. Electrodes used in measuring electrodermal responses. *Amer. J. Psychol.* 1936, **40:**658–662.

Greyson, J. Silver-silver chloride electrodes using optical silver chloride crystals. *J. Electrochem. Soc.* 1962, **109:**745–746.

Grundfest, H. Instrument requirements and specifications in bioelectric recording. *Ann. N.Y. Acad. Sci.* 1955, **60:**841–859.

Grundfest, H. Comparative electrobiology of excitable membranes. *Adv. Comp. Physiol.* 1966, **2:**1–116.

Grundfest, H., and B. Campbell. Origin, conduction and termination of impulses in dorsal spino-cerebellar tracts of cats. *J. Neurophysiol.* 1942, **5:**275–294.

Grundfest, H., R. W. Sengstaken, W. H. Oettinger and R. W. Gurry. Stainless steel microneedle electrodes made by electro-pointing, *Rev. Sci. Instr.* 1950, **21:**360–361.

Guld, C. A glass-covered platinum microelectrode. *Med. Elect. Biol. Eng.* 1964, **2:**317–327.

Guzman, C., M. Alcaraz, and A. Fernandez. Rapid procedure to localize electrodes in experimental neurophysiology. *Bol. Inst. Estud. Med. Biol.* (*Mexico*). 1958, **16:**29–31.

Haggard, E. A., and R. Gerbrands. An apparatus for the measurement of continuous changes in palmar skin resistance. *J. Exp. Psychol.* 1947, **37:**92–98.

Hagiwara, S., K. Naka, and S. Chichibu. Membrane properties of a barnacle muscle fiber. *Science.* 1964, **143:**1446–1448.

Hartman, H. B., and E. G. Boettiger. The functional organization of the propus-dactylus organ in the cancer irroratus say. *Comp. Biochem. Physiol.* 1967, **22:**651–663.

Hecht, H., and L. A. Woodbury. Excitation of human auricular muscle and the significance of the intrinsicoid deflection of the auricular electrocardiogram. *Circulation.* 1950, **2:**37–47.

Hecht, H. H. Normal and abnormal transmembrane potentials of the spontaneously beating heart. *Ann. N.Y. Acad. Sci.* 1956–1957, **65:**700–733.

Helmholtz, H. Studien über electrische Grenzschichten. *Ann. Phys. Chem.* 1879, **7:**337–382.

Hendler, E., and L. J. Santa Maria. Response of subjects to some conditions of a simulated orbital flight pattern. *Aerosp. Med.* 1961, **32:**126–133.

Hendley, C. D., and R. Hodes. Effects of lesions on subcortically evoked movement in cat. *J. Neurophysiol.* 1953, **16:**587–594.

Henry, F. Dependable electrodes for the galvanic skin response. *J. Gen. Psychol.* 1938, **18:**209–211.

Herrmann, G. R. New electrode for taking electrocardiograms. *Circ. Res.* 1962, **11**:736–738.

Hermann, L. *Allgemeine Muskelphysik. Handbuch der Physiologie der Bewegunsapparate.* 1879, **1**:1–260.

Hetherington, A. W., and S. W. Ranson. Hypothalamic lesions and adiposity in the rat. *Anat. Rec.* 1940, **78**:149–172.

Hilderbrand, J. H. Some applications of the hydrogen electrode in analyses research and teaching. *J. Amer. Chem. Soc.* 1913, **35**:847–871.

Hill, D. W. In Hales, Marey, and Chauveau, Rept. on 1966 course "Classical Physiology with Modern Instrumentation." NIH Grant Rept. HE 05125, 1967.

Hills, G. J., and D. J. G. Ives. In *Reference Electrodes,* D. J. G. Ives, and G. J. Janz. New York: Academic Press, 1961. 651 pp.

Hisada, M. Membrane resting and action potentials from a protozoan, *Noctiluca Scintillans. J. Cell. Comp. Physiol.* 1957, **50**:57–71.

Hodgkin, A. L. *The Conduction of the Nervous Impulse.* Springfield, Ill.: Charles C. Thomas, 1964. 108 pp.

Hodgkin, A. L. Evidence for electrical transmission in nerve, I and II. *J. Physiol.* 1937, **90**:183–210, 211–232.

Hodgkin, A. L. A note on conduction velocity. *J. Physiol.* 1954, **125**:221–224.

Hodgkin, A. L. The relation between conduction velocity and the electrical resistance outside a nerve fiber. *J. Physiol.* 1939, **94**:560–570.

Hodgkin, A. L. The subthreshold potentials in a crustacean nerve fibre. *J. Physiol.* 1939, **126**:87–121.

Hodgkin, A. L., and A. F. Huxley. Action potentials recorded from inside a nerve fiber. *Nature.* 1939, **144**:710–711.

Hodgkin, A. L., and A. F. Huxley. The components of membrane conductance in the giant axon of the Loligo. *J. Physiol.* 1952, **116**:473–496.

Hodgkin, A. L., and A. F. Huxley. Currents carried by sodium and potassium ions through the membrane of the giant axon of Loligo. *J. Physiol.* 1952, **116**:449–472.

Hodgkin, A. L., A. F. Huxley, and B. Katz. Ionic currents underlying activity in the giant axon of the squid. *Arch. Sci. Physiol.* 1949, **3**:129–150.

Hodgkin, A. L., and P. Horowicz. The differential action of hypertonic solutions on the twitch and action potential of a muscle fiber. *J. Physiol.* 1957, **136**:17*P*–18*P*.

Hodgkin, A. L., and A. F. Huxley. The dual effect of membrane potential on sodium conductance in the giant axon of *Loligo. J. Physiol.* 1952, **116**:497–506.

Hodgkin, A. L., and A. F. Huxley. Movement of radioactive potassium and membrane current in a giant axon. *J. Physiol.* 1953, **121**:403–414.

Hodgkin, A. L., and A. F. Huxley. A quantitative description of membrane current and its application to conduction and excitation in nerve. *J. Physiol.* 1952, **117**:500–544.

Hodgkin, A. L., A. F. Huxley, and B. Katz. Measurement of current-voltage relations in the membrane of the giant axon of *Loligo. J. Physiol.* 1952, **116**:424–448.

Hoff, H. E., and L. A. Geddes. The rheotome and its prehistory. *Bull. Histol. Med.* 1957, **31**:212–347.

Hoff, H. E., and L. A. Geddes. *Experimantal Physiology,* 1st ed. Houston, Tex.: Baylor University College of Medicine, 1958. Library of Congress Card #58-2380.

Hoff, H. E., L. A. Geddes, A. G. Moore, and J. Vasku. The thermode – an electrode for demonstrating the T-wave change due to delayed repolarization of the ventricle. *J. of Electrocardiology* 1970, **3**:333–335.

Hoff, H. E., L. H. Nahum, and B. Kisch. Influence of right and left ventricles on the electrocardiogram. *Amer. J. Physiol.* 1941, **131**:687–692.

van't Hoff, J. H. Die Rolle des osmotisches Durckes in der Analogie zwichen Losungen und Gasen. *Z. Physik. Chem.* 1887, **1**:480–508.

Hoffman, B. L., and E. E. Suckling. Cellular potentials of intact mammalian hearts. *Amer. J. Physiol.* 1952, **170**:375–362.

Hoffman, B. L., and E. E. Suckling. Cardiac cellular potentials: effect of vagal stimulation and acetylcholine. *Amer. J. Physiol.* 1953, **173**:312–320.

Hoffman, B. L., and P. F. Cranefield. *Electrophysiology of the Heart.* New York: McGraw-Hill, Book Company, The Blakiston Division, 1960. 323 pp.

Hollander, P. B., and J. L. Webb. Cellular membrane potentials and contractility of normal rat atrium and the effects of temperature, tension and stimulus frequency. *Circ. Res.* 1955, **3**:604–612.

Hokin, L. E., and M. H. Hokin. The chemistry of cell membranes. *Sci. Amer.* 1965, **213**:78–86.

Holman, M. E. Membrane potentials recorded with high resistance microelectrodes. *J. Physiol.* 1958, **141**:464–488.

Holter, H. Problems of pinocytosis with special regard to amoebae. *Ann. N.Y. Acad. Sci.* 1959, **78**:525–537.

Horsley, Sir V., and R. H. Clarke. The structure and functions of the cerebellum examined by a new method. *Brain.* 1908, **31**:45–124.

Hoshiko, T. Electrogenesis in frog skin. In *Biophysics of Physiological and Pharmacological Actions.* A. M. Shanes, Ed. *Amer. Assoc. Adv. Sci.* 1961, **29**:31–47.

Hoyle, G., and T. Smith. Giant muscle fibers in a barnacle, *Balanus nubilus Darwin. Science.* 1963, **139**:49–50.

Hubel, D. H. Tungsten microelectrode for recording from single units. *Science.* 1957, **125**:549–550.

Hunsperger, R. W., and O. A. M. Wyss. Quantitative Ausschaltung von Nervengewebe durch Hochfrequenzkoagulation. *Helv. Physiol. Acta.* 1953, **11**:283–304.

Huf, E. Versuche über den Zusammenhangzwichen Stoffwechsel Potentialbildung und Funktion der Froschant. *Arch. ges. Physiol.* 1935, **235**:655–673.

Hursh, J. B. Conduction velocity and diameter of nerve fibers. *J. Physiol.* 1939, **127**:131–139.

Huxley, A. F., and R. Stampfli. Direct determination of membrane resting potential in single myelinated nerve fibers. *J. Physiol.* 1951, **112**:476–495.

Huxley, A. F., and R. Stampfli. Evidence for saltatory conduction in peripheral myelinated nerve fibers. *J. Physiol.* 1949, **108**:315–319.

Jaenicke, W., R. P. Tischer, and H. Gerischer. Die anodische Bildung von Silberchlorid-Deckschichten und Umlagerungserscheinungen nach ihrer kathodischen Reduktion zu Silber. *Z. Elektrochem.* 1955, **59**:448–455.

Jahn, H. Über den Dissociationsgrad und das Dissociationsgleichgewicht stark dissociierter Elektrolyte. *Z. Physik. Chem.* 1900, **33**:545–576.

James, W. B., and H. B. Williams. The electrocardiogram in clinical medicine. *Amer. J. Med. Sci.* 1910, **140**:408–421.

Janz, G. J. Silver-silver halide electrodes. In *Reference Electrodes,* D. J. G. Ives and G. J. Janz. New York: Academic Press, 1961. 651 pp.

Janz, G. J., and D. J. G. Ives. Silver-silver chloride electrodes. *Ann. N.Y. Acad. Sci.* 1968, **148:**210–221.

Janz, G. J., and F. J. Kelly. Reference electrodes. In *Encyclopedia of Electrochemistry.* C. A. Hampel, Ed. New York: Reinhold Publishing Company, 1964. 1206 pp.

Janz, G. J., and H. Taniguchi. The silver-silver halide electrodes. *Chem. Rev.* 1953, **53:**397–437.

Jasper, H. H., R. T. Johnson, and L. A. Geddes. The RCAMC electromyograph. *Can. Army Med. Rept.* C6174, 1945.

Jaron, D., S. A. Briller, H. P. Schwan, and D. B. Geselowitz. Nonlinearity of cardiac pacemaker electrodes. *IEEE Trans. Bio-Med. Eng.* 1969, **BME-16:**132–138.

Jeans, J. H. *The Mathematical Theory of Electricity and Magnetism.* 4th ed. Cambridge: Cambridge University Press, 1920. pp. 300–363.

Jenkner, F. A new electrode material for multipurpose biomedical application. *EEG Clin. Neurophysiol.* 1967, **23:**570–571.

Jenks, J. L., and A. Graybiel. A new simple method of avoiding high resistance and overshooting in taking standardized electrocardiograms. *Amer. Heart J.* 1935, **10:**683–695.

Johnson, J. B. Thermal agitation of electricity in conductors. *Phys. Rev.* 1928, **32:**97–109.

Jolliffe, C. B. A study of polarization capacity and resistance at radio frequencies. *Phys. Rev.* 1923, **22:**293–302.

Jones, G., and G. M. Bollinger. The measurement of the conductance of electrolytes. VII. On platinization. *J. Amer. Chem. Soc.* 1935, **57:**280–284.

Jones, G., and G. M. Bollinger. The measurement of the conductance of electrolytes. *J. Amer. Chem. Soc.* 1931, **53:**411–451.

Jones, G., and S. M. Christian. The measurement of the conductance of electrolytes, VI. *J. Amer. Chem. Soc.* 1935, **57:**272–280.

Kado, R. T. Personal communication (December, 1965).

Kado, R. T., and W. R. Adey. Electrode problems in central nervous monitoring in performing subjects. *Ann. N.Y. Acad. Sci.* 1968, **148:**263–278.

Kado, R. T., W. R. Adey, and J. R. Zweizig. Electrode system for recording EEG from physically active subjects. *Proc. 17th Ann. Conf. Eng. Med. Biol.* 1964. Cleveland. p. 5 (129 pp).

Kahn, A. Fundamentals of biopotentials and their measurement. Biomedical Sciences Instrumentation, 1964 (Dallas); *Amer. J. Pharm. Educ.* 1964, **28:**805–814, U.S. Patent #3, 295, 515.

Kahn, A. Motion artifacts and streaming potentials in relation to biological electrodes. *6th Int. Conf. Med. Elect. Biol. Eng.* 1965. Tokyo, Japan. Conference Digest Paper. pp. 562–563.

Kao, C. Y. A method of making prefilled microelectrodes. *Science.* 1954, **119:**846–847.

Katz, B. The electrical properties of the muscle fiber membrane. *J. Physiol.* 1948, **135:**505–534.

Katz, B., and O. H. Schmitt. Electric interaction between two adjacent nerve fibers. *J. Physiol.* 1940, **97:**471–488.

Kennard, D. W. Glass Microcapillary Electrodes. In P. E. K. Donaldson, *Electronic Apparatus for Biological Research.* London: Butterworth's Scientific Publications, 1958, 718 pp. Chapter 35.

Kennedy, J. L., and R. C. Travis. Surface electrodes for recording bioelectric potentials. *Science.* 1948, **108:**183.

King, E. E. Errors in voltage in multichannel ECG recordings using newer electrode materials. *Amer. Heart J.* 1964, **18:**295–297.

Kniskern, P. W. Detergent between electrodes and skin. *Amer. Heart J.* 1961, **61:**427.

Kohlrausch, F. Uber platinirte Elektroden und Widerstandsbestimmung. *Ann. Phys. Chem.* 1897, **60:**315–332.

Kohlrausch, F., and L. Holborn. Das Leitvermogen der Elektrolyte. Leipzig: Verlag von B. G. Teubner, 1898. 211 pp.

Krasno, L. R., and A. Graybiel. A new "plaster electrode" not requiring paste. *Amer. Heart J.* 1955, **49:**774–776.

Krieg, W. J. S. Accurate placement of minute lesions in the brains of albino rats. *Quart. Bull. Northwestern Univ. Med. Sch.* 1946, **20:**199–208.

Krogh, A. Osmotic regulation in the frog (*R. esculenta*) by active absorption of chloride ions. *Scand. Arch. Physiol.* 1937, **76:**60–74.

Kurella, G. A. The difference of electric potentials and the partition of ions between the medium and vacuole of the alga. In *Glass Microelectrodes*. M. Lavallée, O. F. Schanne, and N. C. Hebert, Eds. New York: John Wiley and Sons, 1969. 446 pp.

Lagow, C. H., R. J. Sladek, and P. C. Richardson. Anodic insulated tantalum oxide electrocardiograph electrodes. *IEEE Trans. Bio-Med. Eng.* 1971, **BME-18:**162–164.

Lategola, M. T., J. Naughton, C. M. Brake, and P. Lyne. Use of simultaneous multilead telecardiography for monitoring cardiovascular rehabilitants during exercise. *Aerosp. Med.* 1969, **40:**1258–1264.

Leaf, A., and E. Dempsey. Some effects of mammalian neurohypophyseal hormones on metabolism and active transport of sodium by the isolated toad bladder. *J. Biol. Chem.* 1960, **235:**2160–2163.

Lewes, D. British Patent applications #52253/64 and #44049/65.

Lewes, D. Multipoint electrocardiography without skin preparation. *Lancet.* 1965, **2:**17–18; British patent application #52253/64.

Lewes, D. Electrode jelly in electrocardiography. *Brit. Heart J.* 1965, **27:**105–115.

Lewes, D. Multipoint electrocardiography without skin preparation. *World Med. Elect.* 1966, **4:**240–245.

Lewes, D., and D. Hill. Personal communication (1966).

Lewis, T. Polarisable as against non-polarisable electrodes, *J. Physiol.* 1914–1915, **49:**L–Lii.

Lewis, T. *The Mechanism and Graphic Registration of the Heart Beat,* 3rd ed. London: Shaw & Sons, 1925. 529 pp.

Ling, G., and R. W. Gerard. The normal membrane potential of frog sartorius fibers. *J. Cell. Comp. Physiol.* 1949, **34:**383–396.

Littmann, D. Electrode contact fluid in electrocardiography. *Amer. J. Cardiol.* 1959, **4:**554.

Livingston, L. G., and B. M. Duggar. Experimental procedures in a study of the location and concentration within the host cell of the virus of tobacco mosaic. *Biol. Bull.* 1934, **67:**504–512.

Lopez, A., and P. Richardson. Capacitive electrocardiographic and bioelectric electrodes. *IEEE Trans. Bio-Med. Eng.* 1969, **BME-16:**99.

Lorente de Nó, R. *A Study of Nerve Physiology.* Part 2, Chapter 16. New York: Rockefeller Institute, 1947.

Loucks, R. B., H. Weinberg, and M. Smith. The erosion of electrodes by small currents. *EEG Clin. Neurophysiol.* 1959, **11:**823–826.

Lucchina, G. G., and C. G. Phipps. An improved electrode for physiological recording. *Aerosp. Med.* 1963, **34:**230–231.

Lucchina, G. G., and C. G. Phipps. A vectorcardiographic lead system and physiologic electrode configuration for dynamic readout. *Aerosp. Med.* 1962, **33:**722–729.

Luttgau, H. C., and R. Niedergerke. The antagonism between Ca and Na ions on the frog heart. *J. Physiol.* 1958, **143**:486–505.

Lux, D. Microelectrodes of high stability. *EEG Clin. Neurophysiol.* 1960, **12**:928–929.

Lykken, D. T. Properties of electrodes used in electrodermal measurements. *J. Comp. Psychol.* 1959, **52**:629–634.

MacInnes, D. A. *The Principles of Electrochemistry* New York: Reinhold Publishing Company, Dover Publications, 1961. 478 pp.

MacInnes, D. A., and J. A. Beattie. The free energy of dilution and the transference numbers of lithium chloride solutions. *J. Amer. Chem. Soc.* 1920, **42**:1117–1128.

MacInnes, D. A., and K. Parker. Potassium chloride concentration cells. *J. Amer. Chem. Soc.* 1915, **37**:1445–1461.

MacNichol, E. J., and G. Svaetichin. Electric responses from isolated retinas of fishes. *Amer. J. Ophthalmol.* 1958, **46**:26–46.

Marchant, E. W., and E. W. Jones. The effect of electrodes of different metals on the skin currents. *Brit. Heart J.* 1940, **2**:97–100.

Marmont, G. Studies on the axon membrane. *J. Cell. Comp. Physiol.* 1949, **34**:351–382.

Macleod, A. G. The electrocardiogram of cardiac muscle. *Amer. Heart J.* 1938, **15**:402–413.

Macleod, A. G. The electrogram of cardiac muscle: an analysis which explains the regression or T deflection. *Amer. Heart J.* 1938, **15**:165–186.

Marples, M. J. Life on the human skin. *Sci. Amer.* 1969, **220**:108–115.

Marshall, W. H. An application of the frozen sectioning technique for cutting serial sections thru the brain. *Stain Technol.* 1940, **15**:133–138.

Mason, R. E. and I. Likar. A new system of multiple lead electrocardiography. *Amer. Heart J.* 1966, **71**:196–205.

Maxwell, J. Preparation of the skin for electrocardiography. *Brit. Med. J.* 1957, **2**:942; *Brit. Med. J.* 1958, **1**:41.

Mela, M. J. Microperforation with laser beam in the preparation of microelectrodes. *IEEE Trans. Bio-Med. Eng.* 1966, **BME-13**:70–76.

Merritt, E. Polarization capacity and polarization resistance as dependent upon frequency. *Amer. Phys. Soc.* 1921, **17**:524–526.

Miller, C. W. A direct measurement of polarization capacity and phase angle. *Phys. Rev.* 1923, **22**:622–628.

Mills, L. W. A fast inexpensive method of producing large quantities of metallic microelectrodes. *EEG Clin. Neurophysiol.* 1962, **14**:278–279.

Montes, L. F., J. L. Day, and L. Kennedy. The response of human skin to long-term space flight electrodes. *J. Invest. Dermatol.* 1967, **49**:100–102.

Moore, J. A., Ed. *Physiology of the Amphibia.* New York: Academic Press, 1964. 643 pp.

Moore, J. W., and K. S. Cole. Voltage clamp techniques. In *Physical Techniques in Biological Research,* Vol. VI, W. L. Nastuk, Ed. New York: Academic Press, 1963. 425 pp.

Morrison, N. K. Development of conductive cloth plantar electrode for use in measuring skin resistance. U.S.A.F. Wright-Patterson Air Development Center. 1958, TN-58-284. Astia Document AD 204425.

Mudd, S., M. McCutcheon, and B. Lucke. Phagocytosis. *Physiol. Rev.* 1934, **14**:210–275.

Murdock, C. C., and E. E. Zimmerman. Polarization impedance at low frequencies. *Physics.* 1936, **7**:211–219.

Nastuk, W. The electrical activity of the muscle cell membrane at the neuromuscular junction. *J. Cell. Comp. Physiol.* 1953, **42:**249–272.

Nastuk, W. L., and W. L. Hodgkin. The electrical activity of single muscle fibers. *J. Cell. Comp. Physiol.* 1950, **35:**39–73.

Nernst, W. Die elektromotorische Wirksamkeit der Jonen. *Z. Physik. Chem.* 1889, **4:**129–188.

Nishi, S., H. Soeda, and K. Koketsu. Studies on sympathetic B and C neurons and patterns of preganglionic innervation. *J. Cell. Comp. Physiol.* 1965, **66:**19–32.

O'Connell, D. N., and B. Tursky. Special modifications of the silver-silver chloride sponge electrode for skin recording. U.S. Air Force Office of Scientific Research. Contract AF 49 (638)–728. 37 pp.

O'Connell, D. N., B. Tursky, and M. T. Orne. Electrodes for recording skin potential. *Arch. Gen. Psychiatr.* 1960, **3:**252–258.

Offner, F. F. Electrical properties of tissues in shock therapy. *Proc. Soc. Exp. Biol. Med.* 1942, **49:**571–575.

Oikawa, T., T. Ogawa, and K. Motokawa. Origin of so-called cone action potential. *J. Neurophysiol.* 1959, **22:**102–111.

Osterhout, W. J. V. Physiological studies of single plant cells. *Biol. Rev.* 1931, **6:**369–411.

Osterhout, W. J. V., and S. E. Hill. Action curves with single peaks in *Nitella* in relation to the movement of potassium. *J. Gen. Physiol.* 1940, **23:**743–748.

Ottoson, D., F. Sjostrand, S. Stenstrom, and G. Svaetichin. Microelectrode studies on the emf of the frog skin related to electron microscopy of the dermo-epithelial junction. *Acta Physiol. Scand.* 1953, **29:**611–624.

Pardee, H. E. B. An error in the electrocardiogram arising in the application of the electrodes. *Arch. Int. Med.* 1917, **20:**161–166.

Pardee, H. E. B. Concerning the electrodes used in electrocardiography. *Amer. J. Physiol.* 1917, **44:**80–83.

Parker, T. G. Simple method for preparing and implanting fine wire electrodes. *Amer. J. Phys. Med.* 1968, **47:**247–249.

Parker, T. G. V. A. Hospital, Houston, Tex. Personal communication (December, 1966).

Parsons, R. Electrode double layer. *The Encyclopedia of Electrochemistry, C. A.* Hampel, Ed. New York: Reinhold Publishing Company, 1964, 1206 pp.

Patten, C. W., F. B. Ramme, and J. Roman. Dry electrodes for physiological monitoring. NASA Tech. Note NASA TN D-3414. Washington, D.C.: National Aeronautics and Space Administration, 1966. 32 pp.

Pfeffer, W. *Osmotische Untersuchungen.* Leipzig: von W. Engelmann, Verlag, 1877. 236 pp.

Phillips, C. G. The dimensions of a cortical motor point. *J. Physiol.* 1955, **129:**21P–22P.

Plonsey, R. *Bioelectric Phenomena.* New York: McGraw Hill Book Company, 1969. 380 pp.

Plutchik, R., and H. R. Hirsch. Skin impedance and phase angle as a function of frequency and current. *Science.* 1963, **141:**927–928.

Prosser, C. L., and N. S. Rafferty. Electrical activity in chick amnion. *Amer. J. Physiol.* 1956, **187:**546–548.

Puck, T. T., K. Wasserman, and A. P. Fishman. Some effects of inorganic ions on the active transport of phenol red by isolated kidney tubules of the flounder. *J. Cell. Comp. Physiol.* 1952, **40:**73–88.

Radenovitch, Ch., L. A. Tsaneva, and A. N. Sinyukhin. The varied nature of rhythmic fluctuations in the potential of the cytoplasm and the membrane of plant cells. *Biophysics* (a translation of *Biofizika*). 1968, **13:**319–333.

Rappaport, M. B., C. Williams, and P. D. White. An analysis of the relative accuracies of the Wilson and Goldberger methods for registering unipolar and augmented unipolar electrocardiographic leads. *Amer. Heart J.* 1949, **37:**892–917.

Ray, C. D. A new multicontact electrode for depth brain studies. *Proc. 16 Ann. Conf. Eng. Med. Biol.* 1963, **5:**44–45.

Ray, C. D., R. G. Bickford, L. C. Clark, R. E. Johnston, T. M. Richards, D. Rogers, and W. S. Russert. A new multicontact, multipurpose, brain depth probe: details of construction. *Proc. Staff Meet. Mayo Clin.* 1965, **40:**771–804.

Ray, C. D., and A. E. Walker. Platinized platinum wire brain depth probes: a comparison with electrodes of stainless steel. *Proc. Amer. EEG Soc.* 1965, **19:**90–91.

Rayport, M. Anatomical identification of somatic sensory cortical neurones responding with short latencies to specific afferent volleys. *Fed. Proc.* 1957, **16** (Part 1):104.

Renshaw, B., A. Forbes, and B. R. Morison. Activity of isocortex and hippocampus: electrical studies with microelectrodes. *J. Neurophysiol.* 1940, **3:**74–105.

Richardson, P. C. The insulated electrode: a pasteless electrocardiographic technique. *Proc. 20th Ann. Conf. Eng. Med. Biol.* 1967, **9:**15.7.

Richardson, P. C. Progress in long-term physiologic sensor development. *Proc. Biomed. Eng. Symp., San Diego.* 1967, :39–44.

Richardson, P. C., F. K. Coombs, and R. M. Adams. Some new electrode techniques for long-term physiologic monitoring. *Aerosp. Med.* 1968, **39:**745–750.

Richardson, P. C. and F. K. Coombs. New construction techniques for insulated electrocardiographic electrodes. *Proc. 21st Ann. Conf. Eng. Med. Biol.* 1968, **10:**13A.1.

Robertson, J. D. Granulo-fibrillar and globular structure in unit membrane. *Ann. N.Y. Acad. Sci.* 1966, **137:**421–440.

Robinson, D. A. Electrical properties of metal microelectrodes. *Proc. IEEE.* 1968, **56:**1065–1071.

Robinson, F. R., and M. T. Johnson. Histopathological studies of tissue reactions to various metals implanted in cat brains. *ASD Tech. Rept.* 61–397, 1961. 13 pp. U.S.A.F. Wright-Patterson AFB, Ohio.

Robinson, R. A., and R. H. Stokes. In *Electrolyte Solutions,* 2nd ed. New York: Academic Press, 1959.

Roman, J. Flight research program—III. High impedance electrode techniques. *Aerosp. Med.* 1966, **37:**790–795.

Roman, J., and L. Lamb Electrocardiography in flight. *Aerosp. Med.* 1962, **33:**527–544.

Rose, K. D. Telemetering physiologic data from athletes. *Proc. Int. Telemetering Conf.* 1963, **1:**225–241. McGavock and Assoc., Pasadena, Calif.

Rosenberg, T. On accumulation and active transport in biological systems. *Acta Chem. Scand.* 1948, **2:**14–33.

Roth, I. A self-retaining skin contact electrode for chest leads in electrocardiography. *Amer. Heart J.* 1933–1934, **9:**526–529.

Roth, J., H. M. Cullen, and V. Milstein. The use of carbon electrodes for chronic cortical recording. *EEG Clin. Neurophysiol.* 1966, **21:**611–615.

Rothschild, Lord. The polarization of a calomel electrode. *Proc. Roy. Soc.* (*London*). 1938, **125:**283–290.

Rowland, V. Sample nonpolarizable electrode for chronic implantation. *EEG Clin. Neurophysiol.* 1961, **13**:290–291.

Rowley, D. A., S. Glagov, and P. Stoner. Fluid electrodes for monitoring the electrocardiogram during activity and for prolonged periods of time. *Amer. Heart J.* 1961, **62**:263–269.

Ruch, T. C., and H. D. Patton. *Physiology and Biophysics.* Philadelphia, Pa.: W. B. Saunders Company, 1965. 1242 pp.

Ruhland, W., and C. Hoffmann. Die Permeabilität von *Beggiatoa Mirabilis. Planta.* 1925, **1**:1–83.

Rush, S., E. Lepeschkin, and H. O. Brooks. Electrical and thermal properties of double-barreled ultra microelectrodes. *IEEE Trans. Bio-Med. Eng.* 1968, **BME-15**:80–93.

Rushton, W. A. H. A theory of the effects of fiber size in medullated nerve. *J. Physiol.* 1951, **115**:101–122.

Russell, H. B. The use of Cambridge electrode jelly. *Lancet.* 1935, **2**:1173.

Rylander, H. G. Capacitive electrocardiograph electrodes. M. S. thesis, University of Texas, Austin, 1970. *Electron. Res. Center Rept.* **99**:Sept. 4, 1970.

Sano, T., M. Ono, and T. Shimamoto. Intrinsic deflections, local excitation and transmembrane action potentials. *Circ. Res.* 1956, **4**:444–449.

Sauer, G. L. *Teen Skin.* Springfield, Ill.: Charles C. Thomas Publisher, 1965. 57 pp.

Schanne, O. F., M. Lavallée, R. Laprade, and S. Gagné. Electrical properties of glass microelectrodes. *Proc. IEEE.* 1968, **56**:1072–1082.

Schiebel, M. E., and A. B. Schiebel. Histological localization of microelectrode placement in brain by ferrocyanide and silver staining. *Stain Technol.* 1956, **31**:1–5.

Schmitt, O. H., M. Okajima, and M. Blaug. Skin preparation and electrocardiographic lead impedance. *Dig. IRE Int. Conf. Med. Electron.* 1961, p. 236. Washington, D.C.: MacGregor & Werner, 288 pp.

Schwan, H. P. *Determination of biological impedances in Physical Techniques in Biological Research,* Vol. VIB., W. Nastuk, Ed. New York and London: Academic Press, 1963. 425 pp.

Schwan, H. P. Electrode polarization impedance and measurements in biological materials. *Ann. N.Y. Acad. Sci.* 1968, **148**:191–209.

Schwan, H. P. Electrode polarization in AC steady state impedance of biological systems. *Dig. 6th Int. Conf. Med. Electron. Biol. Eng. (Tokyo).* 1965, Paper 33–1.

Schwan, H. P., and J. G. Maczuk. Electrode polarization impedance; limits of linearity. *Proc. 18th Ann. Conf. Eng. Biol. Med.* Philadelphia, Pa.. Washington, D.C.: McGregor & Werner, 1965. 270 pp.

Scott, R. N. A method of inserting wire electrodes for electromyography. *IEEE Trans. Bio-Med. Eng.* 1965, **BME-12**:46–47.

Seelig, M. C. Localized gangrene following the hypodermic administration of calcium chloride. *J. Amer. Med. Ass.* 1925, **84**:1413–1414.

Sellier, J., and H. Verger. Étude experimentale des fonctions de la couche optique. *C. R. Soc. Biol.* 1903, **55**:485–487.

Sellier, J., and H. Verger. Recherches experimentales sur la physiologie de la couche optique. *Arch. Physiol. (Paris).* 1898, **10**:706–713.

Shackel, B. A rubber suction cup surface electrode with high electrical stability. *J. Appl. Physiol.* 1958, **13**:153–158.

Shackel, B. Skin drilling: a method for diminishing galvanic skin potentials. *Amer. J. Psychol.* 1959, **72**:114–121.

Silver, M. L. Sensitivity to bentonite. *EEG Clin. Neurophysiol.* 1950, **2**:115.

Simons, D. G., W. Prather, and F. K. Coombs. Personalized telemetry medical monitoring and performance data-gathering for the 1962 SAM–MATS fatigue study. SAM-TR-65-17. U.S.A.F. Brooks AFB, Texas (1965).

Smith, E. R., and J. K. Taylor. Reproducibility of the silver-silver chloride electrode. *J. Res. Natl. Bur. Std. (U.S.)*. 1938, **20**:837–847.

Skov, E. R., and D. G. Simons. EEG electrodes for in-flight monitoring. SAM-Tech. Rept. TR-65-18. U.S.A.F. School of Aerospace Medicine, Brooks AFB, Texas (1965).

Sneddon, I. B., and R. M. L. Archibald. Traumatic calcinosis of the skin. *Brit. J. Dermatol.* 1958, **70**:211–214.

Sollner, K. The electrochemistry of porous membranes. *J. Macromol. Sci.-Chem.* 1969, **A3**(1):1–86.

Sollner, K. Ion exchange membranes. *Ann. N.Y. Acad. Sci.* 1954, **57**:177–203.

Spencer, M. P., T. A. Gornall, and T. C. Poulter. Respiratory and cardiac activity of killer whales. *J. Appl. Physiol.* 1967, **22**:974–981.

Spiegler, K. S., and M. R. J. Wyllie. Electrical potential differences. In *Physical Techniques in Biological Research*. Vol. 2. G. Oster, and A. W. Pollister, Eds. New York: Academic Press, 1956. 502 pp.

Stangl, K. G. Unterdruck–Haftelektrode für elektromedizinsche Zwecke. German Patent 1108820, October 16, 1959.

Straub, W. Nadelelektroden bei Elektrokardiographie. *Klin. Wochenschr.* 1922, **1**:1638.

Stern, O. Zur Theorie der elektrolytischen Doppelschicht. *Z. Elektrochem.* 1924, **30**:508–516.

Sullivan, G. H., and G. Weltman. A low mass electrode for bioelectric recording. *J. Appl. Physiol.* 1961, **16**:939–940. U.S. Patent #3,151,619.

Sutter, C. Über die Beeinflussung der EKG-Kurve durch elektrische Eigenschaften der Aufnahmeanordnung. *Cardiologica* 1944, **8**:246–262.

Svaetichin, G. Low resistance microelectrode. *Acta Physiol. Scand.* 1951, 24. Suppl. **86**:3–13.

Szekely, E. G., J. J. Egyed, C. G. Jacoby, R. Moffet, and E. A. Spiegel. High frequency coagulation by means of a stylet electrode under temperature control. *Confin. Neurol.* 1966, **26**:146–152.

Tasaki, I. Conduction of the nerve impulse. In *Handbook of Physiology*, Section 1, *Neurophysiology*, Chap. 3. J. Field, H. W. Magoun, and V. Hall, Eds. Washington, D.C.: American Physiological Society, 1959.

Tasaki, I. *Nervous Transmission.* Springfield, Ill.: Charles C. Thomas Publishing Company, 1953. 164 pp.

Tasaki, I. Properties of myelinated fibers in a frog sciatic nerve and in spinal cord as examined with microelectrodes. *Jap. J. Physiol.* 1952–1953, **3**:73–94.

Tasaki, I., E. J. Polley, and F. Orrego. Action potentials from individual elements in cat geniculate striate cortex. *J. Neurophysiol.* 1954, **17**:454–474.

Taylor, R. M. Cable theory. In *Physical Techniques in Biological Research*, Vol. VIB., W. L. Nastuk, Ed., New York: Academic Press, 1963.

Taylor, C. V. Microelectrodes and micromagnets. *Proc. Soc. Exp. Biol. Med.* 1925, **23**:147–150.

Taylor, J. K., and E. R. Smith. Reproducibility of silver-silver halide electrodes. *J. Res. Natl. Bur. Std. (U.S.)*. 1939, **22**:307–314.

Telemedics, Inc. *Med. Electron. News.* 1961, **1**(4):9.

Terman, F. E. *Radio Engineers Handbook,* 1st ed. New York: McGraw-Hill Book Company, 1943. 1019 pp.

Terzuolo, C. A., and T. Araki. An analysis of intra-versus extracellular potential changes associated with activity of single spinal motoneurones. *Ann. N.Y. Acad. Sci.* 1961–1962, **94:**547–558.

Thompson, N. P., and J. A. Patterson. Solid salt bridge contact electrodes. System for monitoring the ECG during body movement. *Tech. Rept.* 58–453. ASTIA Doc. AD215538. April 1958.

Tomita, T. The nature of action potentials in the lateral eye of the horseshoe crab as revealed by simultaneous intra- and extracellular recording. *Jap. J. Physiol.* 1956, **6:**327–340.

Tomita, T., and A. Kaneko. An intracellular coaxial microelectrode — its construction and application. *Med. Elect. Biol. Eng.* 1965, **3:**367–376.

Tomita, T., M. Murakami, Y. Sato, and Y. Hashimoto. Further study of the so-called cone action potential (*S*-potential), its histological determination. *Jap. J. Physiol.* 1959, **9:**63–69.

Trank, J., R. Fetter, and R. M. Lauer. A spray-on electrode for recording the electrocardiogram during exercise. *J. Appl. Physiol.* 1968. **24:**267–268.

Ungerleider, H. E. A new precordial electrode. *Amer. Heart J.* 1939, **18:**94.

Ussing, H. H. Transport of ions across cellular membranes. *Physiol. Rev.* 1949, **29:**127–155.

Ussing, H. H., and K. Zerahn. Active transport of sodium as the source of electric current in the short-circuited isolated frog skin. *Acta Physiol. Scand.* 1951, **23:**110–127.

Varley, C. F. Polarization of metallic surfaces in aqueous solutions. *Proc. Roy. Soc. (London).* 1871, **19:**243–246.

Vis, V. A. A technique for making multiple bore microelectrodes. *Science.* 1954, **120:**152–153.

von Bonin, G., W. W. Alberts, E. W. Wright, and B. Feinstein. Radiofrequency brain lesions. *Arch. Neurol.* 1965, **12:**25–29.

Waller, A. D. On the electromotive changes connected with the beat of the mammalian heart, and of the human heart in particular. *Phil. Trans. Roy. Soc. (London).* 1889, **180B:**169–194.

Warburg, E. Über das Verhalten sogenannter unpolarisirbarer Elektroden gegen Wechselstrom. *Ann. Phy. Chem.* 1899, **67:**493–499.

Warburg, E. Über die Polarizationscapacitat des Platins. *Ann. Phys. (Leipzig).* 1901, **6:**125–135.

Ware, R. Personal communication. School of Aerospace Medicine, Brooks Air Force Base, Tex.. 1965.

Weale, R. A. A new micro-electrode for electrophysiological work. *Nature.* 1951, **167:**529–530.

Weidmann, S. The electrical contents of Purkinje fibers. *J. Physiol.* 1952, **118:**348–360.

Weinman, J. and J. Mahler. An analysis of electrical properties of metal electrodes. *Med. Elect. Biol. Eng.* 1964, **2:**299–310.

Welch, W. Self-retaining electrocardiographic electrode. *J. Amer. Med. Ass.* 1951, **147:**1042.

West, T. C. Ultramicroelectrode recording from the cardiac pacemaker. *J. Pharmacol. Exp. Therap.* 1955, **115:**283–290.

Whalen, R. E., C. F. Starmer, and D. H. McIntosh. Electrical hazards associated with cardiac pacemaking. *Ann. N.Y. Acad. Sci.* 1964, **111:**922–931.

Whittier, J. R., and F. A. Mettler. Studies on the subthalamus of the rhesus monkey. *J. Comp. Neurol.* 1949, **90:**281–317.

Wien, M. Über die Polarization bei Wechselstrom. *Ann. Phys. Chem.* 1896, **58:**37–72.

Wilson, F. N., F. D. Johnston, A. G. Macleod, and P. S. Barker. Electrocardiograms that represent the potential variations of a single electrode. *Amer. Heart J.* 1934, **9:**447–458.

Wilson, F. N., A. G. Macleod, and P. S. Barker. The distribution of currents of action and of injury displayed by heart muscle and other excitable tissues. Ann Arbor, Mich.: University of Michigan Press, 1933. 59 pp.

Winsbury, G. J. Machine for the fast production of microelectrodes. *Rev. Sci. Instr.* 1956, **27:**514–516.

Wise, K. D., and J. B. Angell. An integrated circuit approach to extracellular microelectrodes. *Proc. 8th Int. Conf. Med. Biol. Eng. 1969* Paper No. 14–5. Chicago, Ill: Carl Gorr Printing Company, July, 1969.

Wise, K. D., and A. Starr. An integrated circuit approach to extracellular microelectrodes. *Proc. 8th Int. Conf. Med. Biol. Eng.* 1969. Sec. 14–5.

Wolbarsht, M. L., E. F. MacNichol, and H. G. Wagner. Glass insulated platinum microelectrode. *Science.* 1960, **132:**1309–1310.

Wolff, I. A study of polarization capacity over a wide frequency band. *Phys. Rev.* 1926, **27:**755–763.

Wolfson, R. N., and M. R. Neuman. Miniature S_i–S_iO_2 insulated electrodes based on semiconductor technology. *Proc. 8th Int. Conf. Med. Biol. Eng. 1969.* Paper No. 14–6. Chicago, Ill: Carl Gorr Printing Company.

Woodbury, J. W., and A. J. Brady. Intracellular recording from moving tissues with a flexibly mounted ultramicroelectrode. *Science.* 1956, **123:**100–101.

Woodbury, J. W., J. Lee, A. J. Brady, and K. A. Merendino. Transmembranal potentials from the human heart. *Circ. Res.* 1957, **5:**179.

Woodbury, J. W., and D. M. McIntyre. Transmembrane action potentials from pregnant uterus. *Amer. J. Physiol.* 1956, **187:**338–340.

Woodbury, L. A., J. W. Woodbury, and H. Hecht. Membrane resting and action potentials of single cardiac muscle fibers. *Circulation,* 1950, **1:**264–266.

Woodbury, L. A., H. Hecht, and A. R. Christopherson. Membrane resting and action potentials of single cardiac muscle fibers of the frog ventricle. *Amer. J. Physiol.* 1951, **164:**307–318.

Young, J. Z. Structure of nerve fibers and synapses in some invertebrates. *Cold Spring Harbour Symp. Quant. Biol.* 1936, **4:**1–6.

Zerahn, K. Oxygen consumption and active sodium transport in the isolated and short-circuited frog skin. *Acta Physiol. Scand.* 1956, **36:**300–318.

Zervas, N. T. Eccentric radio-frequency lesions. *Confin. Neurol.* 1965, **26:**143–145.

Zimmerman, E. E. The influence of temperature on polarization capacity and resistance. *Phys. Rev.* 1930, **35:**543–553.

Appendix: Experiments

INTRODUCTION

Informative and important experiments in the life sciences can be conducted without the use of rare or expensive subjects. History has repeatedly demonstrated that significant discoveries have been made using very primitive living forms. Life abounds in nearly all terrestrial and aqueous environments, and there is an ample supply of subjects suitable for the widest variety of studies. Many of the so-called lower forms of life exhibit the same phenomena seen in man—the organism about which we desire the most information. In choosing a subject for investigation, life scientists use Galen's dictum that the species of choice is that which presents the phenomenon of interest in the most prominent way. In engineering language, this is merely a way of arranging things so that the highest signal-to-noise ratio is obtained for the information channel. Thus the membrane electrophysiologist chooses the giant axon of the squid because of its large diameter (0.5 mm) and its long survival time; similarly, those interested in active transport choose the frog skin and kidney tissue, because they are readily available, survive well, and demonstrate the phenomenon in a way that is accessible to investigation. Many other examples can be cited, but perhaps the animal that has served life scientists best has been the frog. Not only have important life-science discoveries been made with this animal, but it gave birth to current electricity and indeed electrical engineering because of the Galvani-Volta controversy (see* Fulton and Cushing, 1936; Hoff, 1936; Dibner, 1952; Geddes and Hoff, 1971). Because of its widespread availability and long survival time, the frog has been chosen as the subject for many of the experiments that follow.

Pithing a Frog

In many cases, it is necessary to anesthetize experimental animals to eliminate the sensation of pain that accompanies operative procedures. It is pos-

---

* All references cited in the Appendix are listed at the end of the respective section.

sible to render the frog insensitive to pain without using a chemical anesthetic; physiologists merely quickly destroy the brain (where stimuli are perceived) and often the spinal cord (the center for a large number of motor reflexes). Such mechanical destruction is given the term pithing,* and a pithed frog will continue to live for many hours; the heart will continue to beat, and there is adequate exchange of respiratory gases through the skin.

Pithing the brain and spinal cord of the frog is an art that is easily learned. The frog is first grasped in the left hand as shown in Fig. A1-1*A* using the index finger to bend the head downward so that it is approximately at right angles to the axis of the body. (Very slippery or lively frogs can be handled more easily if they are cooled in the refrigerator or wrapped in a paper

---

* The nervous system is likened to the pith in a plant.

Fig. A1-1.   Pithing the brain (*A*,*B*) and spinal cord (*C*) of a frog.

towel). With the frog in this position the foramen magnum (the hole in the skull through which the spinal cord emerges) becomes accessible and the point of a pithing needle (Fig. A1-5c, available from Fisher Scientific, Pittsburgh, Pa. 15219) is easily passed through it. When this is done, as shown in Fig. A1-1B, the spinal cord is severed and the animal's legs will twitch. Then the pithing needle is moved laterally and vertically and in an angular fashion to destroy the brain. Now the animal is insensitive to pain and is called a spinal-cord preparation. If left for a few minutes, the spinal cord reflexes will become active and a pinch on the toe will cause withdrawal of the limb, although no sensation is perceived.

If the spinal cord is destroyed by pithing, the thoracic and lumbar somatic reflexes disappear. The spinal cord is pithed by bending the head forward and advancing the pithing needle down the central canal of the spinal column which houses the spinal cord (Fig. A1-1C). When the pithing needle enters the spinal cord, the limbs will twitch. The pithing needle is easily advanced down the spinal cord, which is then destroyed by an angular motion of the pithing needle. During this procedure the hindlimbs become fully extended (Fig. A1-1C); this is the sign that all the neurones in the cord are being mechanically stimulated. When the spinal cord has been destroyed, all four limbs are completely flaccid and will not move when pinched.

Exposure of the Sciatic Nerves

The sciatic nerves are easily exposed by the method illustrated in Figs. A1-2 and A1-3. A doubly pithed frog is first pinned down and a fold of skin on the back is grasped with a forceps as in Fig. A1-2A. The fold of skin is lifted and a scissor cut is made in the skin between the animal and forceps so that the skin held by the forceps is cut away. The skin on the back will then fall away, leaving an oval opening (Fig. A1-2B) over the urostyle or tail bone. The next step is to cut away the distal muscular attachments of the urostyle (dashed line in Fig. A1-2B). When this has been done, the free end of the urostyle is grasped firmly with a forceps and the urostyle is raised as shown in Fig. A1-2B1. This procedure exposes the two sciatic nerve trunks lying immediately below the urostyle; then the urostyle itself is cut off as far headward as possible, leaving the sciatic nerve trunks clearly exposed. In order to facilitate later manipulation of the nerves, a loose thread should be slipped under each at the point where the nerves leave the spinal cord to form the sciatic trunks.

The sciatic nerve can be exposed in the thigh region in the manner shown in Fig. A1-3. The skin on the midthigh is grasped with a forceps (Fig. A1-3A) and cut away as described previously, leaving an oval exposure revealing the large semimembranous and the triceps femoris muscles (Fig. A1-3B), between which a branch of the sciatic nerve runs. The nerve can then be exposed by applying lateral force to the two muscles (Fig. A1-3C). The nerve runs along

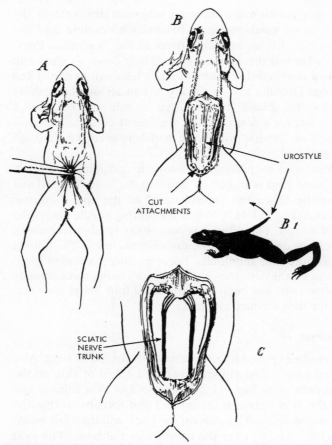

Fig. A1-2. Exposure of the sciatic nerves under the urostyle.

with blood vessels and need not be dissected free of them. A loose thread passed under the nerve will permit easy access to the nerve, and the overlying muscles can be allowed to resume their original position.

Dissection of the Sartorius Muscle

The sartorius is a postural muscle which contains a bundle of almost identical and parallel muscle fibers; for this reason it is frequently used to study the properties of skeletal muscle. It is easily exposed by laying a doubly pithed frog on its back, grasping a pinch of skin in the midthigh region, and cutting it away to leave an oval opening. If the opening is enlarged to expose all the thigh muscles; the sartorius will be visible in about the middle of the exposure (Fig. A1-4). Extend the incision to expose the lower abdominal muscles and cut them free from the pubic arch. Gently advance the point of a pithing needle under the sartorius muscle at the knee. Slip a small (closed) forceps under the tendon end of the muscle; open the jaws of the forceps, grasp a thread, and withdraw the forceps, pulling the thread under the tendon (the tendon is still attached to the knee). Grasp both ends of the thread and raise the muscle slightly. Use the pithing needle to gently tease the muscle away from the underlying muscles and fascia, it may be necessary to slit the fascia parallel to the sartorius muscle. As the muscle is freed from the underlying tissue, advance the loop of thread toward the pelvic end of the muscle. When this dissection is complete, tie the thread around the pelvic end of the muscle. Using the forceps, pass another thread under the muscle as close to the knee as possible and tie it to the muscle. The muscle attachments beyond the threads are then cut, and the muscle is lifted from the animal using the threads.

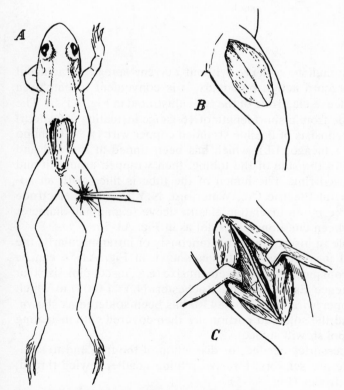

Fig. A1-3. Exposure of the sciatic nerve in the thigh.

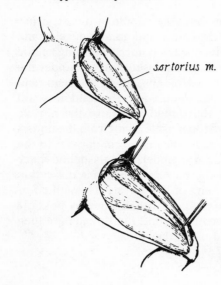

sartorius m.

Fig. A1-4. Exposure of the sartorius muscle.

Electrodes

In order to apply stimuli to a nerve without current spread to adjacent irritable tissue, or to record action potentials, it is convenient to employ a monopolar or bipolar sleeve electrode of the type illustrated in Fig. A1-5*a*. The electrode holder is made from a short length of 16-gauge insulating (spaghetti) tubing. Each electrode consists of flexible stranded copper wire (#8430 phono pickup cable, Belden, Chicago, Ill.), which has been tinned at the end and pushed through a hole in the side of the tubing, then wrapped around it and soldered to form a closed ring. The lumen of the tube is filled with an adhesive (Clearseal, General Electric Co., Waterford, N.Y.), and the electrode is held against the nerve by an insulating elastic sleeve (e.g., small-diameter rubber tubing that has been cut along one side) as in Fig. A1-5*a*.

When it is permissible to insert pins subcutaneously or intramuscularly, the easily constructed and inexpensive electrodes shown in Fig. A1-5*b* can be used. They are made from pins of any convenient size (e.g., insect pins size 0 or 00, Wards Natural Science Establishment, Rochester, N.Y. 14603) to which a flexible stranded copper wire (Belden #8430) has been soldered as shown. The head of the pin and the soldered portion are then covered with insulating varnish (ordinary nail polish will suffice).

In summary, the accessories needed for dissection of the frog and to stimulate nerve and muscle are scissors, forceps, pithing needle, sewing thread, and pin and sleeve electrodes (Fig. A1-5).

The electronic equipment used in the experiments that follow is of a very standard type and is found in most modern laboratories. Basically, the experiments employ a high-quality oscilloscope with a triggerable time base and a stimulator capable of triggering it and delivering an output that is isolated from the chassis and ground. Although the instructions in the experiments are written for a Tektronix storage oscilloscope (model 564 with a 2B67 time base and a 3A9 differential amplifier), a Narco Bio-Systems Mark V stimulator, (in some cases connected to a Grass Model S1U4B or American Electronics Laboratories model 112 radio frequency stimulus isolation unit), any equivalent arrangement of instrumentation can be used. In four of the experiments (3 through 6) a storage oscilloscope, although very convenient,

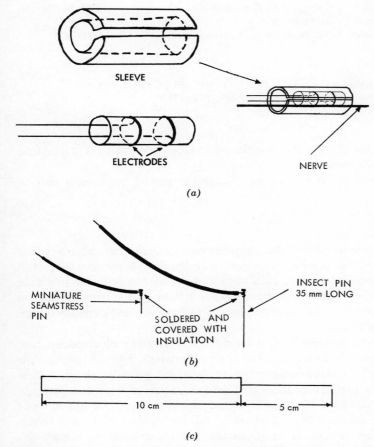

Fig. A1-5. Sleeve and pin electrodes and pithing needle: (*a*) sleeve electrode; (*b*) pin electrodes; (*c*) pithing needle.

is not essential; however, the availability of trace-storage eliminates the need for photography or drawing records on the cathode ray tube with a wax pencil.

In addition to the electronic instruments and supplies described, certain easily constructed pieces of equipment are required; complete instructions for their fabrication are presented in the appropriate experiments.

References

Dibner, B. *Galvani-Volta*. Norwalk, Conn. Burndy Library, 1952. 52 pp.

Fulton, J. F., and H. Cushing. A biographical study of the Galvani and Aldini writings on animal electricity. *Ann. Sci.* 1936, **1**:234–238.

Geddes, L. A., and H. E. Hoff. The discovery of bioelectricity and current electricity (the Galvani-Volta controversy). *IEEE Spectrum* 1971. **8**:38–46.

Hoff, H. E. Galvani and the Pre-Galvani electrophysiologists. *Ann. Sci.* 1936, **1**:157–172.

EXPERIMENT 1   THE ELECTRODE–ELECTROLYTE INTERFACE AND RESISTIVITY

Object

1.   To determine the electrical properties of an electrode-electrolyte interface at various current densities.
2.   To measure the resistivity of an electrolyte placed between a pair of electrodes.

Theory

In order to be able to predict the impedance-frequency characteristics that will appear between the terminals of a pair of electrodes affixed to a subject, it is necessary to know (*a*) the electrical nature of the electrode-tissue interface and (*b*) the electrical properties of the intervening tissue. The former is not too difficult to ascertain; the latter is much harder to describe and depends on the type of intervening tissue.

In nearly all cases of measuring a bioelectric event, the electrodes come into direct contact with electrolytes (body or otherwise). Thus determination of the properties of an electrode-electrolyte interface provides the first step in understanding the impedance measured between a pair of electrodes strategically oriented to detect a bioelectric event. Knowledge of the impedance-frequency characteristics of both electrodes and that of the intervening tissue enables specification of the input impedance of the measuring apparatus, which must be made many times higher than the sum of the three impedances.

Failure to meet this requirement will result in reduction of measured potential, accompanied by waveform distortion (see p. 120).

Electrically, a single electrode-electrolyte interface exhibits both resistive and capacitive properties. The simplest approximation is due to Warburg (1898, 1901) who stated that such an interface could be simulated by a series resistance and capacitance, the reactance of which at any frequency is very nearly equal to the resistance, and both vary nearly inversely as the square root of frequency. This interesting relationship, which may be called Warburg's law, can be tested by measurement of the impedance-frequency characteristic of a variable-length conductivity cell, constructed like a syringe (Fig. E1-1a). Figure E1-1b presents the Warburg series-equivalent electrical

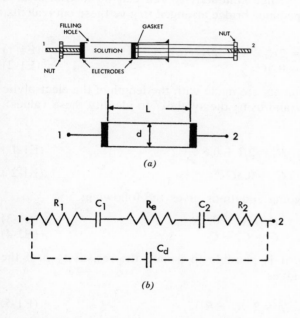

Fig. E1-1. Syringe conductivity cell: (a) conductivity cell; (b) equivalent circuit; (c) simplified equivalent circuit.

circuit in which R_e is the resistance of the electrolytic solution of resistivity ρ; L is the length, and the cross-sectional area A is equal to $\pi d^2/4$. The capacitance C_d, shown between the terminals 1 and 2 represents the capacitance of the cell, considering the electrodes as parallel plates of a capacitor and the electrolyte as the dielectric. If the ratio L/d is large and measurements are made using low-frequency current, C_d can be neglected. Furthermore, if electrodes 1 and 2 are identical, the circuit can be simplified still more to that of Fig. E1-1c, which $R_1 = R_2 = R$, where R is the resistance of a single electrode-electrolyte interface, and $C_1 = C_2 = C$, where C is the capacitance of a single electrode-electrolyte interface.

With the simplification of Fig. E1-1c, the series-equivalent resistance R_s and capacitance C_s of the syringe conductivity cell can be measured at a given frequency using an impedance bridge arranged to give these equivalents. Therefore

$$R_s = 2R + R_e = 2R + \rho L/A \tag{E1-1}$$
$$C_s = 0.5C \tag{E1-2}$$

Now if the same measurements are made with the length of the electrolytic column reduced to one-half (and using the symbol ' to identify these values), the following are obtained:

$$R_s' = 2R + 0.5\,\frac{\rho L}{A} \tag{E1-1'}$$

$$C_s' = 0.5C \tag{E1-2'}$$

Manipulation of the foregoing equations gives the following:

$$R = R_s' - 0.5R_s \tag{E1-3}$$
$$C = C_s' + C_s \tag{E1-4}$$

By another manipulation, it is possible to isolate the resistivity ρ of the electrolyte; performing this gives

$$\rho = 2(R_s - R_s')\,\frac{A}{L} \tag{E1-5}$$

Therefore if ρ is independent of frequency (as it usually is), measurement of the series-equivalent resistance and capacitance at different frequencies, with the syringe conductivity cell set at full length and then at half-length, will permit calculation of R, C, and ρ. The dependence of R and C on current density (as well as frequency) can be determined by repeating the procedure with different currents flowing through the conductivity cell.

The conductivity cell employed to determine the properties of metal electrodes can be easily fabricated from a 1-ml plastic (disposable) syringe as shown in Fig. E1-1a. A circular thin metallic electrode equal to the internal diameter of the syringe is mounted in the needle end of the syringe (electrode 1). An identical electrode is mounted on the end of a threaded rod inside of a stiff plastic tube; immediately behind this electrode is a gasket, which can be expanded by tightening the nut on electrode 2. A small hole is drilled in the side and at the end of the syringe to permit easy filling of the conductivity cell with the desired electrolytic solution.

The impedance bridge used to measure the equivalent series resistance R_s and capacitance C_s appearing between terminals 1 and 2 of the syringe conductivity cell (Fig. E1-1a) appears in Fig. E1-2. In order to ensure that the same current always flows through the conductivity cell, despite its change in impedance with frequency, the bridge is arranged so that a high resistance (50,000 Ω) is placed in series with the conductivity cell and the oscillator; an identical high resistance is placed in series with the bridge-balancing resistors R_s and capacitors C_s. Thus the bridge is a constant-current comparison bridge because the ratio arms Z_1 and Z_2 are equal (50 KΩ). Balancing the bridge is accomplished by detecting the bridge voltage across the diagonal arms using an oscilloscope (e.g., Model 564,* Tektronix, Portland, Ore.) coupled via a bridge-isolation transformer with a 1–1 ratio and low primary-to-secondary capacitance. The bridge is excited by a sine-wave power oscillator (205 AG Hewlett-Packard, Palo Alto, Calif.) or an oscillator connected to a high output impedance power amplifier capable of providing enough voltage to obtain the current required to make measurements at the desired current density. The magnitude of the impedance between the terminals of the syringe conductivity cell can be calculated by using a high-gain, high-input impedance (VTVM) AC voltmeter to measure the oscillator

* The storage feature of this oscilloscope is not used in this experiment; therefore, any high quality oscilloscope can be employed.

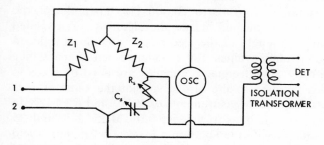

Fig. E1-2.   Impedance bridge for measurement of the series resistive R_s and capacitive C_s equivalents for the conductivity cell connected to terminals 1 and 2. The ratio arms Z_1, Z_2 are 50,000-Ω matched resistors.

voltage and the voltage across the cell terminals. The cell impedance may then be obtained by dividing this voltage by the current, which is equal to the oscillator voltage divided by 50,000.

In summary, in this experiment the electrical properties of a single electrode-electrolyte interface will be investigated by measurement of the capacitance and resistance values appearing between the terminals of an electrolytic conductivity cell at full and at one-half length at various frequencies. Measurements will be made of the series-equivalent values for different current densities and the impedance between the cell terminals will be determined for each frequency and current density. With no additional measurements the resistivity of the electrolyte can also be calculated. At the option of the instructor, the parallel-equivalent values can also be measured by connecting R_s and C_s (Fig. E1-2) in parallel.

Equipment

The following equipment is required: impedance comparison bridge with equal resistive ratio arms and decade resistance (0–1 MΩ) and capacitance (0–10 μF) boxes with resistance steps of 1 Ω and capacitance steps of 0.0001 μF, sine-wave generator (0–10 KHz), oscilloscope, bridge isolation transformer (low primary-to-secondary capacitance), high input impedance AC voltmeter (e.g., VTVM), syringe conductivity cell, 0.6% saline.

Procedure

The procedure consists of filling the syringe conductivity cell with 0.6% saline at room temperature and allowing 10 minutes for the electrode-electrolyte interface to reach equilibrium. The level of the meniscus in the filling hole is made level with the bore of the syringe. (Be sure to eliminate all air bubbles.) The oscillator voltage is set to obtain the appropriate voltage for the desired current density (e.g., 0.05 mA/cm^2), and R_s and C_s are measured in the frequency range extending from 10 to 10,000 Hz. Then the plastic plunger of the conductivity cell is pushed in and the electrolyte is expelled through the filling hole until the length of the electrolytic column is reduced to one-half. The series-equivalent values for resistance and capacitance R_s' and C_s' are measured for the same frequency range. Care must be taken to maintain a constant voltage applied to the bridge so that the same current density is used for all measurements. Measurements of R_s, C_s and R_s', C_s' are to be made for the following current density values: 0.05, 0.25, 2.5, and 25 mA/cm^2 and using 0.6% sodium chloride solutions. For each frequency and current density value, the voltage across the conductivity cell is measured with the VTVM. All measurements are made for one current density, starting with the conductivity cell at full length; then the conductivity cell is reduced to half length and the measurements is repeated, using the same current density. Measuring is to begin with the lowest current density value. Equations E1-3 and E1-4 are used to calculate the series-equivalent values for R and C at each frequency. Equation E1-5 is used to calculate the resistivity of the electrolyte. The magnitude of the cell impedance is ascertained by dividing the cell voltage by the current through it.

It will be necessary to know the areas of the electrodes in the syringe conductivity cell, the full length L, and the position to be taken by the plunger to obtain half this length; these measurements should be made in centimeters.

The data obtained should be presented as follows:

1. A. A log-log plot showing the resistance and capacitance (of a single electrode-electrolyte interface) versus frequency for the lowest current-density measurement.
    B. A log-log plot of the resistance and reactance $(1/2\pi fC)$ versus frequency for a single electrode-electrolyte interface for the lowest value of current density employed.
    C. A log-log plot of the capacitance per square centimeter of electrode area, versus frequency for the lowest value of current density employed.
    D. Calculation of the least-squares (log) representation of the data for $C = Kf^{-\alpha}$ using values found for capacitance C in microfarads per square centimeter of area.
2. A. Graphs (semilog) of the manner in which R nd C vary with increasing current density for 100, 200, 500, and 1000 Hz.
    B. Identify the locus for 20% decrease in R and 20% increase in C as found in part 2 A.
3. Log-log plots of R and X versus frequency, using current density as the parameter.
4. On semilog paper plot p (resistivity) versus frequency for the various current density values employed.
5. A semilog plot of cell impedance versus frequency for the various current density values investigated.
6. Do the data obtained in this experiment support Warburg's law?

References

Fricke, H. The theory of electrolyte polarization. *Phil. Mag.* 1932, **14**:310–318.

Geddes, L. A., and L. E. Baker. The specific resistance of biological material, a compendium of data for the biomedical engineer and physiologist. *Med. Biol. Eng.* 1967, **5**:271–293.

Geddes, L. A., C. P. da Costa, and G. Wise. (1971) The impedance of stainless-steel electrodes. *Med. Biol. Eng.* In press.

Schwan, H. P., and C. F. Kay. The conductivity of living tissues. *Ann. N.Y. Acad. Sci.* 1956–1957, **65**:1007–1013.

Schwan, H. P., and C. F. Kay. The specific resistance of body tissues. *Circ. Res.* **4:**664–670.

Warburg, E. Über das Verhalten sogenannter unpolarisirbarer Elektroden gegen Wechselstrom. 1899, *Ann. Phys. Chem.* **67:**493–499.

Warburg, E. Üver die Polarizationscapacitat des Platins, *Ann. Phys.*, 1901, **6:**125–135.

EXPERIMENT 2   THE POTENTIAL FIELD SURROUNDING A DIPOLE IN A VOLUME CONDUCTOR

Object

1. To map the isopotential lines surrounding a current dipole in a volume conductor.
2. To demonstrate waveform distortion when amplifier input impedance is reduced.
3. To investigate the dipole representation for excitation and recovery.

Theory

Because many bioelectric generators are likened to a dipole source in a volume conductor, mapping the isopotential lines surrounding a current-driven dipole permits visualization of the manner in which potential is distributed around such a bioelectric generator. Although in the biological situation the environmental volume conductor is neither uniform nor infinite in extent, the many isopotential line studies indicate that often the distortion due to these factors is not excessive. Another important difference between the biological and physical situations relates to the position of the dipole, which changes as a result of the propagation of activity (depolarization) and recovery (repolarization) of the living cells; therefore, a temporal sequence of isopotential lines describes the activity of a bioelectric generator. Knowledge of the distribution of the isopotential lines surrounding a bioelectric generator permits prediction of the instantaneous potential difference appearing between a pair of electrodes located anywhere in the known potential field; the potential is merely the difference between the potentials at the two electrodes. If the temporal sequence of the potential field is known, the time course of the bioelectric event measured by the electrodes can be derived.

A practical method of establishing a potential field consists of exciting a pair of small, closely-spaced spherical electrodes (the dipole) in an electrolytic environment (0.6% saline) in a large vessel. Wilson et al. (1933) showed that the potential V_p at any point P is given by

$$V_p = \frac{IL\rho}{4\pi} \left(\frac{\cos \alpha}{r^2} \right)$$

where I is the current, L is the distance between the centers of the spherical electrodes, ρ is the resistivity of the volume conductor material, α is the angle that the line, r units long and extending to P, makes with the axis of the dipole. Figure E2-1 illustrates the dipole and its potential field.

Passage of a steady current through an electrolyte will result in electrolysis and polarization of the electrodes. It is possible to minimize these factors by exciting the dipole with a short-duration rectangular pulse of low-intensity

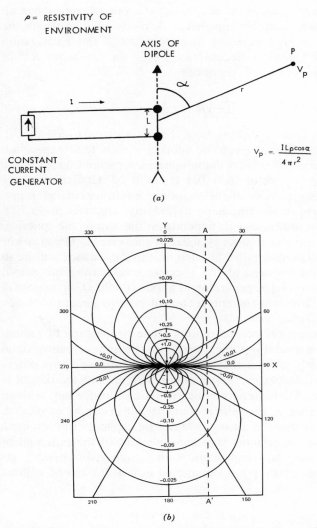

Fig. E2-1. (*a*) The dipole; (*b*) isopotential line distribution around dipole in (*a*).

current; this will assure that the potential field will be established only while current is flowing, disappearing when the current flow ceases.

Application of a rectangular pulse of voltage to the dipole electrodes will not produce the flow of a rectangular pulse of current because the electrode-electrolyte impedance is reactive. However, if a constant-current generator is employed, the current waveform will be rectangular and the potential picked up by a pair of electrodes in the volume conductor will also be rectangular. A simple method of converting a constant-voltage generator to one that delivers a constant current consists of increasing the output impedance so that it is many times higher than the impedance Z_L of the load to which the generator is connected. In this case the load consists of the polarization impedance of each of the spherical electrodes Z_p, plus the resistance R_e* of the environmental volume conductor; therefore

$$Z_L = 2Z_p + \frac{\rho}{2\pi} \left(\frac{1}{R} - \frac{1}{L} \right)$$

where ρ is the resistivity of the fluid comprising the environment (0.6% NaCl), R is the radius of each spherical electrode (cm), and L is their separation (cm). In a practical situation, if 0.5-cm diam. spheres separated by 1 cm are placed in an 0.6% saline conductor ($\rho = 103$ Ω-cm at 20° C, Experiment 1 item 4, p. 299), the resistance of the electrolytic environment, as represented by the second term in the foregoing expression, amounts to 49.2 Ω. The electrode polarization impedance Z_p depends on the size of the spherical electrodes, the current, and the frequency used to measure it. Using a low current density and a rectangular pulse of current, this impedance will be in the neighborhood of a few hundred ohms. Therefore, increasing the output impedance of the constant-voltage generator to about 10,000 Ω, or more will make it approximate a constant-current rectangular-pulse generator when connected to the "dipole" electrodes.

Two other important practical considerations relate to the type of rectangular pulse generator to be used. First, the output must be isolated from ground and chassis so that the only current that flows in the electrolytic tank enters and leaves via the dipole electrodes. A stimulator with an isolated output is adequate for this purpose. Alternately, a bridge-isolating transformer, with an adequate frequency response to pass the rectangular pulse can be employed. The second requirement is that the pulse generator be capable of providing a triggering pulse. This feature permits easy measurement of the potential of the exploring electrode using an oscilloscope with a triggered sweep.

A convenient experimental setup is diagrammed in Fig. E2-2. The volume

---

* Jeans (1920) showed that $R_e = \frac{\rho}{2\pi} \left(\frac{1}{R} - \frac{1}{L} \right)$, provided L is much greater than R.

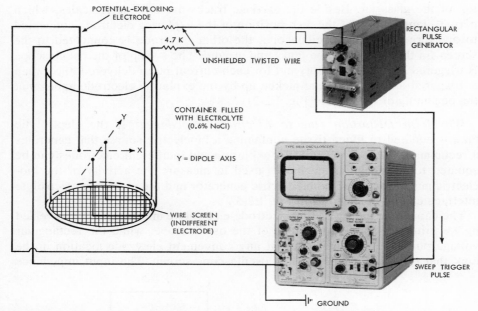

Fig. E2-2. Equipment for mapping the isopotential lines surrounding a dipole source in a volume conductor.

conductor consists of a glass jar (9 3/4 in. diam., 10 in. deep*) filled with 0.6% sodium chloride solution. The dipole consists of two small spheres (5 mm diam.) mounted on two insulated rods protruding from a plastic disk, slightly smaller than the inside diameter of the jar. Well-insulated wires, connected to the rods, are brought out of the solution and connected to two 4.7 KΩ resistors, which are in turn connected to the pulse generator (E & M Physiograph Stimulator Mark V, Narco Bio-Systems, Houston, Texas).

The potential-measuring electrodes consist of a circular metal-screen (reference electrode)† on top of the circular plastic disk that supports the dipole, and an exploring electrode—a 1-mm rod exposed only at the tip and mounted in a carriage that can be moved in two directions at right angles across the

* Cylindrical jar, Kimax - serial #32600, 2.5 gal, Curtin Co. #12355-4G.

† An alternate method of providing an indifferent electrode involves connecting two equal resistors (e.g., 10 KΩ) directly across the terminals of the dipole electrodes and using the midpoint as a point of zero potential for measurement of the dipole field. Thus the ground side of the oscilloscope will be connected to this zero reference point and the potential-exploring electrode in the tank is connected to the other (ungrounded) terminal of the oscilloscope. This technique is particularly useful when it is desired to plot a potential field that has been distorted by the presence of a material having a conductivity different from that of the volume conductor.

top of the glass jar. Beside the carriage tracks are millimeter scales, which permit identification of the *X-Y* position of the exploring electrode connected to one terminal of the oscilloscope; the other electrode is connected to the screen (in the bottom of the tank) and ground. The sweep of the oscilloscope is triggered by the stimulator, and for each current pulse delivered, the sweep is triggered and the potential picked up by the exploring electrode appears at the beginning of the sweep (Fig. E2-2).

Waveform Distortion Due to Electrode Impedance   If the electrolytic tank is equated to living tissue containing a bioelectric source that generates a rectangular pulse of current, the exploring and indifferent electrodes can be equated to extracellular electrodes used to measure the activity of the bio-electric generator; thus the bioelectric generator and the electrode-electrolyte interface can be modeled as in Fig. E2-3*a*.

The capacitive nature of this electrode-electrolyte interface can be revealed by lowering the input impedance of the oscilloscope, while the rectangular voltage pulse is being measured at any convenient electrode location in the tank that produces a pulse on the oscilloscope screen. The input impedance

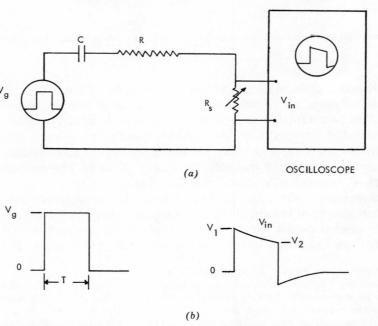

(a)   OSCILLOSCOPE

(b)

Fig. E2-3.   The capacitive nature of an electrode-electrolyte interface revealed by reducing amplifier input impedance: (*a*) approximate equivalent circuit; (*b*) waveforms recorded on the oscilloscope.

R_s can be lowered by connecting a decade resistance box across the input terminals V_{in} of the oscilloscope and reducing the resistance (refer to Fig. E2-3).

If the generator voltage is a pulse of amplitude V_g Fig. E2-3b), then due to the exponential nature of an RC circuit, the envelope of V_{in} will be described by $\left(\dfrac{V_g R_s}{R + R_s}\right) e^{-t/(R+Rs)C}$ as in Fig. E2-3b.

If the pulse is of duration T (Fig. E2-3b), the following relationships arise:

$$V_1 = \frac{V_g R_s}{R + R_s}$$

$$V_2 = \left(\frac{V_g R_s}{R + R_s}\right) e^{-T/(R+R_s)C}$$

Or equivalently

$$R = R_s \left(\frac{V_g}{V_1} - 1\right)$$

$$C = \frac{T V_1}{R_s V_g \ln (V_1/V_2)}$$

The Dipole Representation of Excitation and Recovery   Because active tissue is negative to resting regions and because the area of activity is propagated, it is customary to represent propagated excitation and recovery as traveling dipoles. Therefore, if the tissue is in a volume conductor, a nearby monopolar electrode will detect the propagation of excitation by an initial positivity followed by negativity; the passage of recovery will be indicated by negativity followed by positivity. The type of waveform obtained can be illustrated in this experiment by allowing the dipole to remain in a fixed location and moving the monopolar exploring electrode along a line parallel to the dipole axis. Figure E2-1b shows that when the electrode is moved from A to A', isopotential lines of increasing voltage are encountered; then the potential detected decreases to zero as the exploring electrode is opposite the center of the dipole. As the electrode continues to move toward A', the potential increases with the opposite polarity, reaches a maximum, and starts to decrease. In summary, a diphasic wave (positive followed by negative) occurs when the exploring electrode is moved along the line (from A to A') parallel to the dipole axis.

Equipment

The following equipment is required: rectangular-pulse generator with isolated output, electrolytic tank with centrally mounted dipole, reference electrode (screen) and exploring electrode mounted to calibrated positioning device, oscilloscope with a good low-frequency response, (preferably to dc), decade resistance box (0–1 MΩ), 0.6% sodium chloride solution (14 liters), two 4.7-KΩ resistors (one-half watt).

Procedure

1. Isopotential Lines   Set up the equipment in accordance with Fig. E2-2; with this arrangement the stimulator triggers the sweep of the oscilloscope. Set the stimulus duration to 2 msec, the frequency to 25/sec, and the output to approximately 100 V and, while current is being delivered to the dipole in the electrolytic tank, measure the stimulator output, the voltage across each of the 4.7 KΩ resistors, and the voltage across the dipole. Calculate the current *I* flowing and estimate the impedance of the dipole-electrolytic tank circuit.

Connect the oscilloscope ground to the indifferent (screen) electrode in the bottom of the tank and connect the potential-exploring electrode to the input of the oscilloscope. Place the tip of the exploring electrode at the same depth as the dipole. Move the exploring electrode to different positions in the tank and adjust the VOLTS/DIV control of the oscilloscope to obtain a voltage pulse at the beginning of the sweep of the oscilloscope. Note that the amplitude depends on distance and orientation of the exploring electrode with respect to the dipole axis; refer to Fig. E2-1*b* to obtain an idea of the voltage to be expected for different electrode locations.

Check the orientation of the dipole with respect to the calibrated *X*-*Y* scales beside the carriage that allows movement of the exploring electrode. Be sure that the scales are adjusted so that one scale (*Y* axis) is coincident with the dipole axis; the other (*X*) will, of course, be at right angles.

Move the exploring electrode around in the electrolytic tank until a deflection of +20 mV is obtained on the oscilloscope screen; read the *X* and *Y* coordinates for this point and plot the point on graph paper. Keeping *X* constant, now move the exploring electrode across the tank along this line, which is parallel to the dipole axis, and plot the coordinates for the other +20-mV and the two −20-mV points. (Refer to Fig. E2-1 for guidance). Note that the exploring electrode will cross the same isopotential line twice on the positive side and twice on the negative side of the dipole as the exploring electrode is moved along a line parallel to the dipole axis.

Now move the exploring electrode to a new *X* value 5 to 10 mm away from the previous location and again move it along the *Y* axis (parallel to the dipole axis), and plot the points where ± 20 mV are found. Repeat the procedure until enough points have been obtained to draw the smooth ± 20-mV isopotential curves surrounding the two poles of the dipole. In a similar manner determine and draw the isopotential curves for ±10, ±5, and ±2 mV.

Using the current flowing through the dipole, calculate and plot the ±20-mV isopotential lines on the dipole map. If Experiment 1 has been performed, use the value of resistivity found for 0.6% saline; if not, a value of 103 Ω-cm at 20° C can be used. Compare the measured and calculated values for the 20-mV isopotential line on the dipole axis.

Calculate the resistance R_e of the electrolytic conductor between the dipole electrodes, using the following expression:

$$R_e = \frac{\rho}{2\pi} \left(\frac{1}{R} - \frac{1}{L} \right)$$

where R is the radius (cm) of each of the dipole electrodes and L is their (center-to-center) separation (cm). Compare this value with that measured for the sum of the dipole electrode impedances plus the resistance of the electrolytic tank. Note that the value of the series resistance (9.4 KΩ) is much higher than the impedance between the dipole electrode terminals.

2. Waveform Distortion With the exploring electrode at any convenient location in the tank to obtain a large amplitude rectangular wave on the oscilloscope, measure the amplitude of the generated rectangular wave V_g, and the duration of the rectangular wave T. Connect a 1-MΩ decade resistance box across the input terminals of the oscilloscope (R_s in Fig. E2-3). Starting with the maximum resistance, reduce the resistance and note the increasing distortion of the waveform. Not only is amplitude lost, but electrical differentiation of the waveform occurs. (This phenomenon is studied further in Experiment 4).

Set R_s the decade resistance box, to 100 Ω. Measure V_1 and V_2, the initial and final values, respectively, at the top of the distorted pulse. (Fig. E2-3) Now calculate the equivalent resistive and capacitive components of the electrode-electrolyte interface from the following equations:

$$R = R_s \left(\frac{V_g}{V_1} - 1 \right)$$

$$C = \frac{TV_1}{R_s V_g \ln (V_1/V_2)}$$

Measure the area A of the exposed tip of the exploring electrode. Calculate the capacitance per unit area C/A of the electrode-electrolyte interface.

3. The Dipole Representation of Excitation and Recovery. Place the monopolar exploring electrode on a line 15 mm away from the dipole axis and at one side of the electrolytic tank. Move the electrode along this line and measure the potential for each 5-mm increment; plot the values obtained on a graph which also identifies the location of the dipole. Additional points should be taken to obtain a smooth curve.

Results

From the measurements obtained in this experiment, obtain the following data:

1.   Measured distance (Y) for 20 mV isopotential line_____cm (on the dipole axis).

2.   Calculated distance (Y) for 20 mV isopotential line_____ cm (on the dipole axis).

3.   Resistance R_e of the electrolytic tank_____ Ω.

4.   Resistance R of exploring electrode_____ Ω.

5.   Capacitance C of exploring electrode _____ μF.

6.   Capacitance per square centimeter of exploring electrode _____ μF/cm².

References

Geddes, L. A., and L. E. Baker. The relationship between input impedance and electrode area in recording the ECG. *Med. Biol. Eng.* 1966, **4**:439–450.

Geddes, L. A., L. E. Baker, and M. McGoodwin. The relationship between electrode area and amplifier input impedance in recording muscle action potentials. *Med. Biol. Eng. 1967*, **5**:561–569.

Jeans, J. H. *The Mathematical Theory of Electricity,* 4th Ed. Cambridge: Cambridge University Press, 1920, Chapter X.

Wilson, F. N., A. G. Macleod, and P. S. Barker. *The Distribution of the Currents of Action and of Injury Displayed by Heart Muscle and Other Excitable Tissues.* Ann Arbor, Mich.: University of Michigan Press, 1933. 39 pp.

EXPERIMENT 3   THE ELECTRICAL AXES OF THE HEART

Object

1. To simulate use of the limb leads to locate the electrical axes reflecting excitation and recovery of the cardiac chambers.

2. To locate the electrical axes of the ventricle of the frog and the ventricles of a human subject.

Theory

Because the human heart is almost at the center of an equilateral triangle with the apices at the points where the right and left arms and left leg join the trunk, it is customary to use the potentials recorded from electrodes on the right and left arms and left leg to locate the electrical axes of excitation and recovery of the atria and ventricles. The principle employed can be easily simulated by the use of an electrolytic tank containing a current-driven dipole to imitate the electrical activity of the heart. The potentials appearing between the limb leads can be simulated by placing, in the tank, three electrodes located at the apices of an equilateral triangle with the current-driven dipole at its center. The orientation of the dipole can then be determined from the potentials appearing between any two pairs of electrodes. This method, which was described by Cantrell et al. (1969), is employed in this experiment.

The polarity convention adopted for the three standard limb leads, from which the frontal-plane components of the cardiac potentials can be obtained, appears in Fig. E3-1*a*. This convention was originated by Willem Einthoven (1908), the Dutch physician who, by his development of the string galvanometer (1903), initiated clinical electrocardiography. The triangular aspect of these three leads became known as the Einthoven triangle; and despite the obvious inaccuracy in geometry and inhomogeneity in the electrical components constituting the torso, the Einthoven triangle provides valuable diagnostic information. The lead configuration employed is as follows: lead I = right arm to left arm (RA–LA), lead II = right arm to left leg (RA–LL) and lead III = left arm to left leg (LA–LL); the left leg electrode is sometimes designated F, for foot.

Einthoven showed that the equilateral triangle furnishes a reasonably good representation of the anatomical situation because of the agreement between one of its properties and the potentials measured by the limb electrodes. This relationship, which later became known as Einthoven's law, can best be understood by examination one of the "cardiac vectors" E at the center of the triangle in Fig. E3-1*b*. The components of this potential, which are the voltages appearing between the three limb leads, are given by the projections on the corresponding sides of the triangle. The important property of this particular representation is that the amplitude of the voltage that is known as lead II is equal to the algebraic sum of the voltages that are called leads I and III.

The three components of the "cardiac vector" E, which is described by its orientation angle α measured with reference to lead I, are leads I, II, and III and are therefore

$$\text{lead} \quad \text{I} = E \cos \alpha$$
$$\text{lead} \quad \text{II} = E \sin (\alpha + 30°)$$
$$\text{lead III} = E \sin (\alpha - 30°)$$

Einthoven's law states that the amplitude recorded in lead II equals the algebraic sum of the amplitudes in leads I and III. Writing this statement mathematically:

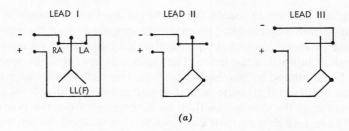

(a)

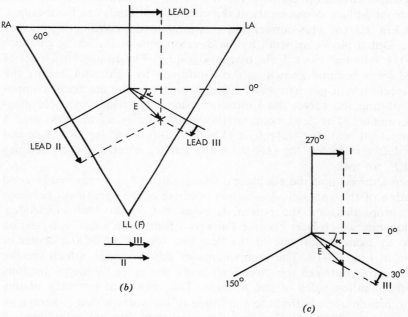

(b)

(c)

Fig. E3-1. Location of the electrical axis of the heart: (*a*) the three standard leads (polarities are for upward deflection of recorder); (*b*) the Einthoven triangle; (*c*) location of electrical axis by using leads I and III.

$$II = I + III$$
$$E \sin (\alpha + 30°) = E \cos \alpha + E \sin (\alpha - 30°)$$
$$E[\sin \alpha \cos 30° + \cos \alpha \sin 30°] = E[\cos \alpha + \sin \alpha \cos 30° - \cos \alpha \sin 30°]$$
$$E\left[\frac{\sqrt{3}}{2} \sin \alpha + \frac{\cos \alpha}{2}\right] = E\left[\cos \alpha + \frac{\sqrt{3}}{2} \sin \alpha - \frac{\cos \alpha}{2}\right]$$
$$\frac{E}{2}\left[\sqrt{3} \sin \alpha + \cos \alpha\right] = \frac{E}{2}\left[\sqrt{3} \sin \alpha + \cos \alpha\right]$$

This simple trigonometric proof shows that, in the particular case of an equilateral triangle reference system, the amplitude designated as the lead-II component of the vector E, having any orientation angle α, is equal to the algebraic sum of the amplitudes of the components designated as leads I and III. It cannot be concluded that the same holds true for any other reference system. As stated previously, despite the obvious errors in the practical application of this concept, clinical electrocardiographic practice has proven that the errors are small enough to be ignored in most instances.

The current-driven dipole in the electrolytic tank (9 3/4 i.d. × 10 in. high),* which is used to simulate the heart and to illustrate Einthoven's law, appears in Fig. E3-2. The measuring electrodes RA, LA, LL consist of 1/4-in. brass rods (7 1/2 in. long), which have been threaded at one end for a distance of 1 1/2 in. to accommodate a 1/4-in. −20 nut. The electrodes are mounted to a circular disk (12 in. diam.) of 3/4-in. plywood, which forms the top of the electrolytic tank. The 1/4-in. rods are mounted at the apices of an equilateral triangle having 8 1/4 in. sides and centrally located on the circular wooden

* Cylindrical 2 1/2 gal jar; Kimax, serial 32,600; Curtin Co. #12355-4G.

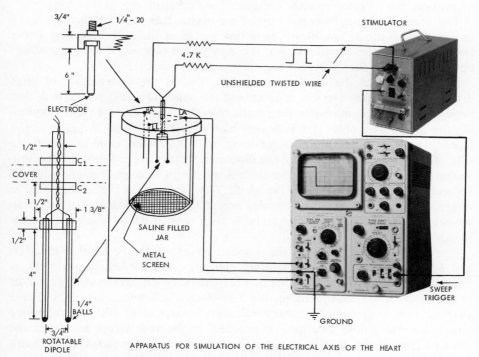

Fig. E3-2. Apparatus for simulation of the electrical axis of the heart.

disk. The electrode holes on the tank side of the cover should be countersunk to accommodate the 1/4-in. −20 nuts as in Fig. E3-2. The electrodes are held firmly in place by one nut and washer below and one nut and washer above the circular wooden cover. A third nut is used to fasten the wire that is connected to the oscilloscope. The rods are covered with shrink-tubing so that only their exposed ends are in contact with the saline that is used to fill the glass jar.

The poles of the rotatable dipole (Fig. E3-2) are two 1/4-in. brass balls soldered to the ends of two lengths of stranded, insulated hookup wire. These wires are passed up the lumens of two pieces of 1/4-in. plastic tubing (4 1/2 in. long), cemented with a center-to-center spacing of 3/4-in. to a block of plastic 1/2-in. thick, 3/4-in. wide, and 1 3/8 in. long, which is in turn cemented to a piece of 1/2-in. plastic tubing, which constitutes the axle passing through the wooden cover of the electrolytic tank. Two adjustable plastic collars C_1, C_2, one above and one below the tank cover, are adjusted to maintain the dipole electrodes in the same plane as the ends of the three rod electrodes. The brass ball electrodes are secured in the plastic tubes by service cement (General Cement Co., Rockford, Ill.); this cement also provides a fluid-tight seal to prevent the electrolyte (0.6% saline) from entering the plastic tubes.

The wires emerging from the tops of the plastic tubes are fed through holes in the sides of the 1/2-in. plastic tube that serves as the axle, emerging at the end above the upper circular collar; these wires are connected to the constant-current generator that provides the current for the dipole.

In the center of the equilateral triangle (i.e., at the intersection of lines drawn from the apices to the midpoints of the opposite sides), a hold is drilled large enough to accommodate the plastic tube that serves as the shaft supporting the dipole assembly; this hole acts as a bearing to permit rotation of the dipole. On the shaft is placed a retaining collar C_2. Then, from the electrode side of the cover, the wires from the dipole electrodes are fed through the hole in the circular wooden cover, the shaft is advanced through the hole, and the second collar C_1 is placed on the shaft. The setscrew in the lower collar is tightened when the dipole electrodes are at the same level as the potential-measuring electrodes, the upper collar is then slipped down, and the setscrew is tightened.

The electrode terminals on the wooden cover should be labeled RA, LA, and LL, to correspond to the right arm, left arm, and left leg, respectively. A line parallel to the line joining RA and LA should be drawn through the center of the dipole shaft; this identifies the zero-degree reference point. The dipole shaft is then rotated so that the dipole axis is coincident with the reference line when the dipole electrode connected to the red wire (identifying the positive pole of the dipole) is under the zero mark. A prominent index mark is then placed on the upper collar and a scale (0–360° clockwise in 15° increments) is marked on the cover to enable setting or determining the orientation of the dipole. Just prior to use, the glass tank is three-quarters filled with 0.6% saline.

The Human ECG   Action potentials, that can be recorded from body-surface leads, precede contraction and relaxation of the atria and ventricles. The four-chambered (two atria and two ventricles) mammalian heart contracts in an orderly sequence and derives its excitation and basic rate from the pacemaker, the sino-atrial (S-A) node. Thus the events start here and a wave of excitation travels over the atria (resulting in atrial systole). Lying in the septal tissue between the atria and ventricles is another specialized tissue, the atrioventricular (A-V) node. The excitation wave propagated from the atria activates the A-V node, which in turn propagates the excitation down a bundle of specialized fibers (bundle of His) to the ventricles via a second, widely dispersed fiber net (the Purkinje fibers). Propagation of the excitation wave to the ventricular musculature results in ventricular systole. Briefly, this is the sequence of the excitatory events of the cardiac cycle.

In a typical record using the standard leads (Fig. E3-3*a*), there are usually three clearly recognizable waves accompanying each cardiac cycle. The first one, a small upward wave, is called the *P* wave and is the potential preceding atrial systole. Following it is usually a large upright triangular wave having small downward waves at its base; this complex is known as the *QRS* wave, *R* being the apex of the triangle. The *QRS* wave is the potential preceding ventricular systole. The third recognizable wave is dome-shaped and can be upward or downward; it is called the *T* wave; and occurs just before ventricular diastole. It is the characteristic repolarization wave of ventricular recovery. A similar recovery wave occurs after atrial systole but it is of low amplitude and is usually obscured by the ventricular *QRS* wave. When seen, it is designated the T_p or T_a wave.

Of paramount importance in deriving useful information from the ECG are the temporal relationships among the various waves along with the amplitudes and the types of waveforms occurring in the various leads. The designation for the waveforms was established by Einthoven and adopted by all electrocardiographers; the scheme is illustrated in Fig. E3-3*a*. The clinical aspects of electrocardiography are well described by Lepeschkin (1951).

The Frog ECG   The ECG of the frog can be recorded using limb electrodes connected to a recording instrument having about three times the sensitivity of conventional electrocardiographs; that is, a recording sensitivity of about 3 cm for 1 mV is required. When such a system is employed, the ECG (Fig. E3-3*b*) obtained has many of the same components found in the human ECG.

The frog heart has more contractile chambers than the human heart. In the

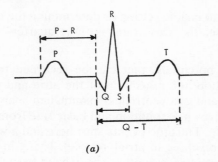

(a)

P-R Interval –   Measured from the beginning of P to the beginning of Q
                  (or if no Q, to the first part of R or S, whichever is first)
                  and represents the conduction time from the beginning of
                  excitation of the atria to the beginning of ventricular
                  excitation.

QRS Interval –   Measured from the beginning of Q to the end of S and
                  represents the time occupied by the excitation wave pas-
                  sing over the ventricles.

Q–T Interval –   Measured from the beginning of Q to the end of T and
                  represents the total time of ventricular electrical systole.

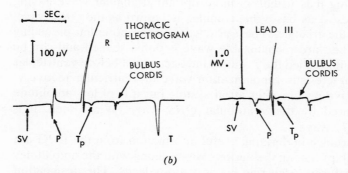

(b)

Fig. E3-3. The electrocardiogram: (*a*) conventional nomenclature; (*b*) frog ECG.

frog there is a sinus venosus (SV), two atria, a single ventricle, and a bulbus
cordis; the sequence of excitation is in this order. The heart beat in the frog
originates in the sinus venosus and is initiated by the SV action potential
(Fig. E3-3*b*); the recovery potential of the sinus venosus is difficult to record
without electrodes placed directly on it. Excitation of the sinus venosus is
propagated to the atria, which give rise to the *P* wave, after which atrial
contraction occurs. During recovery of the atria, excitation is propagated

to the ventricle, giving rise to the R wave—the action potential that precedes its contraction. Often in the frog, recovery of the atria can be identified by the T_p wave occurring just after the R wave. Immediately before ventricular recovery, the ventricular T wave occurs. In the frog heart, the outflow tract of the single ventricle, the bulbus cordis, is also contractile and gives rise to action and recovery potentials; the action potential occurs after the QRS wave, the recovery potential is not often seen in records made using electrodes distant from the heart. Thus excitation of the contractile chambers of the frog heart, as signaled by the ECG, starts with the SV action potential, followed by the P wave, then the R wave and then the bulbus cordis (BC) action potentials. Recovery of the sinus venosus is usually not seen; recovery of the atria (identified by the T_p wave) occurs very close to the R wave. Recovery of the bulbus cordis is rarely recorded, but ventricular recovery, as signaled by the T wave, is readily apparent. In an experimental situation, the various waveforms can be more clearly identified by cooling the frog to slow the heart and to prolong the propagation of both excitation and recovery. Additional data on the electrophysiology of the frog heart can be found in the reports by Bein (1948–1949) and Moore (1964).

Cardiac Vectors   From the foregoing it should be obvious that, associated with the contraction of each of the cardiac chambers, are two bioelectric phenomena that can be represented as cardiac vectors (e.g., vectors of excitation and recovery). In addition, because the amplitudes signaling the same bioelectric event do not reach their maxima and minima at the same instant in each lead, the cardiac vector reconstructed from two limb leads will describe a loop (as shown for the QRS in Fig. E3-4A) rather than a straight line as in Fig. E3-1b. The frontal-plane loop describing ventricular excitation (QRS) is made by projecting the instantaneous amplitudes of leads I and III, as in Fig. E3-4A; the point of intersection gives the magnitude and direction of the vector of ventricular activation at that instant. Repetition of the same procedure for other instants in ventricular excitation gives the corresponding temporal vectors. Joining the tips of the instantaneous vectors produces a loop whose major axis describes the magnitude and direction of the vector. This procedure is illustrated in Fig. E3-4A.

In clinical practice it is customary to visually estimate the cardiac vectors from the standard (and augmented) limb leads. Although some error accompanies this practice, the information derived is of genuine clinical value. The vectors so obtained are called the mean cardiac vectors, and the method (which ignores time differences) involves subtracting the downward deflections from the upward ones. The net amplitudes are then plotted on the Einthoven triangle to obtain the resultant mean vector, as performed in Fig. E3-4B for the QRS wave recorded from leads I and III. Note that the mean

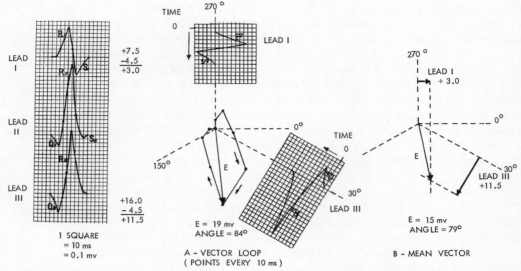

Fig. E3-4. Methods of determining cardiac vectors: (*a*) vector loop (points every 10 msec); (*b*) mean vector.

QRS vector is not very different from that which represents the major axis of the vector loop.

Equipment

To simulate the electrical activity of the heart, it is convenient to employ the equipment illustrated in Fig. E3-2; for the most part it is similar to that employed in Experiment 2 and illustrated in Fig. E2-2. It consists of a stimulator with an isolated output (e.g., Mark V stimulator, Narco Bio-Systems, Houston, Texas) and capable of delivering a triggering pulse to initiate the oscilloscope sweep. Via two 4.7-kΩ resistors, the stimulator delivers a rectangular pulse of current to the dipole immersed in a volume conductor (cylindrical jar containing 0.6% saline). A storage oscilloscope (Tektronix model 564 with a 2B67 time base and a 3A9 preamplifier or equivalent) is employed for viewing the potentials presented to the electrodes in the electrolytic tank or those provided by a frog or human subject. A maximum sensitivity of 3 cm for 1-mV peak and a bandwidth extending from 0.1 to 10 kHz will provide an adequate reproduction of the square pulse and the ECGs (the latter require only a high-frequency response extending to 100 Hz).

Needle electrodes (Fig. A1-5*b*), plate electrodes (Fig. 2-7), and electrode jelly, if human ECGs are to be taken, are also necessary, as well as a pithing needle and a frog.

Procedure

1. Simulation of the Electrical Axis of the Heart Set up the apparatus as in Fig. E3-2 and three-quarters fill the glass jar with 0.6% saline. Set the stimulator duration to 2 msec, the output to zero volts and the frequency to 25/sec. Connect the stimulator X-TRIG to the oscilloscope sweep so that the sweep is triggered by the stimulator, and verify that the stimulator is triggering the sweep by momentarily varying the stimulator frequency. Set the oscilloscope display controls in the NON-STORE position and set the oscilloscope sweep to 2 msec/cm. Connect the oscilloscope vertical amplifier (3A9) to the *RA* and *LA* electrodes in the electrolytic tank and place the dipole axis parallel to this lead (i.e., the dipole reference mark on the plastic collar is at zero degrees). Set the controls of the 3A9 differential preamplifier to VOLTS and DC and the AMPLIFIER-3db FREQUENCY UPPER control to 10 KHz and the LOWER control to 0.1 Hz; this will provide differential amplification with a bandwidth adequate to reproduce the 2-msec pulse. Connect the positive terminal of the stimulator to the red wire connected to the dipole electrode; connect the negative terminal to the other dipole electrode. Set the stimulator output to 100 V and the duration to 2 msec. Adjust the VOLTS/DIV control to obtain a 4-cm-high pulse at the beginning of the oscilloscope sweep. Now rotate the dipole slowly and measure the amplitude of the pulse for each 15° of rotation. Repeat the procedure using lead II (*RA–LL*) and lead III (*LA–LL*). On the same sheet of graph paper, plot the amplitudes obtained for the various angular positions of the dipole. Verify Einthoven's law (lead II = lead I + lead III). When the dipole is parallel to any lead, is the potential difference appearing between the electrodes consistent with the data given in Experiment 2?

2. Location of the Electrical Axes of the Heart The axes of excitation and recovery of the frog and human heart can be located by the use of limb electrodes. In the frog, needle electrodes inserted subcutaneously can be employed; in the human, plate electrodes applied with electrode jelly are customary.

THE FROG ECG Lay a doubly pithed frog (see Appendix Introduction) down on its back on an insulating surface and insert needle electrodes into all four limbs. Connect the right hindlimb electrode to the oscilloscope ground terminal and the right forelimb electrode to the −INPUT terminal and the left hindlimb electrode to the + INPUT terminal. (This is the connection for recording lead II.) Set the VOLTS/DIV control to 0.2 mV and obtain a recording of the frog ECG. If the amplitude is not large enough, increase it by turning the VOLTS/DIV control clockwise. Adjust the sweep (2B67) TIME/DIV control to obtain a sweep speed slow enough to identify all the waves in one cardiac cycle. When a satisfactory record has been obtained, set the oscilloscope controls in the STORAGE mode, obtain a single stored record of the frog ECG, and make a sketch of it. Label all the waveforms and include amplitudes and time calibrations on the sketch. Repeat the procedure for leads I and III. From any two of the three limb lead records, determine the mean QRS vector. The vectors for the other waves should be determined if the amplitudes are sufficiently large.

THE HUMAN ECG   Choose areas on the surfaces of the forearms and legs (near the ankles) for applying metal-plate electrodes. Rub electrolytic jelly into these sites, place one metallic-plate electrode on each, and secure it with a wide rubber band or masking tape. Using the same oscilloscope setting as for the frog, connect the right leg electrode to the oscilloscope ground terminal of the differential amplifier and the right arm and left leg electrodes to the −INPUT and +INPUT terminals, respectively; this arrangement permits recording lead II. Adjust the VOLTS/DIV control and TIME BASE controls to obtain a large amplitude display of the $PQRST$ complex for a single cardiac cycle. When a satisfactory display has been obtained, place the oscilloscope controls in the STORE node so that a stored record of the human ECG will be obtained. Make a sketch of the recording and include amplitude and time calibrations. Repeat the procedure for leads I and III. Identify the various waveforms and for any two limb leads determine the mean vector for the QRS wave. The vectors for the P and T waves can also be determined. (The mean electrical axis for the ventricles in the normal human subject is about 60°; however, the normal range is quite wide.

Additional Experiments

With the equipment employed in this experiment, several other interesting and informative experiments can be performed. For example, the thorax of the frog can be opened and the heart exposed. The ventricle can be moved to the right or left while the ECG is being recorded, and the effect on the amplitude of the R wave can be seen clearly. In addition, if the heart rate is slowed by cooling the frog, extra beats (extrasystoles) can be initiated by pricking an atrium or the ventricle; how such evoked excitations reveal their presence in the ECG will depend on the site of the stimulus. Delivery of repetitive stimuli (pacemaking) permits establishing a faster rate in the chamber stimulated.

Electrodes can be placed directly on the heart chambers to record their electrical signals. The heart can be removed from the frog and placed in 0.6% saline, where it will continue to beat for a short time. Electrodes placed in the saline environment will detect the electrical activity.

If two differential amplifiers are available for use with the oscilloscope, the vector loops can be displayed. To do so merely requires connecting the X-axis amplifier to lead I $(RA-LA)$ and the Y-axis amplifier to lead $V-LL$ (F)

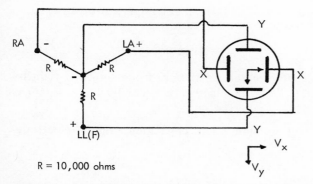

Fig. E3-5. Arrangement of connections for recording the vector loop using the Wilson central terminal to obtain V_y. The polarities deflect the oscilloscope beam in the directions indicated. Calibration: V_x 1 mV = 1 unit; V_y, 1 mV = 1.7 units.

that is, VF, observing the polarity conventions in Fig. E3-5. A discussion of vectorcardiographic patterns can be found in the reports by Cronvich et al. (1950) and Burch et al. (1953).

References

Bein, H. J. Automatisme et conduction physiologiques dans le sinus in situ du coeur de grenouille. *Arch. Int. Physiol.* 1948–1949, **56**:370–379.

Burch, G. E., J. A. Abildskov, and J. A. Cronvich. *Spatial Vectorcardiography*. Philadelphia, Pa.: Lea & Febiger 1953, 173 pp.

Cantrell, G., A. G. Moore, J. D. Bourland, H. E. Hoff, and L. A. Geddes. A training device for visualization of the meaning of leads in electrocardiography. *J. Electrocardiol.* 1969, **2**:407–413.

Cronvich, J. A., J. P. Conway, and G. E. Burch. Standardizing factors in electrocardiography. *Circulation.* 1950, **2**:111–121.

Einthoven, W. Ein nueues Galvanometer. *Ann. Phys. (Leipzig)* 1903, 12 Supp. **4**:1059–1071.

Einthoven, W. Weiteres über das Elektrokardiogramm. *Arch. Ges. Physiol.* 1908, **122**:517–584.

Einthoven, W., and A. de Waart. On the direction and manifest size of the variations of potential in the human heart and on the influence of the position of the heart on the form of the electrocardiogram. Transl. H. E. Hoff, and P. Sekely. *Amer. Heart J.* 1950, **40**:163–211.

Lepeschkin, E. *Modern Electrocardiography*. Baltimore: The Williams and Wilkins Company, 1951. Chapter IV.

Moore, J. A., Ed. *Physiology of the Amphibia*. New York: Academic Press, 1964. 654 pp.

Valentinuzzi, M. E., L. A. Geddes, H. E. Hoff, and J. Bourland. Properties of the 30° hexaxial (Einthoven-Goldberger) system of vectorcardiography. *Cardiovasc. Res. Cen. Bull.*, 1970, **9**:64–73.

EXPERIMENT 4   ELECTRODES: THE EFFECT OF REDUCED
AMPLIFIER INPUT IMPEDANCE

Object

1.   To demonstrate the capacitive nature of the electrode-subject circuit by reducing the input impedance of the bioelectric recorder while the ECG is being recorded.

2.   To estimate the equivalent resistance and capacitance of the electrode-subject interface.

Theory

In Experiment 1 it was shown that an electrode-electrolyte interface can be equated to a series or parallel combination of resistance and capacitance.* When two electrodes are employed to measure a bioelectric event, the event is separated from the measuring instrument by the impedance of the two electrodes; Fig. E4-1a and E4-1b show how the ECG is recorded. The impedances of electrodes A and B have been equated to a parallel combination of resistance and capacitance. It is important to recall that the values of R_a, R_b, C_a, and C_b are frequency dependent. The resistors r and R constitute a voltage divider to account for the loss in amplitude due to shunting by the body tissues and fluids and the remoteness of electrodes A and B from the heart. Although the illustration oversimplifies the situation, it provides an instructive way of viewing the recording of many bioelectric events with extracellular electrodes.

When the ECG is recorded with a local small-area (active) electrode B, paired with a much larger (reference) electrode A, (Fig. E4-1a), the impedance of the latter can be neglected. A further simplification of the circuit (Fig. E4-1c) can be made by obtaining the Thévenin equivalent for r and R, represented by the "frog resistance" R_f in Fig. E4-1c.

One very informative means of revealing the capacitive nature of the electrode-subject circuit is recording a bioelectric event (e.g., the ECG) with a bioelectric recorder having a very high input impedance. The records obtained can be assumed to be faithfully reproduced. If the input resistance (R_i, Fig. E4-1c) of the bioelectric recorder is reduced by placing different values of noninductive resistors across it, there will occur a loss of amplitude and an alteration in waveform; the latter is dramatically revealed by increasing the gain of the amplifier to restore the loss in amplitude. When this procedure is carried out, the electrode-subject impedance becomes a dominant part of the

---

* The electrode potential need only be considered when direct-coupled recording is employed.

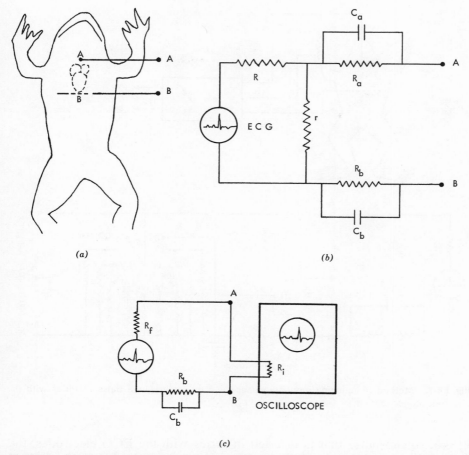

Fig. E4-1. Method of recording the frog ECG: (*a*) electrode location; (*b*) approximate equivalent circuit; (*c*) simplified equivalent circuit.

input circuit, and electrical differentiation of the bioelectric waveform occurs, manifesting itself as a loss of low-frequency components.

The behavior of the electrode-subject circuit can further be demonstrated by insertion of a square pulse in series with the bioelectric event and electrode-subject circuit. With this technique (Fig. E4-2), it is easy to demonstrate the differentiating property of the electrode-subject circuit. (This interesting phenomenon, which relates to waveform distortion, electrode area and input impedance, is discussed in Chapter 3.)

In Fig. E4-3, which illustrates the equivalent electrode-subject circuit, if

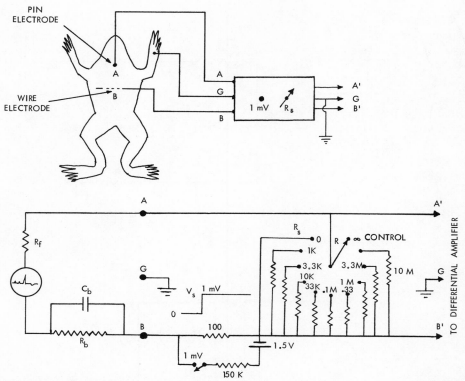

Fig. E4-2. Method of reducing input impedance and inserting square pulse in series with the electrode-subject circuit.

V_s is a square pulse that is inserted in series with the ECG electrodes, the voltage V appearing across the amplifier input terminals, which are shunted by R_i, will consist of the ECG riding on a baseline that is displaced by the square pulse. Immediately after insertion of the square pulse, the baseline displacement will diminish exponentially owing to the differentiating action of C_b and the resistances in the circuit. The initial instantaneous displacement V_i of the baseline is given by $V_s R_i/(R_i + R_f)$. Because V_s and R_i are known and V_i can be measured on the oscilloscope screen, R_f can be calculated. The terminally sustained (final) displacement of the baseline V_f is given by $V_s R_i/(R_i + R_f + R_b)$. Because R_i and V_s are known, V_f can be measured from the oscilloscope screen, and R_f has just been calculated, it is now possible to calculate the electrode-subject resistance R_b.

Measurement of the constant t of the exponential decay (to 37%) of the displacement of the baseline from V_i to V_f will permit determination of C_b (the equivalent electrode-subject capacitance); this time constant is equal to $R_b C_b (R_i + R_f)/(R_i + R_f + R_b)$. All the resistance values are now known, and therefore the value of C_b can be calculated.

In summary, this experiment reveals the nature of the electrode-subject circuit via the distortion imposed on the frog ECG and on a square pulse inserted in series with the ECG, as recorded with a small-area active electrode and a large-area reference electrode. This distortion is demonstrated by successively decreasing the resistive input impedance of the bioelectric recorder.

Equipment

The following equipment is required: Tektronix model 564 storage oscilloscope with model 3A9 preamplifier (or equivalent), variable shunt resistance

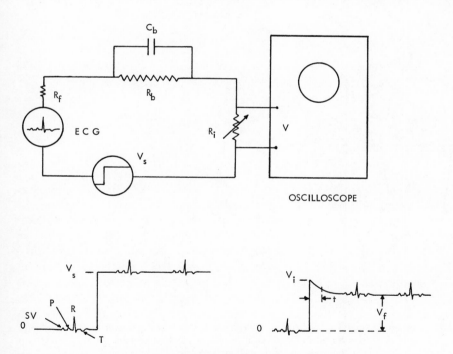

Fig. E4-3. Simultaneous recording of the ECG and square pulse V_s.

box with 1-mV series-calibrating voltage, #32-gauge enameled copper wire, 22-gauge hypodermic needle (spinal type), two pin electrodes. cork-backed frog board, 4 pins, frog (preferably cooled to about 45° F).

Procedure

1. Capacitive Nature of the Electrode-Subject Circuit.*   In this experiment a pithed frog is to be pinned (back down) to the frog board. Intrathoracic electrodes are used to detect the electrocardiographic voltages.

* It is advisable to become familiar with the oscilloscope instruction manual.

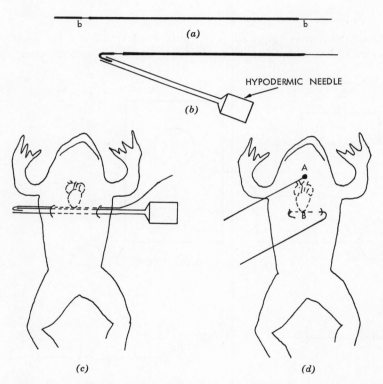

Fig. E4-4. Method of applying electrodes for recording the frog electrocardiogram. (*a*) #32 enameled copper wire (b = bare); (*b*) method of preparing electrode for insertion; (*c*) insertion method; (*d*) location of pin electrode (A) and small area electrode (B).

Approximately 2 cm from the end of a length (about 20 cm) of #32 enameled copper wire, scrape off the insulation to expose circumferentially about 2 to 3mm of copper. In addition, remove about 2 cm of insulation from the opposite end to permit making electrical contact with the recording apparatus (Fig. E4-4*a*). Place a length of about 3-mm of the insulated end of the wire in the tip of the hypodermic needle and bend the wire back 180° as in Fig. E4-4*b*. With the wire in the hypodermic needle, insert the needle laterally through the thorax of the frog just beyond the apex of the ventricle (Fig. E4-4*c*). (Notice the relation between the apex of the heart and the rear of the forelimbs.) Withdraw the hypodermic needle, leaving the enameled copper wire protruding beyond the frog. Adjust the wire so that the bared portion *b* is just opposite the apex of the ventricle. Anchor the electrode in this position by folding both ends of the wire back over the thorax (Fig. E4-4*d*). Insert a pin electrode vertically and midline into the thorax and just anterior to the atria (Fig. E4-4*d*). Insert another pin electrode into a leg of the frog to ground the animal, and connect the electrodes surrounding the heart to the variable resistance box as shown in Fig. E-4-2*a*; set the shunt resistance R_s to infinity.

With the oscilloscope differential amplifier VOLTS/DIV control set to 0.5 mV/cm, set the upper and lower AMPLIFIER-3 dB, frequency controls to 3 KHz and 0.1 Hz, respectively and place the AC-GND-DC lever switches in the DC position. (This will permit obtaining a control ECG with a capacitively coupled amplifying system having a bandwidth extending from 0.1 to 3000 Hz). Place the 2B67 time base MODE selector in the NORM position and set the oscilloscope DISPLAY controls to obtain a NON-STORE display. Set the TIME/DIV control to 0.5 sec/div. Adjust the differential amplifier POSITION control to obtain an ECG located in the center of its screen. (It may be necessary to reposition electrode *B* to obtain a sufficiently large ECG, i.e., about 1 mV).

Now verify that single sweeps can be obtained by placing the time base MODE control in the SINGLE SWEEP position and depress the control to the RESET position. Place the DISPLAY controls in the STORE position, depress the LOCATE button, and adjust the TIME BASE position control to locate the beam at the left-hand edge of the screen. Now adjust the POSITION control of the differential amplifier to place the beam about 2 divisions from the top of the screen. Release the display LOCATE button and depress the time base MODE control to the RESET position to obtain a single stored trace of the ECG on the oscilloscope screen. Note the sinus venosus, *P*, *R*, and *T* waves (Fig. E3-3*b*).

Place the shunt resistance switch in the 100-KΩ position, depress the

DISPLAY LOCATE button, and adjust the differential amplifier POSITION control to place the spot about 4 divisions from the top of the screen. Release the LOCATE button and depress the time base MODE control to obtain another single stored sweep with the reduced input impedance. Repeat this procedure with the shunt resistor in the 10-KΩ position and with the sweep located about 6 divisions from the top of the oscilloscope screen. The display should now show the differentiating action of the electrode-subject interface as manifested by the loss of low-frequency components in the electrocardiographic waves. In particular, compare the control record with those taken with reduced input impedance and notice the difference in the relative amplitudes of the positive and negative deflections. Make a sketch of the waveforms appearing on the oscilloscope screen.

2. Estimation of Equivalent Resistance and Capacitance of the Electrode-Subject Interface. Erase the display on the oscilloscope screen, place the shunt resistor in the infinity position, and short circuit terminals *A* and *B* (Fig. E4-2). Set the LOWER-3 dB frequency control in the DC OFFSET position and place the time base MODE switch in the NORM position. The sweep is now running and should be brought on to the screen by careful manipulation of the DC OFFSET controls. Set the time base MODE switch to the SINGLE position and erase the display. Obtain a single sweep and, when the beam is halfway across the screen, depress the 1-mV button. Measure the amplitude on the screen; this is the value for V_s. Remove the short circuit connected across the frog (terminals *A* and *B*) and position the sweep on the screen using single sweeps while manipulating the DC OFFSET—COARSE control. Place the shunt resistor R_s in the 100-KΩ position, erase the display, depress the DISPLAY LOCATE button and, using the DC OFFSET control, position the trace to the center of the screen. Obtain a single sweep and insert the 1-mV pulse near its beginning as before. Measure peak (transient) deflections of the baseline V_i and the final sustained deflection V_f. Repeat the procedure to ensure that the injected pulse is not coincident with one of the waveforms of the ECG. Enter V_s, V_i, V_f, and R_s in Table E4-1.

To accurately measure the time constant *t*, increase the sweep speed to 0.1 sec/div. and repeat the procedure just described; enter the value for the time constant in the table. Calculate the correct value of R_i, which is equal to the parallel combination of the shunt resistance R_s employed and 1 MΩ, the input resistance of the oscilloscope. Enter R_i in the table and calculate R_f, R_b, and C_b.

Remove the active electrode *B* from the frog, measure the length and diameter of the exposed portion of the copper wire, and calculate the area. Now calculate the equivalent capacitance per unit area.

From the measurements obtained in this experiment, complete Table E4-1.

Table E4-1 Table of Data

Measured Quantities		Calculated Quantities	
V_s	_____ mV	R_f	_____ Ω
V_i	_____ mV	R_b	_____ Ω
V_f	_____ mV	C_b	_____ μF
t	_____ sec	C_b/cm^2	_____ μF/cm²
R_{shunt}	_____ Ω		
R_{scope}	_____ 1.0 MΩ		
R_i	_____ Ω		
Area of active electrode	_____ cm²		

References

Geddes, L. A., and L. E. Baker. The relationship between input impedance and electrode area in recording the ECG. *Med. Biol. Eng.* 1966, **4**:439–450.

Geddes, L. A., L. E. Baker, and M. Mc Goodwin. The relationship between electrode area and amplifier input impedance in recording muscle action potentials. *Med. Biol. Eng.* 1967, **5**:561–569.

EXPERIMENT 5 THE ACTION POTENTIAL AND THE EFFECT OF INJURY

Object

1. To record the diphasic action potential of skeletal muscle with extracellular electrodes.

2. To investigate the development of an injury potential.

Theory

In many circumstances of recording bioelectric events, one or both electrodes are placed directly on an irritable cell. The type of potential change recorded during excitation and recovery depends on the type of cell and the orientation and spacing of the electrodes. With surface electrodes, the potential detected is the instantaneous algebraic sum of the changes in membrane potential "seen" by the two electrodes. If the irritable tissue is long and cylindrical and the amount of tissue occupied by excitation and recovery is small, a

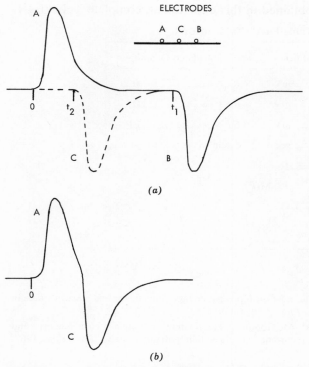

Fig. E5-1. Action potentials appearing on the surface of a long strip of irritable tissue: (*a*) under widely spaced (*A–B*) electrodes; (*b*) under closely spaced (*A–C*) electrodes.

widely spaced pair of surface electrodes will detect a negative (upward) excursion, followed at a time t_1 by a positive (downward) excursion in potential (Fig. E5-1*a*). If the electrodes are moved closer together, the propagation time t_2 between the two phases of the action potential will be shorter and may indeed overlap each other, giving rise to a diphasic action potential (Fig. E5-1*b*).

In order to record the diphasic action potential as described, it is necessary to employ a long cylindrical cell in which excitation and recovery occupies a short distance. Although a squid giant axon could be employed, such a preparation is not readily available. It is more convenient to use the frog sartorius muscle, which consists of a bundle of long cylindrical contractile fibers, almost identical in length and properties.

In this experiment one end of a sartorius muscle is stimulated, using two closely spaced electrodes connected to a stimulator equipped with an isolated output circuit. The evoked action potentials are recorded from electrodes

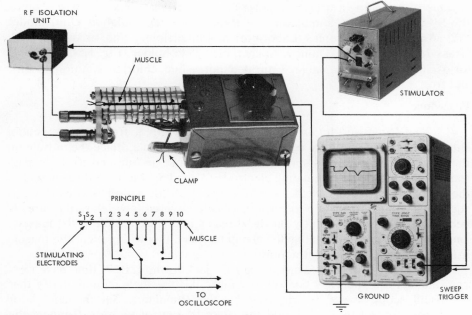

Fig. E5-2. Electrode assembly for frog sartorius muscle.

located at different distances along the muscle. A rotary switch is used to connect the various electrodes instantly to the potential-measuring apparatus.

The holder for the sartorius muscle (Fig. E5-2) consists of a series of electrodes insulated from each other and the holders that support them. The two electrodes at the end ($S1$ and $S2$ in the inset) are connected to the stimulator via the RF insolation unit. Each of the other electrodes (1–10) is connected to a rotary switch that permits connecting the potential-measuring instrument between electrodes 1 and 2, 1 and 3, and so on; in this way the distance between the measuring electrodes can be increased incrementally while the muscle is being stimulated.

The effect of local injury on the action potential can be demonstrated by crushing the muscle under one electrode or depolarizing its membranes by the local application of concentrated potassium chloride solution.

Equipment

The following equipment is required: storage oscilloscope with triggerable time base (e.g., Tektronix 564 with 2B67 time base generator and 3A9 pre-amplifier, Tektronix, Portland, Ore.), stimulator with isolated output and capable of providing a triggering pulse for the oscilloscope sweep (e.g., Narco Bio-Systems Mark V stimulator and Radiofrequency Isolation Unit, model 112, American Electronics Laboratories, Philadelphia, Pa., or model RF1 Grass Instrument Co., Quincy, Mass.), electrode holder (Fig. E5-2) stimulator-oscilloscope trigger cable, saturated potassium chloride solution (5 ml), and frog.

Procedure

1. The Diphasic Action Potential   Doubly pith a frog (see Appendix Introduction) and carefully remove one sartorius muscle. Place the muscle on the electrode assembly as shown in Fig. E5-2 and tie one end to the plastic frame adjacent to the stimulating electrodes S_1 and S_2. Pass the threads tied to the other end of the muscle between the last electrode and the switch box and squeeze the alligator clamp to open its plastic-covered jaws. Pass the threads between the jaws, pull gently on the threads, applying tension to the muscle, and allow the jaws of the alligator clamp to close; this will prevent the muscle from shortening during contraction.

Connect the stimulator (Mark V) x-trig jack to the (2B67) time base ext-trig and set the time/div control to 2 msec. Using single stimuli, verify that the oscilloscope sweep is triggered by the stimulator. Set the stimulator duration control to 1 msec and the output control to zero. Connect the radio-frequency isolation unit to the stimulator and to the stimulating electrodes on the muscle holder in accordance with Fig. E5-2.

Connect the shielded electrode cable from the muscle holder to the oscilloscope positive and negative input (+ −) terminals and the shield to the oscilloscope ground terminal. Ground the oscilloscope and the metal box supporting the muscle holder. Set the volts/div control on the (3A9) differential amplifier to 10 mV and the upper and lower amplifier − 3 dB controls to 10 KHz and 0.1 Hz, respectively. Place the lever switches in the volts and ac positions; these settings will provide a recording sensitivity of 10 mV/cm and a bandwidth extending from 0.1 to 10,000 Hz. Place the electrode selector switch in position 2 and connect the electrode selector unit to the oscilloscope, as in Fig. E5-2. With the oscilloscope controls in the non-store position and while delivering single stimuli, increase the stimulus intensity until there is no further increase in amplitude of the action potential recorded on the oscilloscope screen. Note the relationship between the upward and downward phases of the action potential. Place the oscilloscope display switch in the store position. Depress the locate button and position the beam approximately one-fifth the way down from the top of the screen. By using a single stimulus, obtain a single stored sweep of the action potential on the screen. Depress the locate button and position the beam to approximately two-fifths the way down from the top of the screen. Place the electrode selector switch in position 3. By using a single stimulus, obtain another single stored sweep on the screen.

In a similar manner, continue to reposition the beam and obtain single stored sweeps with the electrode selector switch in positions 4, 5, and so on. Calculate the propagation velocity (m/sec) by dividing the distance (mm) between a pair of electrodes by the time (msec) between the beginning of each phase of the action potential for the selector switch position that gives the clearest separation between the upward and downward phases of the action potential.

2. The Effect of Injury under One Electrode Set the electrode selector switch to the position that gave a clear separation between the upward and downward phases of the action potential; erase the display. With the oscilloscope controls returned to the STORE position and while delivering single stimuli, apply the wooden end of the pithing needle to the muscle over the distal electrode. After each stimulus, increase the force applied to the muscle to crush it against the electrode, noting the gradual decrease in amplitude and the virtual disappearance of the second phase of the action potential. After a satisfactory recording has been obtained, erase the display.

While delivering single stimuli, place the electrode selector switch in the position that records from the electrode just proximal to the site of injury and note the reappearance of a second phase of the action potential. Now, while delivering single stimuli, rotate the selector switch to record from electrodes distal to the injured area; note that a small-amplitude second phase can be recorded, indicating either that excitation is still being propagated by uninjured fibers in the body of the muscle or that some recovery has occurred.

Return the selector switch to permit recording from the electrode at the site of injury and wait about 10 minutes before delivering single stimuli. When this time has elapsed, a second phase of the action potential may be recordable, because some recovery may have occurred at the site of injury.

When the muscle has recovered adequately (or by using the other sartorius muscle), set the electrode selector switch to obtain clear upward and downward phases of the action potential in response to single stimuli; obtain a stored trace of these events. Note the position of the electrode selector switch; then place a drop of saturated potassium chloride solution on the muscle at this electrode location. Note the disappearance of the second phase of the action potential. Rotate the electrode selector switch to record from electrodes beyond the site of potassium chloride depolarization; no action potentials are recordable because the salt solution has penetrated into the muscle. Place the electrode selector switch to record from a site ahead of the point of application of the potassium chloride; note the presence of the second phase of the action potential. Wait several minutes and deliver single stimuli, observing the gradual disappearance of the second phase of the action potential as the potassium chloride diffuses along the muscle fibers.

Summary

1. Make a calibrated sketch of the upward and downward phases of the action potential.
2. Calculate the propagation velocity (m/sec).
3. Sketch the action potential when injury was produced under one electrode.

EXPERIMENT 6 PROPAGATION* VELOCITY OF THE ACTION POTENTIAL IN FROG NERVE

Object

1. To record the action potential of skeletal muscle.
2. To measure the velocity of propagation of the nerve impulse.
3. To record the action potential of nerve.
4. To demonstrate that a nerve fiber propagates in both directions.
5. To create a nerve block by the use of a ligature.

Theory

If electrodes are placed on nerve or muscle during activity, action potentials can be recorded; their form is dependent on the characteristics of the particular tissue and the electrode arrangement. If bipolar electrodes are placed on the uninjured surface of a single fiber of nerve of skeletal muscle (or on a bundle of identical fibers), the action potential recorded is diphasic as the impulse passes. As the advancing impulse appears under the proximal electrode, it becomes negative with respect to the distal electrode. As the impulse passes onward toward the distal electrode, for a brief period it will be between the recording electrodes. During this period the tissue under the proximal electrode will have recovered and no potential difference will be presented to the two electrodes. Similarly, when the impulse reaches the distal electrode, it becomes negative with respect to the proximal electrode. As the impulse passes the distal electrode, the membrane recovers its polarity, hence no difference of potential is measurable between the electrodes. Thus, in summary, the action potential can be thought of as a traveling wave of negativity, and recording electrodes placed on the uninjured surface of nerve or muscle will exhibit a diphasic electrical wave. The first (negative) phase is completed as the impulse passes under the proximal electrode and the second phase is completed as the impulse passes under the distal electrode. If the electrodes are closely spaced, the temporal separation between the phases will be short and a sinusoidal-like waveform will be recorded. If the area under one electrode is injured, however, only one phase will be recorded.

When the tissue under investigation is a structure with varied characteristics,

---

* The term "conduction" velocity is usually employed by physiologists.

the action potential may be complex in form. For example, if electrodes are placed on a nerve trunk containing a population of fibers of many diameters, hence different propagation rates, a complex polyphasic action potential may be recorded consisting of the monophasic or diphasic action potentials of the A_α, A_β, A_γ, B, and C fibers. In a mixed nerve and with surface electrodes, it is seldom possible to identify more than a few of these components because of the large differences in propagation times and their relative distance from the recording electrodes on the surface of the nerve.

Skeletal muscle, being relatively homogeneous tissue, usually exhibits a simple diphasic action potential. Pin electrodes inserted into a muscle are generally adequate to record the action potential. Because the pins cause local injury to some of the muscle fibers, the recorded action potential may not be symmetrically diphasic. To avoid injury in recording nerve action potentials, it is convenient to use a sleeve electrode. Often, however, the roughness and pressure of the sleeve electrode against the nerve are enough to cause slight injury, resulting in an action potential that is not fully diphasic in form.

Measurement of the propagation velocity of nerve is most easily accomplished by employing an *in vivo* nerve-muscle preparation. For a given stimulating site on a nerve, the time between the stimulus and the muscle action potential consists of (*a*) the time necessary to depolarize the nerve and propagate the impulse to the motor end plate, (*b*) the end-plate delay, and (*c*) the time to propagate the muscle action potential to the recording electrodes. If a position on the nerve nearer to the muscle is stimulated, the muscle action potential will occur earlier, because the length of the nerve between the stimulating electrode and muscle is shorter. Since (*b*) and (*c*) are constant, the time difference between the action potentials recorded corresponds to the time taken for the nerve impulse to traverse the distance between the two stimulating positions. This length difference divided by the time difference is the propagation velocity. It is usually expressed as meters per second at a specified temperature. Helmholtz (1850) (see Hoff and Geddes, 1960) employed this principle to measure the velocity of the nerve impulse, which he determined without measuring action potentials, because at that time there were no rapidly responding bioelectric recorders. Surprisingly enough, he obtained 30 m/sec for the frog sciatic nerve—a value that is repeatedly confirmed by more elegant apparatus!

Equipment

Animal Supplies The following animal supplies are required: frog, frog board (5 × 9 in.), thread, pithing needle, pins, forceps, scissors, pin electrodes (3), sleeve electrodes (two bipolar, two monopolar), wax pencil, celluloid centimeter scale.

Electronic The following electronic supplies are required: storage oscilloscope* with triggerable sweep and high gain amplifier (e.g., Tektronix model 564, model 2B67 time base and model 3A9 differential amplifier), stimulator with isolated output and output for triggering oscilloscope sweep (e.g., Narco Bio-Systems Mark V) single-pole double-throw switch with a neutral position, ground wire, and shielded stimulus-trigger cable.

Procedure

Pin a doubly pithed frog (see Appendix Introduction) down to a board. Expose the sciatic nerves in the back and in the thigh region (Figs. A1-2 and A1-3). Apply a monopolar sleeve electrode (Fig. A1-5a) to the sciatic nerve trunk in the back, as far headward as possible. Apply a second monopolar sleeve electrode to the sciatic nerve in the thigh on the same side of the animal. Connect these electrodes to the switch box as shown in Fig. E6-1. The pin electrodes (separated by about 5 mm) are employed to detect the action potential of the gastrocnemius muscle and to pin it to the board to prevent movement when the muscle is stimulated by its nerve. Connect the muscle electrodes directly

---

* A nonstorage oscilloscope (e.g., Tektronix 561A) may be used.

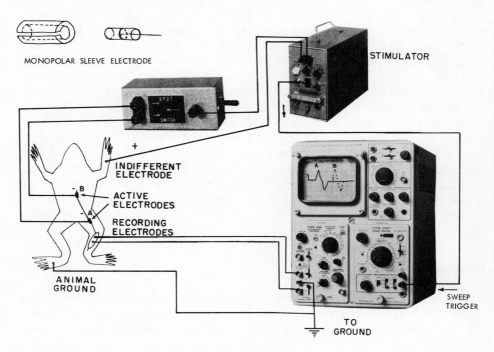

Fig. E6-1. Apparatus for recording frog muscle action potential.

to the positive and negative input terminals of the 3A9 differential amplifier. Connect a pin electrode placed in the opposite hindlimb to the ground terminal of the 3A9 preamplifier. Set the lever switches on the preamplifier to VOLTS and AC. Place the UPPER and LOWER—3 dB FREQUENCY controls in the 10 KHz and 1 Hz positions, respectively (this provides a bandwidth extending from 1 to 10,000 Hz). Set the VOLTS/DIV control to 5 mV.

Connect the negative terminal of the stimulator to the SPDT (single-pole double-throw) switch (Fig. E6-1) and the positive terminal to a pin electrode placed in the right forelimb of the animal.

1. Muscle Action Potential   Set the stimulator output to zero, the duration to 0.1 msec, and the frequency to 2/sec. Place the CONT-OFF-SINGLE control in the CONT position; this causes the stimulator to trigger the oscilloscope sweep at 2/sec. Set the oscilloscope TIME/DIV control to 1 msec/div. Verify that the time base is operating satisfactorily using the nonstore operation of the oscilloscope.

Connect the positive and negative inputs terminals of the 3A9 amplifier directly to the pin electrodes. Place the SPDT switch in position *A* (Fig E6-1) and increase the stimulus output voltage slowly until the gastrocnemius muscle twitches visibly. At this point place the SPDT switch in the *A* position and operating with the oscilloscope controls in the NON-STORE mode, adjust the VOLTS/DIV control so that an adequately large muscle action potential is seen on the oscilloscope screen. Do not stimulate for too long, because the myoneural junctions fatigue easily and the muscle action potentials will be lost. (If this occurs, the procedure can be carried out on the opposite leg.) Adjust the stimulator output so that a slight increase does not increase the amplitude of the muscle action potential. Place the oscilloscope controls in the STORE position and, using single stimuli, obtain a stored display of the amplitude and time course of the muscle action potential. (If a nonstorage oscilloscope is used, the action potential can be sketched with a wax pencil on the face of the cathode-ray tube. Make a calibrated sketch of the amplitude (mV) and time course (msec) of the muscle action potential.

2. Velocity of Propagation of the Nerve Impulse   Using the same settings as before, place the SPDT switch in position *B* to stimulate the main sciatic nerve and obtain a stored tracing. (It may be necessary to increase the stimulus output voltage.) Note that the time between the beginning of the sweep and the muscle action potential is longer. Now move the SPDT switch from position *B* to *A* and note the temporal displacement of the muscle action potential. Measure this temporal movement, which is equal to the time taken for the nerve impulse to travel the distance from *B* to *A* on the nerve. Measure the time difference and divide it into the distance along the nerve between electrodes *A* and *B*; the result is the propagation velocity (m/sec) of the fastest fibers in the sciatic nerve.

3. Nerve Action Potential Remove the monopolar stimulating electrode at *A* and in its place apply a bipolar sleeve electrode; connect this electrode to the positive and negative input terminals. Set the oscilloscope sweep velocity to 0.2 msec/cm. Place the SPDT switch in the *B* position in order to deliver stimuli to the sciatic nerve. Increase the gain of the oscilloscope (and the stimulus intensity, if necessary) until a satisfactory nerve action potential is recorded with the oscilloscope controls in the NON-STORE position. Because nerve does not fatigue easily, the stimulus frequency can be increased to obtain a brighter display of the nerve action potential on the screen. After a satisfactory record is obtained, place the oscilloscope controls in the STORE position and obtain a stored tracing of the nerve action potential. Make a sketch of the amplitude (mV) and duration (msec) of the nerve action potential as it appears on the oscilloscope screen.

If the trailing edge of the stimulus can be seen clearly at the beginning of the sweep, the approximate nerve propagation velocity can be determined by measuring the time between the trailing edge of the stimulus and the onset of the nerve action potential. To this time must be added the duration of the stimulus (0.1 msec). The length of nerve which provides this latency is that measured between the stimulating electrode *B* and the nearest electrode in the bipolar sleeve. Measure this length of nerve and divide it by the propagation time. Practical uncertainties usually prevent the value obtained from coinciding with the value found in Section 2 of this experiment.

4. Bidirectional Propagation in Nerve Bidirectional propagation is an interesting and important property of nerve fibers (axons) that can be demonstrated by interchanging the location of the stimulating and recording electrodes. Referring to Fig. E6-1, apply a bipolar sleeve electrode to the distal sciatic branch *A* (the negative pole of the bipolar stimulating electrode at *A* should be closest to site *B*) and connect it to the stimulator. Connect the other bipolar sleeve electrode (located at *B*) to the input terminals of the oscilloscope. With the oscilloscope operating in the NON-STORE mode, deliver single stimuli to *A* and adjust the stimulus intensity and oscilloscope gain controls to obtain a satisfactory nerve action potential. Obtain a stored tracing of this event and calculate the nerve propagation velocity by dividing the distance between the nearest stimulating and recording electrodes by the time between the onset of the stimulus and initial rise (or fall) of the action potential.

5. Nerve Block Place a thread loosely under the sciatic nerve trunk between the bipolar stimulating and recording electrodes on the sciatic nerve. With the oscilloscope controls in the STORE mode, while delivering single stimuli to the nerve and recording nerve action potentials, make a loose knot in the thread, then tighten the knot gradually, and observe disappearance of the nerve action potential on the oscilloscope, indicating that the impulse has been blocked by mechanical injury.

Enter all data in Table E6-1.

Table E6-1   Table of Data

Muscle Action Potential
    Amplitude (peak-to-peak)   ——————— mV
    Duration   ——————— msec

Nerve Action Potential
    Amplitude (peak-to-peak)   ——————— mV
    Duration   ——————— msec

Propagation Velocity of Nerve Action Potential
    By using the nerve-muscle preparation   ——————— msec
    By recording nerve action potentials   ——————— msec
    By recording propagation in the
        opposite direction   ——————— msec

References

Helmholtz, H. Note sur la vitesse de propagation de l'agent nerveux dans les nerfs rachidiens. *C. R. Acad. Sci. (Paris)*. 1850, **30**:204–206

Hoff, H. E., and L. A. Geddes. Ballistics and the instrumentation of physiology: The velocity of the projectile and of the nerve impulse. *J. Hist. Med. Allied Sci.* 1960, **15**:133–146.

Index